STALLCUP'S GENERATOR, TRANSFORMER, MOTOR AND COMPRESSOR BOOK

National Fire Protection Association
Quincy, Massachusetts

Written by James Stallcup, Sr.
Edited by James Stallcup, Jr.
Design, graphics and layout by Billy G. Stallcup

Notice Concerning Liability: Publication of this work is for the purpose of circulating information and
opinion among those concerned for fire and electrical safety and related subjects. While every effort has
been made to achieve a work of high quality, neither the NFPA nor the authors and contributors to this
work guarantee the accuracy or completeness of or assume any liability in connection with the
information and opinions contained in this work. The NFPA and the authors and contributors shall in no
event be liable for any personal injury, property, or other damages of any nature whatsoever, whether
special, indirect, consequential, or compensatory, directly or indirectly resulting from the publication, use
of or reliance upon this work.

This work is published with the understanding that the NFPA and the authors and contributors
to this work are supplying information and opinion but are not attempting to render engineering or other
professional services. If such services are required, the assistance of an appropriate professional
should be sought.

® Registered Trademark National Fire Protection Association, Inc.

NFPA No.: SGM05
ISBN: 0-87765-669-X
Library of Congress Card Catalog No.: 2004109215

Printed in the United States of America
09   08   07   06   05   5   4   3   2   1

Introduction

While this book is intended for all who are interested and work in a daily capacity with these subjects, it is also designed to help the student in his search for learning. For this reason, the book is profusely illustrated to help visualize for the reader the points brought out in the text while joining theory and practice into a closer relationship.

For user friendly and easy study, Stallcup's Generator, Transformer, Motor and Compressor Book has been divided into three parts, and they are as follows:

Part One: Generators

Part Two: Transformers

Part Three: Motors, Controllers and Compressors

Review questions have been provided at the end of each chapter, and for instructional purposes, transparency masters are available upon request. Answers to the review questions can be found in the Instructor's Guide/ Answer Key.

Note: When "must" is used it carries the same meaning as "shall" when used as a requirement pertaining to rules in the NEC.

Table of Contents

Part Three. MOTORS

Generators

Part One

From the small standby unit to the largest hydroelectric plant, today's world demands speed, light, and power. The **generator** is the device that converts mechanical energy into electrical energy and supplies power where needed. The generator can be designed and installed to provide the starting point for power to an electrical service as well as the backup power when things go wrong.

Engine-driven generators fueled by diesel, gasoline or natural gas are commonly used to produce and provide an alternative source of emergency or standby power when normal power systems fail. Gas-turbine generators are also used to generate such power.

A facility with engine-driven generators provides the necessary power for human safety as well as the protection of property, while maintaining continuous operation of specific types of equipment.

The design requirements for selecting an on-site generator differ depending upon the generator's use. For example, generators can be utilized as emergency systems, standby power systems, or other power sources when used in health-care facilities. A portable generator can provide power for construction, remodeling, maintenance, or making repairs on equipment.

Part One covers the theory of generators, the various types, and the rules and regulations of the NEC pertaining to the their design and installation.

Magnetism and Electromagnetism

Magnetism is one of the fundamental forces involved in the use of electricity. Therefore, it is imperative that electricians, technicians, and maintenance personnel obtain a good understanding and knowledge of the subject.

THEORY OF MAGNETS

A magnet is an object that attracts such magnetic substances as iron or steel, by producing an external magnetic field that reacts with a magnetic substance. A permanent magnet maintains an almost constant magnetic field without the application of any magnetizing force. As an example, for many years, some magnetized substances show practically no loss of magnetic strength and therefore maintain such strength.

MAGNETIC FIELDS

A magnetic field is assumed to consist of invisible lines of force that leave the north pole of a magnet and enter the south pole. The direction of this force is used only to establish rules and references for such operation.

This action is indicated, for example, by the fact that a north pole will repel another north pole and be attracted by a south pole and vice versa.

NATURAL MAGNETS

A natural magnet is called a lodestone, or "leading stone." The natural magnet gets its name from being used by early navigators to determine direction.

When a lodestone is freely suspended, one end always points in a northerly direction. Because of this action, one of the lodestones is called the "north-seeking" and the other the "south-seeking" end. The terms are better known as the north and south poles respectively. The reason that a freely suspended magnet assumes a north-south position is that the earth is a large magnet and its magnetic field exists over its entire surface.

For example, the magnetic lines of force leave the earth at a point near the south pole and enters near the north pole. Therefore, since the north pole of a magnet is attracted to the south pole of another magnet and repels another north pole, one can understand that the magnetic pole near the geographic north pole of the earth is actually a south pole, and that the pole near the geographic south pole of the earth is actually a north pole. **(See Figure 1-1)**

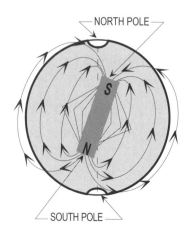

PERMANENT MAGNET AND EARTH

Figure 1-1. The above illustrates the magnetic field of a magnet and its relationship to the earth's magnetic field.

PERMANENT MAGNETS

Certain metallic alloys such as hard steel have the ability to retain magnetism and are able to do so due to the fact they are difficult to magnetize. Hard steel is more difficult to magnetize than soft iron because of the internal friction among the atoms. If such a substance is placed in a strong magnetic field and struck with a hammer, the atoms become aligned with the field. When the substance is removed from the magnetic field, it will retain its magnetism and becomes a permanent magnet.

ELECTROMAGNETS

A bar magnet can be pushed into a coil of wire (solenoid), and current will flow in a certain direction as the magnet moves into the coil. However, as soon as the magnet stops

moving, the current flow will stop. When the magnet is withdrawn, the current reverses and the current will flow in the opposite direction. The current induced in the coil is caused by the field of the magnet as it cuts across the turns of wire in the coil.

Theory Tip: If a piece of soft iron is placed in the magnetic field of a permanent magnet, it will take on the same characteristics as the permanent magnet and becomes magnetized. **(See Figure 1-2)**

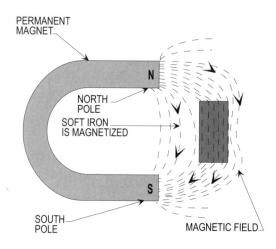

SOFT IRON IN A MAGNETIC FIELD

Figure 1-2. The above illustrates a permanent magnet, magnetizing a piece of soft iron that is placed in its magnetic field.

Figure 1-3(a) illustrates that the north pole of the coil is adjacent to the north pole of the bar magnet and opposes the insertion of the magnet into the coil. However, the instant that the magnet begins to move out of the coil, current induced in the coil changes to the opposite direction. This is due to the field of the coil being reversed. Note that the south pole of the coil field is now adjacent to the north pole of the bar magnet and opposes the withdrawal of the magnet as shown in **Figure 1-3(b)**.

MAGNETIC FIELD

The field of force existing between the poles of a magnet is called a magnetic field. The lines of force of this field may be demonstrated by placing a stiff paper over a magnet and sprinkling iron fillings on the paper.

See Figure 1-4 for a detailed illustration of the lines of force for magnetic circuits. A magnetic force, known as magnetic flux, travels from north to south in invisible lines that cannot be seen.

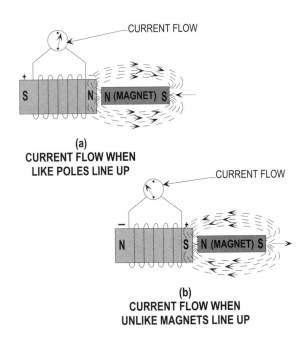

Figures 1-3(a) and (b). The current flows in a magnet when magnets are induced by a changing magnetic field.

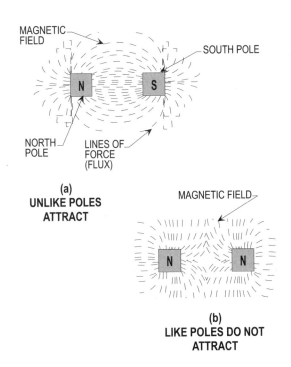

Figure 1-4. The magnetic lines of force between the pole of a magnet is called a magnetic field.

For example, if a soft iron bar is placed across the poles of a magnet, almost all the magnetic lines of force (flux) go through the bar, and the bar becomes magnetized. **(See Figure 1-2)**

ELECTROMAGNETICS

An electric current flowing through a conductor creates a magnetic field around the conductor. When a wire is grasped in the left hand with the thumb pointing from negative to positive poles, the magnetic field around the conductor is in the direction that the fingers are pointing. Note that this can be easily demonstrated by the use of the left-hand rule, which is based upon the true direction of current flow. **(See Figure 1-5)**

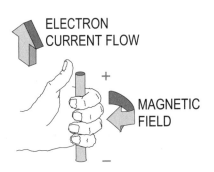

Figure 1-5. When a wire is grasped in the left hand, the thumb points from negative to positive poles, one finger points in the direction of the magnetic field and the index finger in the direction of movement.

When a current-carrying conductor is formed into a loop, the loop takes on the properties of a magnet. For example, one side of the loop will be a north pole and the other side will be a south pole. Where a soft-iron core is placed in the loop, the magnetic lines of force will magnetize the iron core and it becomes a magnet. When a wire is formed into a coil and connected to a source of power, the fields of the separate turns join and travel through the entire coil. **(See Figure 1-6)**

> **Theory Tip:** When a coil is grasped in the left hand with the fingers pointing in the direction of current flow, that is, from negative to positive, the thumb will point toward the north pole of the coil.

ELECTROMAGNETIC INDUCTION

The transfer of electric energy from one circuit to another without the aid of electric connections is known as induction. When electric energy is transferred by means of a magnetic field, it is known as electromagnetic induction.

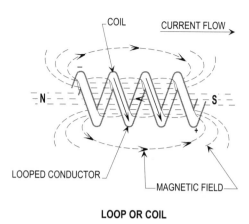

LOOP OR COIL

Figure 1-6. When a conductor is formed into a coil and connected to a source of power, the magnetic fields of each separate turn add and travel through the entire coil.

Electromagnetic induction occurs whenever there is a relative movement between a conductor and a magnetic field, that is, when the conductor is cutting across magnetic lines of force and is not moving parallel to them.

Note that this relative movement may be accomplished in two ways (1) by using a stationary conductor and a moving field and (2) by using a moving conductor with a stationary field. A moving field may be created by a moving magnet or by changing the value of the current in an electromagnet. **(See Figure 1-7)**

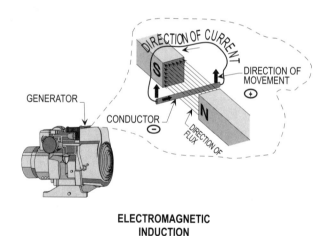

ELECTROMAGNETIC INDUCTION

Figure 1-7. The above illustrates the transfer of electric energy by means of a magnetic field and conductor. This process is called electromagnetic induction.

GENERATOR ACTION

Figure 1-8 illustrates the basic principle action of a generator. As the conductor moves through the field, a voltage is induced in it. Note that the same action occurs if the conductor is stationary and the magnetic field is moved.

The direction of the induced voltage depends on the direction of the field and can be verified by applying the left-hand rule for generators. **(See Figures 1-9(a) and (b))**

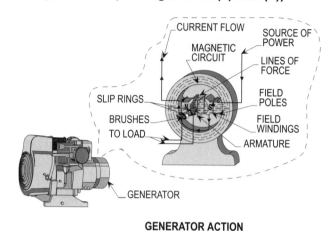

GENERATOR ACTION

Figure 1-8. The above illustrates the basic action that takes place in generators to produce electricity.

Theory Tip: When using the left-hand rule for generators, extend the thumb, forefinger and middle finger of the left-hand so that they are at right angles to one another. Then turn the hand so that the index finger points in the direction of the magnetic field and the thumb points in the direction of the conductor movement. The middle finger will then be pointing in the direction of the induced voltage and flow of current.

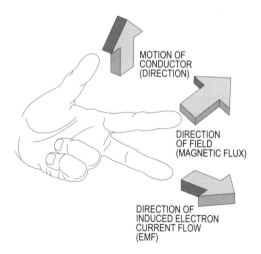

LEFT HAND RULE FOR GENERATORS

Figure 1-9. The above illustration represents the left hand rule for generators.

TYPES OF POWER

All electric power supplied by generators or batteries may be divided into two general groups as follows:

   (1) Alternating current (AC) unit
   (2) Direct current (DC) unit

ALTERNATING CURRENT POWER

AC power sources feed electricity directly to the electrical network to which the lights, motors, appliances and other equipment are connected. Generators are usually used between the power source and the line instead of batteries. Although in most cases, a battery is used for engine-starting purposes only.

Generators and batteries with additional equipment can be utilized to produce the following voltage levels.

(1) 600 volts or less
- Single-phase
  - 120 volts, two-wire
  - 120/240 volts, three-wire
  - 240 volts, two-wire
- Three-phase
  - 240 volts, three-wire
  - 120/208 volts, four-wire
  - 120/240 volts, four-wire
  - 480 volts, three-wire
  - 277/480 volts, four-wire
  - 600 volts, three-wire

(2) Over 600 volts
- Three-phase
  - 2,400 volts
  - 4,160 volts
  - 12,470 volts
  - 13,200 volts or 13,800 volts

See Figures 1-10(a) and (b) for a detailed illustration of the different voltages supplying electrical systems.

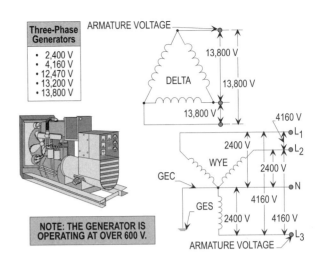

Figure 1-10(b). The above illustrates the most popular voltages used to supply loads requiring voltages rated over 600 volts.

DIRECT CURRENT POWER

DC power sources feed electricity directly to the line to which the lights, motors, appliances, and other equipment are connected. There are limitations to DC power sources. For example, they will only operate ordinary light bulbs, motor and appliances that are designed for DC, and motors of the universal type. DC power may also feed fluorescent bulbs, but only if they are installed with a special converter. Because DC power sources cannot operate any equipment designed for AC, they are not usable as a standby for most emergency systems. (**See Figure 1-11**)

> **Generator Tip:** DC power sources are normally used to supply special types of equipment including accessories, etc.

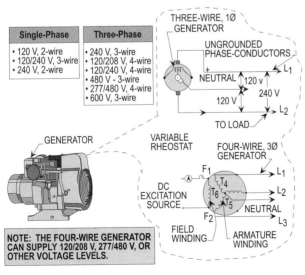

THREE OR FOUR-WIRE GENERATORS

Figure 1-10(a). The above illustrates the most popular voltages used at 600 volts or less.

See Figure 1-11 on next page >>>>

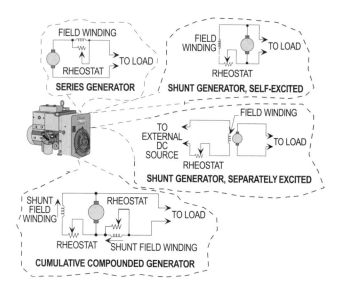

TYPES OF DC GENERATORS

Figure 1-11. The above illustrates the different types of DC generators. For more details, see Chapter 2.

Chapter 1: Magnetism and Electromagnetism

Section   **Answer**

_____   T   F   **1.** A magnetic field is assumed to consist of invisible lines of force that leave the north pole of a magnet and enter the south pole.

_____   T   F   **2.** Hard steel is less difficult to magnetize than soft iron because of the internal friction among the atoms.

_____   T   F   **3.** An electric current flowing through a conductor creates a magnetic field around the conductor.

_____   T   F   **4.** When a current-carrying conductor is formed into a loop, the loop takes on the properties of a magnet.

_____   T   F   **5.** The transfer of electric energy from one circuit to another without the aid of electric connections is known as electromagnetic induction.

_____ _____   **6.** A magnet is defined as an object that _____ magnetic materials such as iron or steel.

_____ _____   **7.** A natural magnet is called a _____ or _____.

_____ _____   **8.** The magnetic lines of force between the poles of a magnet are called a _____ field.

_____ _____   **9.** When electric energy is transferred by means of a magnetic field, it is known as _____ induction.

_____ _____   **10.** All electric power supplied by generators or batteries may be _____ or _____ current.

Generator Principles

The conversion of mechanical to electrical energy occurs in the generator by the rotation of a magnetic field that intersects the windings and induces a voltage. In generators, the growth and collapse of the magnetic field is accomplished by physically moving or revolving the fixed field, called the primary winding, past the conductors (secondary winding). Naturally, the stronger the field, the higher the voltage and the weaker the field, the lower the generated voltage. The contents of this chapter deal with the basic operation of generators.

BASIC OPERATION OF GENERATORS

The basic AC generator consists of a loop of wire that is free to rotate in a magnetic field. The loop of wire is called the armature, and the magnetic field is called the field. The armature is turned by an element called the prime mover. The prime mover can be water, steam, or wind turbines, an engine, or an electric motor depending on the application and use.

Note that the terms *armature* and *field* are electrical terms. Electrically, the armature windings are those windings that are connected to the load. The field windings are those windings that are used to create the magnetic field. The rotor always rotates, and the stator is always stationary.

The armature loop is connected to slip rings. Such slip rings have an electrical conducting brush that slips over the surface of the ring as the armature rotates through the field. **(See Figure 2-1)**

The armature of an AC generator making a complete turn through the magnetic field.

• When the armature reaches position 2. The armature (loop of wire) is moving perpendicular to the magnetic field; therefore, they are cutting the maximum number of lines per second.

• As the loop rotates past position 2, the voltage drops off since the loops are not perpendicular to the magnetic field, therefore cutting few lines of flux.

• As the armature reaches position 3, its motion is again parallel to the field and the output voltage is once more zero. The same as position one.

• As the armature rotates from position 3 to 4, the voltage again reaches a maximum value.

• When the armature completes its turn past position 4, the voltage drops to zero again. No lines of flux cut.

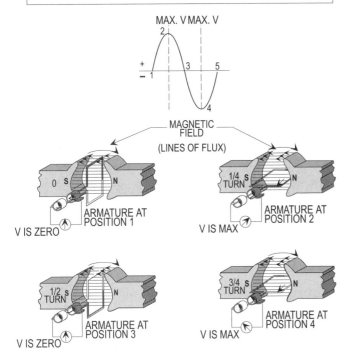

THE ARMATURE (LOOP OF WIRE) ROTATING
THROUGH THE MAGNETIC FIELD.

Figure 2-1. The above illustrates an armature rotating through the flux of a magnetic field and completing a full turn with an entire output of voltage.

As the armature rotates in the field, a voltage is generated that can be utilized to supply a transformer and a switchgear that can be used to step the voltage up or down. Loads are then supplied by the switchgear voltage or the transformer voltage. **(See Figure 2-2)**

Note that AC generators are usually referred to as alternators. Alternators generate most of the electrical power used in modern-day electrical systems.

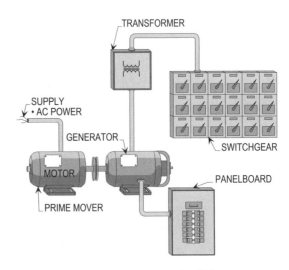

**GENERATOR SUPPLYING
TRANSFORMER OR PANELBOARD**

Figure 2-2. The above illustrates a generator driven by a motor and such generator is supplying a transformer and panelboard. **Note:** The transformer can be used to step up or step down the voltage to serve the switchgear.

BASIC OPERATION OF DC GENERATORS

By replacing the slip rings on a basic AC generator with two semicylindrical segments called a commutator and connecting two stationary brushes on opposite sides of the commentator, a basic DC generator is obtained. The brushes are so mounted that each brush contacts each segment of the commutator, which revolve simultaneously with the loop.

Note that the rotating parts of a DC generator, the coil and two piece commutator, are called an armature.

The switching action of the commutator segments makes the output of the DC generator produce direct current with no part of the output current going in reverse direction, as would occur in an AC generator. At the instant each brush contacts two segments of the commutator, a direct short-circuit is produced. If an EMF were generated, a high current would flow in the short circuit, which would cause an arc and thus could damage the commutator. To prevent this from happening, the brushes must be placed in the exact position where the short will occur when the generated EMF is zero. In a DC generator, this position is called the neutral plane. **(See Figure 2-3)**

BRUSHES

Brushes ride on the surface of the commutator and form the electrical contact between the armature coil and the external circuit. Brushes are made of high grade carbon and are held in place by brush holders. The brushes are insulated from the frame and are free to slide up and down in their holders so that they can follow the surface of the commutator. The pressure of the brushes may be varied, and their position on the commutator as well may be adjusted for neutral plane position.

The loop of a DC generator making one revolution through the magnetic field

- When the loop is in position 1, (0°) the voltage is zero. No lines of flux cut.

- After the loop has rotated 90° from position 1 and is passing through position 2, the black coil side is moving downward and the white side is moving upward. Both sides are cutting a maximum number of flux lines and the voltage is maximum.

- As the loop moves through an angle of 180° in position 3, the coil sides are again cutting no flux lines. The generated EMF or voltage is zero.

- As the loop moves through an angle of 270°, as in position 4, the coil sides are cutting a maximum number of flux lines. The generated voltage is at a negative maximum.

- The next 90° turn of the loop completes a 360° revolution and the generated voltage falls to zero.

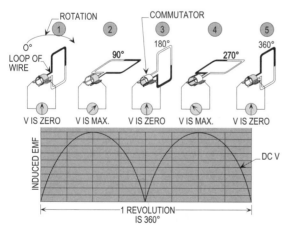

THE LOOP WIRE OF A DC GENERATOR PASSING THROUGH THE FLUX FIELD

Figure 2-3. The above illustrates a loop of wire rotating through the flux field of a magnetic field and completing one revolution of 360 degrees.

COMMUTATOR

A commutator is a mechanical rectifier that is nothing more than a slip ring split into segments. The ends of the rotating armature coil are attached to each segment of the commutator.

Figure 2-3 shows the commutator cutting through the magnetic field with current flowing in one direction toward the commutator. Note that current is flowing from the unshaded side of the armature to the shaded side and the cross-hatched brush is touching the shaded section of the commutator while the other brush is touching the unshaded section of the commutator.

By this operation, current flows out of the exciter armature through the cross-hatched brush to the main generator field winding and returns through the other brush to complete the circuit.

As the armature continues to turn through 180 degrees, current flows in the opposite direction in the armature. The current flow is now flowing from the shaded section of the armature to the unshaded side and the unshaded section of the commutator is now touching the cross-hatched brush. It is by this action of the commutator that DC voltage/current is produced.

COMMUTATION

As an armature revolves in a DC generator, the armature coil cuts through the magnetic lines of force (magnetic flux) and a voltage is induced in them that appears at the brushes. As the commutator segments (to which the coils are connected) pass the brushes, current is drawn from the segments. This is due to voltage being induced when the coil passes the field poles. Coils that are in the interpole spaces are shorted momentarily, and the connection to the coils are reversed to allow DC current flow. **(See Figure 2-4)**

ARMATURE REACTION

There is an EMF generated in a moving armature that opposes the magnetic field used to produce the electrical output. The neutral plane of the armature is perpendicular to the lines of force, or flux field, when there is no current in the armature.

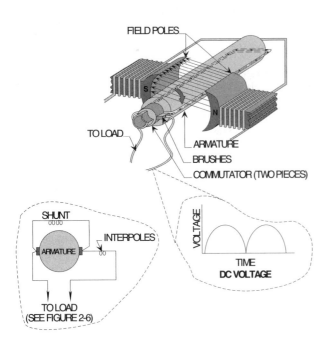

Purpose Of Armature

• To produce electricity, the armature must be mounted between *field coils,* so that the magnetic force (flux) generated by the electromagnet will be cut by the rotating armature.

Purpose Of Brushes

• *Brushes* make sliding contact so that they may contact the commutator and carry generated electricity to the load.

Purpose Of Commutator

• The commutator acts as a reversing switch as the armature rotates in the different fields.

Purpose Of Switching Action

• As a result of the switching action, the current output is a series of maximums and minimums with current flowing in only *one direction.*

Figure 2-4. The above illustrates the main purpose and use of the armature brushes, and commutator, and their relationships in a generator to produce DC voltage.

The effect of armature reaction can be minimized or overcome by shifting the brush assembly as follows:

    (1) By using chamfered poles
    (2) By using commutating poles
    (3) By using pole face windings
    (4) By any combinations of the above

Basically, the procedure for eliminating such shift in the neutral plane is actually nullifying the change.

For example, the entire brush assembly can be adjusted to bring the brushes in line with the shifted neutral plane. Because the neutral plane shifts with the load, this means shifting the brushes every time the load changes. Note that this method is not practical.

A more practical method is that the poles be slightly chamfered. In other words, the radial distances between the pole face and the armature is increased slightly at the edges of the poles.

The effect of this procedure produces an increase in the air gap at the edges of the poles, which offsets to some extent the tendency of the field to shift due to the armature reaction. **(See Figure 2-5)**

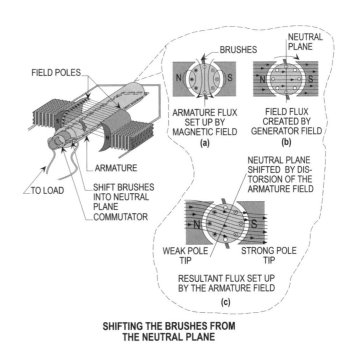

SHIFTING THE BRUSHES FROM THE NEUTRAL PLANE

Figure 2-5. The above illustrates the problems of armature reaction by shifting the position of the brushes so that they are in the neutral plane when the generator is producing its normal load current; the generator operates properly under a fairly constant load.

Another method is to place the commutating poles in the interpolar spaces; such poles are smaller and narrower than the main field poles, and their winding is in series with the armature. They are so connected that their field opposes the field created by the armature reaction.

The final method is to design the faces of the main field pole so they are slotted longitudinally with the windings placed in the slots. These windings are then connected so that their field opposes the field created by the armature.

See Figure 2-6 for a detailed illustration of using interpoles and windings to correct armature reaction.

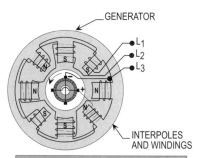

The Effect Of Interpole Windings

- The windings of the interpoles are in series with the load, therefore, the effect of the interpole is proportional to the load.

- The polarity of the interpoles is such that their effect is opposite to that of the armature field.

- Basically, each interpole is of the same polarity as the next field pole in the direction of the rotation.

- With the aid of such polarity, the interpole pulls the generator field into the correct position.

INTERPOLE WINDINGS USED TO CORRECT ARMATURE REACTION

Figure 2-6. The above illustrates the use of interpoles and windings to correct the problems due to the armature reaction of generators.

GENERATED VOLTAGE

The voltage output of generators can be generated for single-phase or three-phase use. The frequency of the generated power is directly related to the speed of the generator, which in turn is directly related to the prime mover speed.

The voltage output will have a sine-wave pattern in the field poles that turn at a constant speed. The sinusoidal voltage is sinusoid because of the field flux that intersects the windings produces a voltage that grows and collapses with each rotation of the field poles. (See **Figure 2-7**.)

SINGLE-PHASE OUTPUT

Single-phase output is obtained by having one set of armature windings in the stator. A two-pole, single-phase generator consists of a North pole and a South pole and conductors that are part of a continuous armature conductor (winding) that fills the slots in the stator.

Note that the stator slots are separated mechanically and electrically by 180 degrees. When the flux from the North pole intersects the A (1) side of the conductor in Figure 2-7, the flux returning to the South pole intersects the A (2) side of the conductor, resulting in generation of a peak voltage between A (1) and A (2). When the North and South poles are perpendicular to the plane of the A (1) and A (2) conductors, no lines of force are intersecting the conductors and the voltage difference between A (1) and A (2) is zero. One complete revolution of the rotor through 360 degrees is considered one cycle.

See Figure 2-7 for a detailed illustration of a two-pole generator producing a single-phase output.

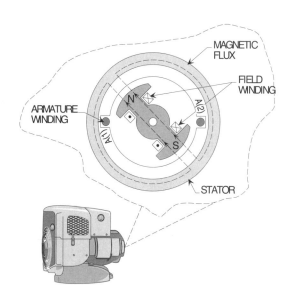

Figure 2-7. The above illustrates a single-phase generator that is used to supply single-phase power of 120/240 volts.

THREE-PHASE OUTPUT

Three-phase output can be produced in a rotating field having two or four poles as shown in Figure 2-8. As illustrated, the rotating field is equipped with one North and one South pole. Note that there are three sets of conductors, A (1) and A (2), B (1) and B (2), and C (1) and C (2). Each set of conductors is located 120 degrees apart, with each group of conductors generating a single-phase voltage. Since the groups are spaced 120 degrees apart, the single-phase voltage of each group is electrically spaced 120 degrees from the other two. The total output of the three single-phase voltages produce a three-phase output.

A four pole generator requires two North poles and two South poles on the rotor with a three-group set of conductors on the stator.

See Figure 2-8 for a detailed illustration of a two-pole and four-pole generator producing a three-phase output.

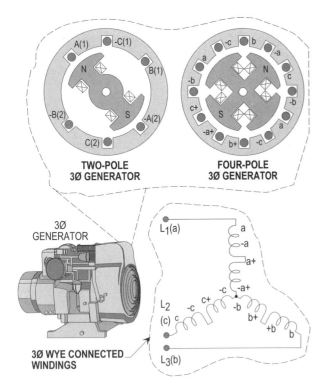

TWO- AND FOUR-POLE 3ϕ GENERATORS

Figure 2-8. The above illustrates two- and four-pole generators with the windings of the four-pole generator connected in a wye configuration.

THE DIFFERENCE BETWEEN A GENERATOR AND MOTOR

In comparing a generator to a motor, there are similarities that must be pointed out and discussed. For example, a generator is a machine that converts the mechanical power of a prime mover into electrical energy, expressed in kilowatts (kW). In comparison, a motor is a machine that converts electrical energy into mechanical energy and delivers this energy in the form of horsepower to the shaft of a driven load.

TYPICAL SYNCHRONOUS GENERATOR

A typical synchronous generator consists of field windings supplied by DC voltage that are mounted on a rotor and rotated inside of a stationary winding called the armature. The generator shaft is turned by a mechanical prime mover. As the generator shaft turns, the magnetic field is rotated, causing flux to intersect the armature winding and induce an electromotive force (EMF). The rotation of the field causes the induced EMF to increase and decrease, which produces a voltage at the terminals of the armature winding. By connecting the terminals of the armature winding to an electrical load, an alternating current will flow. **(See Figure 2-9)**

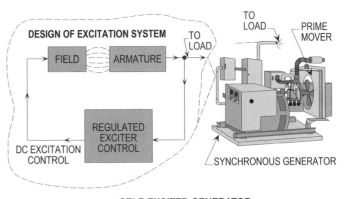

SELF-EXCITED GENERATOR

Figure 2-9. The above illustration shows an excitation diagram and synchronous generator.

TYPICAL SYNCHRONOUS MOTOR

A typical synchronous motor consists of the same elements as the generator. For example, a rotating magnetic field stator is mounted inside of a stationary armature winding with an additional motor that has an induction winding that is used for starting and it is mounted on the surface of the rotor. During motor start-up, no DC current is applied to the field winding rotor; instead, an alternating current (AC) is supplied to the terminals of the armature winding, which creates a magnetic field in the winding. Because this field is served by AC, it travels around the armature winding at the same frequency as the supplying current. The rotating armature field induces a current in the winding on the surface of the rotor, which develops a torque that causes the rotor to turn and the motor to start as an induction motor. When the speed is near the synchronous speed of the motor, DC current is then applied to the rotating field and the motor is brought up to synchronous speed. **(See Page 15-14.)**

GENERATOR EXCITERS

The value of the AC voltage generated by a synchronous machine is controlled by varying the current in the DC field windings, while frequency is controlled by the speed of rotation.

Power input is controlled by the torque applied to the generator shaft by the driving engine. It is by this procedure that the synchronous generator controls the power in which it generates.

Synchronous generators normally use a brushless exciter, which is nothing more than a small AC generator mounted on the main shaft. The AC voltage generated is rectified by a three-phase rotating rectifier assembly, also on the shaft.

This DC voltage is applied to the main generator field, which is also mounted on the main shaft. A voltage regulator controls the exciter field current, which controls the field voltage. In this manner, a well controlled generator output can be obtained. **(See Figure 2-10)**

DC EXCITERS

This type of exciter operates on the principle of an AC voltage being induced in a coil that is rotating in a magnetic field. A commutator added to the output connection makes this device a DC generator. The process by which AC voltage is induced and by which the AC voltage is then converted to DC voltage is called rectification. **(See Figure 2-10)**

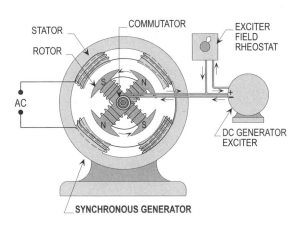

Figure 2-10. The above illustrates a synchronous generator with a DC exciter.

IN COMPARISON

In selecting a generator for various applications, synchronous generators are normally applied instead of induction generators because induction generators operate at a fixed power factor. As a result of the characteristic, induction generators must always operate in parallel with synchronous machines or capacitors to correct power factor. However, a synchronous generator is capable of correcting power factor while delivering a constant frequency with an over adjustment of the field current and power.

TYPICAL INDUCTION GENERATOR

A typical induction generator is essentially the same as that of an induction motor in that they both have a squirrel-cage rotor and wound stator. When this machine is driven above its designed synchronous speed, it becomes a generator. When operated at less than synchronous speed, it functions as a motor. Because induction generators do not have an exciter, they must operate in parallel with the utility. This outside power source provides the reactive power for generator operation. Note that its frequency is automatically locked in with the utilities.

An induction generator is also a popular choice for use when designing and installing cogeneration systems, which operate in parallel with the utility.

ADVANTAGES

This type of induction generator offers several advantages over a synchronous generator, and they are as follows:

(1) Voltage and frequency are controlled by utilities.

(2) Regulators are not required.

(3) Construction of generator allows high reliability and requires little maintenance.

(4) Only a minimum of protective relays and controls are necessary.

DISADVANTAGES

The major disadvantage of an induction generator is that it is difficult to operate alone as a standby or emergency generator. **(See Figure 2-11)**

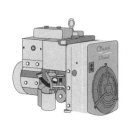

Advantages Of An Induction Generator
• Utilities control voltage and frequency.
• Regulators are not needed.
• Provides high reliability and needs very little maintenance.
• Relays and controls are easy to service and maintain.
Disadvantages Of An Induction Generator
• It is hard to use as an emergency or standby unit.

ADVANTAGES OF AN INDUCTION GENERATOR

Figure 2-11. The above lists the advantages and disadvantages of an induction generator.

TYPES OF ENGINES

Fuel availability determines the type of the engine-generator set to be used. If a certain type fuel is already in use at the site, a generator set using the same fuel is usually selected and utilized.

If the generator set is located in an area where public utilities are not available, LPG or diesel fuel are normally the types of fuel sources used.

GASOLINE ENGINES

Gasoline engines are economical up to about 100 kW. Initial cost are comparatively low, and they have reliable starting ability.

DIESEL ENGINES

Diesel engines are very popular due to their reliability, ruggedness, low-maintenance, economical operation, and the low initial cost for larger units. For industrial and commercial applications, diesel engines are built in sizes up to about 2000 kW. For the needs of a prime power installation, they come in sizes up to 20,000 HP or greater.

GASEOUS FUEL ENGINES

Gaseous fuel engines are comparable to diesel engines, except that the normal gas supply is subject to interruption in the event the supply line is damaged. To compensate for this problem, an on-site propane gas tank is usually used to provide an alternate fuel supply should the normal supply be lost.

GAS TURBINE ENGINES

Gas turbine engine sets have had success as on-site power sources for heavy loads ranging from about 500 kW or greater. They are small in weight, and due to their lack of vibration, they can be installed on floors or on roofs.

They have the ability to burn a wide variety of fuels, either liquid or gas. If needed, they come as dual-fuel sets, such as natural gas and diesel, or natural gas and liquid petroleum gas.

See Figure 2-12 for a detailed illustration of the various kinds of engine-generator sets.

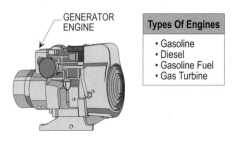

TYPES OF ENGINES THAT ARE MOST USED

Figure 2-12. The above illustrates the types of engines that are most used, based upon the types of fuel used.

For the same reason above, synchronous motors are selected over induction motors for applications where constant speed is necessary. Induction motors will decrease in speed as a mechanical load is applied, and this decrease in speed causes a decrease of counter electromotive force, which allows more current to be supplied by the source. Because of a separate DC supply, a synchronous motor always runs at synchronous speed, even if the load is increased.

SINGLE-PHASE GENERATORS

A single-phase generator consists of a rotating magnet called the field, which is inside a stationary winding called an armature. The rotating magnet is generally an electromagnet that is wound on a cylindrically shaped shaft called the rotor. The rotor is elongated on one end, and to this end a coupling is attached to connect the generator to a prime mover. The stator core is contained in the generator frame, and the bearings are mounted on end plates, which are called bearing brackets. The armature winding exits the generator frame through insulated terminals known as bushings. These terminals are attached to the generator frame in a compartment called a terminal box. The power leads from the load are connected in the box to the terminals loads of the armature. **(See Figure 2-13)**

THREE-PHASE GENERATORS

A three-phase generator has three separate windings, which are placed in the slots of the stator core. The windings are arranged so that three voltages are produced that are 120 electrical degrees apart. In a two-pole generator, each phase winding is divided into two parallel groups, and in a four-pole generator, each phase is divided into four parallel phase groups. These phase groups are connected to the main and neutral leads by parallel rings. These rings are located at the stator winding and positioned at the collector end for proper continuity. **(See Figure 2-14)**

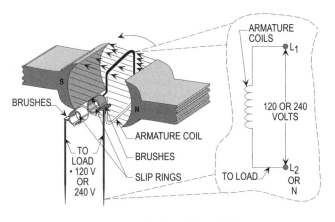

SINGLE-PHASE GENERATOR

Figure 2-13. The above illustrates a single-phase generator supplying either 120 volts or 240 volts to specific loads.

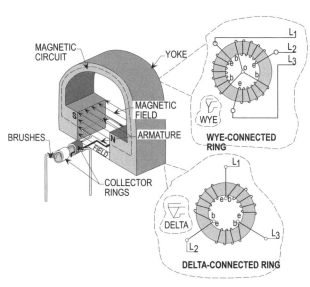

TWO OR FOUR-POLE GENERATORS CAN BE CONNECTED WYE OR DELTA

Figure 2-14. Windings are shown connected in wye and delta configurations on rings to demonstrate how easily they can be connected.

OPERATION

Three-phase generators are much more efficient than comparably sized single-phase generators. For example, as the magnetic field rotates across the armature winding, an electromotive force (EMF) is induced in the armature winding. The rotation of the field causes this induced EMF to increase and decrease at the terminals of the armature winding. As the rotor spins, three sets of AC voltages are generated in the stator windings as the rotor turns through the magnetic field. These voltages are equal in amplitude, but they are shifted in phase by 120 electrical degrees from each other. When the terminals of the armature winding are connected to complete a circuit through a load, such as the primary winding of a three-phase transformer, an alternating current will flow and supply the load.

COMPONENTS OF AN AC GENERATOR

The following components are considered major elements of an AC generator that electrical personnel must understand:

(1) Stator,
(2) Rotor,
(3) Cooling system,
(4) Exciters, and
(5) Commutator.

STATOR

The following items are the most pertinent elements to consider about the stator components of an AC generator:

(1) Mechanical components, bearings, shafts, etc.
(2) Wye-connected, and
(3) Delta-connected windings.

MECHANICAL COMPONENTS

The most important mechanical components of an AC generator are as follows:

(1) The frame,
(2) The core,
(3) Termination box, and
(4) The winding (coils).

FRAME

The stator frame, called the housing, is fabricated from steel plates and bars electrically welded into a rigid box section. A short piece of duct work is provided on the bottom of the frame through which ventilating air can be discharged. Holes drilled and tapped around the edges of the duct provide a means for attaching the necessary duct work. Port holes with a removable glass serve as windows for the inspection of the end windings during operation.

CORE

The stator core is basically built up of low-loss segmental silicon steel laminations and assembled on bars that span the entire length of the core. Both sides of the laminations are treated with an insulating material to prevent short-circuiting the laminations. Vent spacers are built in with the laminations at intervals to provide radial passages through the core for the ventilating air.

Adequate pressure is applied at intervals during the stacking operation to produce a tight core. Heavy end plates and nonmagnetic finger plates are used at the ends of the core to maintain adequate pressure at all times.

TERMINATION BOX

This component of the generator frame is located either on the bottom or the top of the frame. Contained in the termination box are the six lead bushings. These bushings are designed to serve two purposes. They provide a gas-tight penetration in the generator frame for the three line leads and the three neutral leads that make up both ends of the three phases of the stator winding. Secondly, such bushings are used to insulate the high-voltage leads from the generator frame, which is at ground potential. (**See Figure 2-15**)

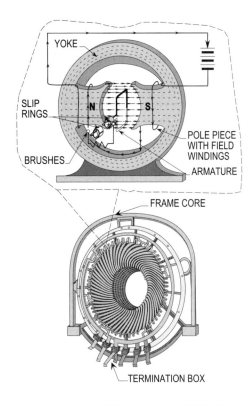

MAJOR ELEMENTS OF AN AC GENERATOR

Figure 2-15. The above illustration shows the main components that are necessary to operate an AC generator.

WYE-CONNECTED SYSTEMS

Generator stator windings are typically connected in a wye configuration. Line leads numbered T_1, T_2, and T_3 are the line leads connected to the hot conductors, and T_4, T_5, and T_6 are the neutral leads tied together and usually connected to ground. (**See Figure 2-16**)

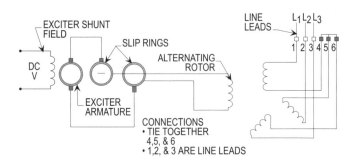

WYE CONNECTED GENERATOR

Figure 2-16. The above illustrates a generator that has its windings connected in a wye configuration.

DELTA-CONNECTED SYSTEMS

The delta connection is made by connecting terminals 1 to 6, 2 to 4, and 3 to 5. Line leads are then connected to terminals 1, 2, and 3 accordingly. The generator windings when connected in a delta configuration reduce the line-to-line voltage. However, the available line current is increased. (**See Figure 2-17**)

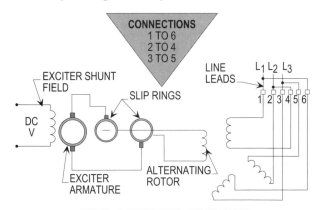

DELTA-CONNECTED GENERATOR

Figure 2-17. The above illustrates a generator that has its windings connected in a delta configuration.

TYPES OF ROTORS

To impose the magnetic field on the rotor, poles are utilized that consist of stacked magnetic iron laminations (to reduce eddy currents) with copper conductors wrapped around the iron. These poles are excited by a DC current. Poles built in this manner reduce the problem created by eddy currents. The rotor poles must be arranged in pairs with a minimum arrangement of one pair of poles. The pairs are located 180 electrical degrees apart. As the North pole of the magnetic field of the rotor intersects one phase group of the stator winding, the South pole of the rotor intersecting the diametrically opposite portion of the same phase winding. The construction of the rotor is primarily determined by the speed of operation.

To accomplish this purpose, two basic rotor constructions for generators are available. They are salient poles, which have projecting poles with concentrated windings and cylindrical poles, which are equipped with distributed windings.

The rotor selected is based on the characteristics of the generator. For example, large, low-speed generators usually are designed with salient pole rotors. However, generators operating at 1800 rpm (4/pole) or 3600 rpm (2/pole) use cylindrical rotors.

SALIENT POLE ROTORS

Salient pole rotors can be built with either laminated poles or solid poles. The constructions of such poles are as follows:

LAMINATED POLES

This type of construction is more efficient because the magnetic lines of flux travel through the laminated core in a perpendicular direction to the field winding. This reduces iron losses and provides a more efficient magnetic coupling to the laminated core of the stator winding.

SOLID POLES

This type of construction is utilized where the rating of the generator requires less concentrated flux density.

CYLINDRICAL POLES

This type of construction is utilized in high speed generators. These rotors generally carry higher field ratings and are more rigid. The surface of these rotor bodies is grooved to reduce surface currents and to increase heat transfer to the cooling medium of air or hydrogen. Radial slots for the field windings are machined in the rotor body, and the field coils are imbedded in slots. (**See Figures 2-18(a) and (b)** for the types of rotors)

COOLING SYSTEMS

The types of cooling systems normally used for AC generators are air-cooled, air-to-water heat exchanger, and gas-to-water heat exchanger. Each type is explained as follows:

AIR-COOLED

A natural, air-cooled generator uses outside air, at ambient temperature, as a cooling medium. Such air is circulated through the stator and rotor by propeller-type blowers on both ends of the rotor. The warm air exiting the stator and rotor is exhausted back outside the generator in a complete cycle. In other words, air passes through only one time. (**See Figure 2-19**)

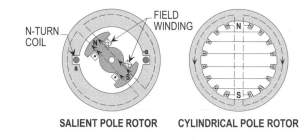

TYPES OF ROTORS

SALIENT POLE ROTOR   CYLINDRICAL POLE ROTOR

Figure 2-18(a). The above illustrates the difference between a salient pole rotor and cylindrical pole rotor.

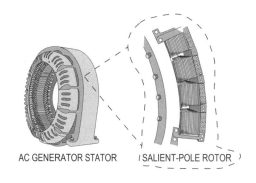

AC GENERATOR STATOR   SALIENT-POLE ROTOR

AC GENERATOR WITH A SALIENT POLE ROTOR

Figure 2-18(b). The above is a detailed illustration of an AC generator with a salient pole rotor.

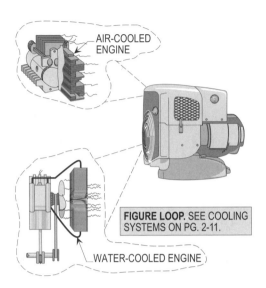

TWO TYPES OF COOLING ENGINES

Figure 2-19. The above shows a generator engine cooled by either air or water.

Generator Tip: This type of cooling system sometimes requires air filters on the intake to minimize the contaminants that can get into the generator. Such filters can get dirty and can thus restrict air flow. To prevent this problem, proper maintenance must be provided.

AIR-TO-AIR HEAT EXCHANGER

A generator with an air-to-air heat exchanger is different from the natural cooled type due to the heat exchanger constantly recirculating the same air through the stator and rotor. Note that such circulation keeps the generator windings cleaner than a system that does not recirculate the same air.

Generator Tip: This type of air circulation eliminates the need for a filter system, but does require additional secondary air-cooling equipment.

AIR-TO-WATER HEAT EXCHANGER

A generator with an air-to-water heat exchanger is different from the air-to-air heat exchanger type. The warmer air coming out of the stator and rotor is circulated across a cooler that consists of a number of copper tubes with circular fins around the outside diameter of the tubes. Water circulated through the tubes removes the heat from the air being passed over the outside of the tubes. Note that a source of cooling water must be pumped through these coolers.

Generator Tip: This type of air circulation reduces the problem of contaminants getting into the generator because the same air is being recirculated through the generator constantly.

GAS-TO-WATER HEAT EXCHANGER

This system has many advantages over other types because it uses hydrogen as a cooling means. For example, hydrogen has lower density and better thermal conductivity to reduce windage loss as well as increasing heat transfer output per unit volume.

Generator Tip: The benefit of having a closed gas system is that it reduces dirt and moisture combination in the machine and achieves quieter operation.

OTHER EXCITERS

The commutator illustrated in Figure 2-3 requires a great deal of maintenance and has been replaced primarily by the following devices:

STATIC EXCITATION

Static excitation uses power from the main generator output, which is fed back to the voltage regulator through an excitation transformer to produce the DC field current for the field windings on the generator rotor. The direct current is then connected to the field windings of the generator's rotor through collector rings. A closed-loop feedback circuit allows the voltage regulator to monitor and regulate the output of the generator.

BRUSHLESS EXCITATION

The brushless excitation method eliminates the inherent inefficiencies of slip rings and brushes. The exciter in the brushless excitation system consists of an alternating current generator with a rotating armature and a stationary magnetic field. The alternating current generated in the rotating armature is converted to direct current by a rectifier, which is mounted on the same shaft as the armature. Note that the brushless exciter uses a permanent magnet generator as a pilot exciter to supply power to the voltage regulator. **(See Figure 2-20)**

REGULATOR

The function of a regulator is to use the feedback signals from the voltage potential transformers (PTS) and current transformers (CTS - connected in series) to keep the generator voltage at the desired usable level. Its function also protects from sudden load swings or voltage spikes by tripping and taking the generator off the line if it becomes necessary. **(See Figure 2-21)**

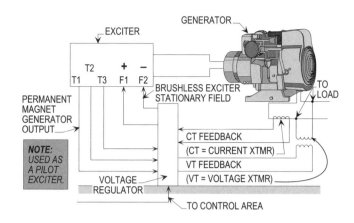

REGULATOR WITH BRUSHLESS EXCITATION

Figure 2-20. The above illustrates a generator with a voltage regulator and a brushless excitation system.

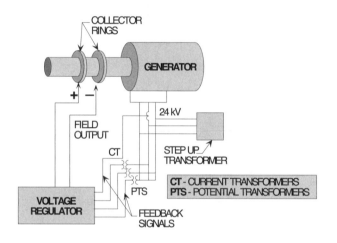

VOLTAGE REGULATOR

Figure 2-21. The above illustrates a voltage regulator receiving signals and using such signals to regulate the desired voltage level.

MANUAL SYNCHRONIZATION OF GENERATORS

A simple device capable of monitoring instantaneous voltage between generators and recognizing a voltage difference is a neon lamp. A common circuit arrangement that uses three neon lamps to monitor and indicate voltage differences between two three-phase generators is often used.

The circuit in generator 1 is supplying current to the load and generator 2 is about to be synchronized and connected in parallel with generator 1. The three neon lamps are con-

nected across the open contacts of the tie breaker to monitor and indicate voltage differences between the respective phases for the two generators. When the lamps indicate that the generators are matched or synchronized, the tie breaker is closed and the generators are connected in parallel.

When the neon lamps are continuously dark, the generators are usually in exact and continuous synchronization. **(See Figure 2-22)**

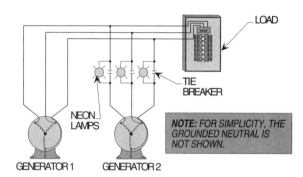

Figure 2-22. The above shows the procedure for manually synchronizing two generators supplying a load.

For installations where the voltages of the generators are greater than 480 volts, step-down potential transformers must be used to reduce the voltage to the neon lamps to a safe level. Note that when transformers are utilized for this purpose, care must be exercised to properly connect the secondary leads of the potential transformers before energizing the generators.

AUTOMATIC SYNCHRONIZATION OF GENERATORS

The difference between automatic and manual synchronization is that automatic synchronization uses some form of logic circuitry to automatically monitor, adjust and connect the generators. The logic circuitry automatically controls the generator to speed it up or to slow it down and to regulate the generator field current to increase or decrease the voltage. Note that a special interlock circuit is used for the tie circuit breaker to prevent it from closing until synchronization has been established.

Automatic synchronization is mainly used for permanent installations where the phase rotation of the generator's lines are known to be correct. Just like (similar) the manual process of synchronization, the automatic process controls adjust the parameters of the generator that is to be connected in parallel with the generator that is already on-line. **(See Figure 2-23)**

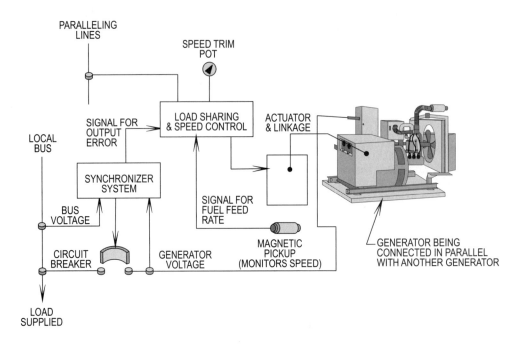

Figure 2-23. The above illustrates the procedure for automatic synchronization of two generators supplying a load.

Chapter 2: Generator Principles

Section	Answer	

_____   T   F   **1.** The basic AC generator consists of a loop of wire that is free to rotate in a magnetic field.

_____   T   F   **2.** Brushes ride on the surface of the commutator and form the electrical contact between the armature coil and the external circuit.

_____   T   F   **3.** The output of generators generate three-phase voltage only.

_____   T   F   **4.** The three-phase output of a generator is obtained by having one set of armature windings in the stator.

_____   T   F   **5.** A four-pole, three-phase generator requires two North poles and two South poles on the rotor with a three-phase set of conductors on the stator.

_____   T   F   **6.** An induction generator always operates in series with the utility.

_____   T   F   **7.** Fuel availability determines the type of engine-generator set to be used.

_____   T   F   **8.** A three-phase generator has three separate windings which are placed in the slots of the stator core.

_____   T   F   **9.** The generator windings, when connected in a delta configuration, increase the line-to-line voltage.

_____   T   F   **10.** DC exciters operate on the principle of a DC voltage being induced in a coil that is rotating in a magnetic field.

_____ _____   **11.** The basic AC generator consists of a loop of wire called the _____.

_____ _____   **12.** The commutator acts as a _____ switch as the armature rotates in the different fields.

_____ _____   **13.** The windings of the interpoles are in _____ with the load, therefore, the effect of the interpole is proportional to the load.

_____ _____   **14.** A single-phase output is obtained by having one set of armature windings in the _____.

_____ _____   **15.** A three-phase output can be produced in a rotating field having _____ or _____ poles.

_____ _____   **16.** A generator converts mechanical power of a prime mover into electrical energy measured in _____.

_____ _____   **17.** A motor converts electrical energy into mechanical energy and delivers this output, which is measured in _____.

_____ _____   **18.** Synchronous generators usually use a _____ exciter.

_____ _____   **19.** Utilities control the _____ and _____ when using an induction generator.

_____ _____   **20.** A single-phase generator consists of a rotating magnet called the _____.

_____ _____   **21.** As the armature rotates in the field of a basic generator, a voltage is generated that can be utilized to supply a transformer for:

    (a) Stepping down the voltage   **(c)** All of the above

    (b) Stepping up the voltage   **(d)** None of the above

_____ _____   **22.** The effect of armature reaction can be minimized or overcome by shifting the brush assembly and by:

    (a) Using chamfered poles   **(c)** Using pole face windings

    (b) Using commutating poles   **(d)** All of the above

_____ _____   **23.** A typical synchronous generator consists of a magnetic field that is mounted and rotated inside of a stationary winding called the:

    (a) Armature   **(c)** Stator

    (b) Commutator   **(d)** Rotor

_____ _____   **24.** For industrial and commercial applications, diesel engines are built in sizes up to about:

    (a) 1000 kW   **(c)** 3000 kW

    (b) 2000 kW   **(d)** 5000 kW

_____ _____   **25.** Gas turbine engines have had success as on-site power sources for heavy loads ranging from about:

    (a) 500 kW and greater   **(c)** 1000 kW and greater

    (b) 750 kW and greater   **(d)** 2000 kW and greater

26. Generator stator windings are typically connected in a:

    (a) Delta configuration   **(c)** Delta-wye configuration

    (b) Wye configuration   **(d)** Wye-delta configuration

27. Salient pole rotors can be built with which type of poles?

    (a) Laminated poles   **(c)** None of the above

    (b) Cylindrical poles   **(d)** All of the above

28. Which type of cooling system is used for AC generators?

    (a) Air-cooled   **(c)** Gas-to-water heat exchanger

    (b) Air-to-water heat exchanger   **(d)** All of the above

29. AC voltage induced in a coil that is rotating in a magnetic field is accomplished by using which type of exciter?

    (a) DC exciters   **(c)** Brushless excitation

    (b) Static excitation   **(d)** Regulators

30. Inherent inefficiencies of slip rings and brushes are eliminated by using:

    (a) DC exciters   **(c)** Brushless excitation

    (b) Static excitation   **(d)** Regulators

Generators and The NEC

In addition to the requirements of **Article 445**, generators must comply with the requirements of other sections of the NEC, most notably are **Articles 215, 230, 250, 700, 701, 702, and 705**.

Articles 215 and 230 deal with generators when they are used to supply service equipment and feeder circuits. **Article 250** addresses the special grounding techniques based upon where the generator is installed and used. **Article 700** contains the rules for generators that are utilized when supplying power to emergency systems. **Articles 701 and 702** pertain to generators that serve legally required and optional standby systems. **Article 705** is used when generators are connected in parallel with the utility power sources to serve as an interconnected electric power production source.

LOCATION OF GENERATORS
445.10

One of the first requirements is that the generator be suitable for the location where it is installed. Basically, standard-type generators are designed to operate indoors in dry places. The requirements of **430.14** must be met to help protect the operation of generators. If generators are installed in hazardous locations or used to supply special equipment, the requirements of **Articles 500 through 503, 505, 510 through 517, 520, 525, 530, 665, and 695** must also be complied with.

NAMEPLATE MARKINGS
445.11

To aid designers and electrical personnel, every generator must have a nameplate that contains the following information:

(1) The manufacturer's name and operating frequency,
(2) Number of phases and power factor,
(3) The subtransient and transient impedances,
(4) The rating in kW's and kWA's,
(5) Rating in kVA or kW, with the corresponding volts and amperes,
(6) The rated revolutions per minute,
(7) Insulation type, ambient temperature, and time rating.

Such information must be used when designing, installing, and maintaining generators in residential, commercial, and industrial applications.

OVERCURRENT PROTECTION FOR GENERATORS
445.12

Constant-voltage generators, except for AC generators and exciters, are protected from excessive current by circuit breakers or fuses or by inherent design. (AC generators are exempt from the need of overcurrent protection.) This is due to impedance, which limits the short-circuit current to a value that is not damaging to their windings. All generator exciters are usually separately excited. In most installations, DC as well as AC units are normally operated without overcurrent protection. **(See Figure 3-1)**

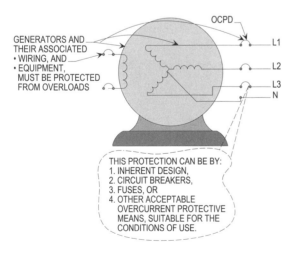

PROTECTION OF GENERATORS
NEC 445.12

Figure 3-1. Generators and their elements must be protected from overloads, short-circuits, and ground-faults.

CONSTANT-VOLTAGE GENERATORS
445.12(A)

The basic rule requires DC generators to have overcurrent protection. However, AC generators may be so designed that on a high overload the voltage of the generator falls off, thereby reducing the overload current to a safe value. For this reason, the NEC does not always require overload protection for all AC generators.

There are installations where overload protection can be omitted. In some cases, it is considered better to risk damage to the exciter rather than have the generator shut down through operation of an exciter overcurrent protection device. **(See Figure 3-2)**

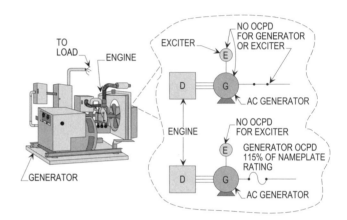

OVERCURRENT PROTECTION FOR GENERATORS
NEC 445.12(A)

Figure 3-2. The above illustrates when overcurrent protection is required for generators.

> **Generator Tip:** Generators that produce a constant voltage (most commonly used generators are of this type) are required to be protected from overloads. This may be accomplished by using overcurrent protective devices such as fuses, circuit breakers, etc. or by inherent design.

TWO-WIRE GENERATORS
445.12(B)

Two-wire DC generators are only required to have overcurrent protection in one wire, if the overcurrent protection device is activated by the entire current and not the current in the shunt coil. However, the overcurrent device must not, under any circumstances, open the shunt coil.

> **Generator Tip:** The NEC does not allow an overcurrent protection device in the generator's positive lead only, because an overcurrent device in the positive lead would

not always be actuated by the entire current generated. Note that an overcurrent protection device is not permitted for the shunt field. If the shunt field circuit were to open and the field was at full strength, a dangerously high voltage would be induced, that might damage the generator. (**See Figure 3-3**)

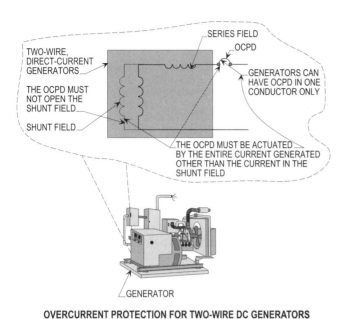

OVERCURRENT PROTECTION FOR TWO-WIRE DC GENERATORS
NEC 445.12(B)

Figure 3-3. The above illustrates overcurrent protection for two-wire DC generators.

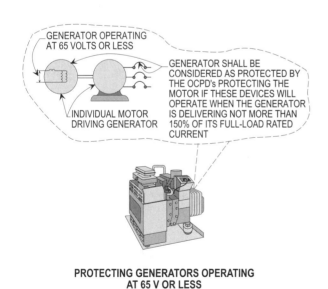

PROTECTING GENERATORS OPERATING
AT 65 V OR LESS
NEC 445.12(C)

Figure 3-4. The above illustrates the protection of generators operating at 65 volts or less.

GENERATORS OPERATING AT 65 VOLTS OR LESS
445.12(C)

A generator operating at 65 volts or less, and driven by an individual motor, is considered adequately protected by the motor overcurrent protection device, where such OCPD will open the circuit, if the unit is delivering not more than 150 percent of the generator's full-load current. (**See Figure 3-4**)

Generator Tip: If the fuse(s) or circuit breaker protecting the motor is set to operate when the generator is 50 percent or less overloaded, no protection is required in the generator leads. However, if the generator voltage is above 65 volts, it is the intent of the NEC to require separate overcurrent protection for the generator.

BALANCER SETS
445.12(D)

Balancer sets consist of two smaller DC generators used with a larger two-wire generator. The two balancer generators are connected in series across the two-wire main generator lines. A neutral tap is brought out from the midpoint connection between the two balancer generators. Note that each of the two balancer generators carries about one-half of any unbalanced load condition.

With such an arrangement, where there is a heavy unbalance in the load, the balancer generators may become overloaded, while there is no overload on the main generator. The balancer generators must be equipped with an overload device that will actuate the main generator disconnect if the balancer generators should become overloaded. (**See Figure 3-5**)

Generator Tip: Balancer sets must be equipped with overload devices that disconnects the three-wire system in case of an excessive unbalanced condition. Three-wire direct current generators must be provided with overcurrent protection devices, one in each armature lead arranged to disconnect the three-wire circuit in case of heavy overloads or extreme unbalanced current conditions.

THREE-WIRE, DC GENERATORS
445.12(E)

As in two-wire generators, the overcurrent protection device protecting a three-wire generator must be capable of taking the full generator current. When equalizer leads are provided, the overcurrent protection device, if not properly installed in the circuit, might take only a part of the generator current. To help solve this problem, three-wire DC generators operating in parallel are equipped with two equalizer leads. The overcurrent protection devices must be so placed in the circuit that they will take the full generator current without tripping open the circuit.

A two-pole breaker placed ahead of the junction of the main and equalizer leads will provide such protection. However, a four-pole circuit breaker with two poles for the main leads and two poles for the equalizer leads can also be used, provided such CB is actuated by the full current flow of the generator. **(See Figure 3-6)**

> **Generator Tip:** These generators, which are either shunt wound or compound wound, must be provided with overcurrent protection devices in each armature lead. Such devices must sense the entire armature current and be multi-pole devices that open all the poles in the event of an overcurrent condition.

EXCEPTION TO (A) THROUGH (E)
445.12(A) THROUGH (E), Ex.

There are cases where a generator fails and it is less of a hazard than disconnecting it when an overcurrent condition occurs. In such instances, the AHJ may allow the generator to be connected to a supervision panel with an annunciator or alarm, instead of requiring overcurrent protection.

AMPACITY OF CONDUCTORS FROM GENERATORS
445.13

Ungrounded phase conductors from a generator must be sized at no less than 115 percent of the nameplate current value. Neutrals can be computed and sized according to **220.61**. (**Table 310.16**, or the over 2000 volt Tables.) Phase-conductors must be capable of carrying ground-fault currents and are required to be sized in accordance with **250.24(C)**. (Also, see **310.15(B)(4)**)

For example, a generator with 100 amp output must have conductors with an ampacity of at least 115 amps respectively (100 A x 115% = 115 A). **(See Figure 3-7)**

PREVENTING OVERLOAD CONDITIONS
445.13, Ex.

The conductors can be protected at 100 percent of the rated generated current (100 amps). However, to do so, the design or operation of the generator must be such as to prevent overloading. Therefore, an ampacity of 100 percent loading is all that is permitted.

For example, a generator with 100 amp output must have conductors with an ampacity of at least 100 amps (100 A x 100% = 100 A) and the load limited to this value. **(See Figure 3-8)**

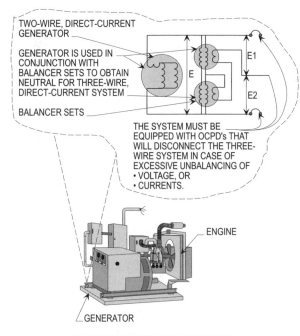

**GENERATOR USING BALANCER SETS
NEC 445.12(D)**

Figure 3-5. The above illustrates balancer sets used with a generator to disconnect the system if an excessive unbalanced current condition should occur.

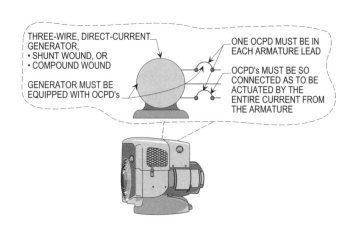

**PROTECTING A THREE-WIRE DC GENERATOR
NEC 445.12(E)**

Figure 3-6. The above illustrates the overcurrent protection requirements for three-wire DC generators.

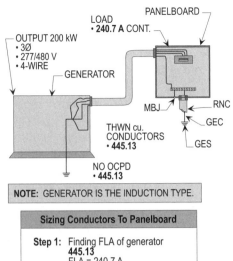

NOTE: GENERATOR IS THE INDUCTION TYPE.

Sizing Conductors To Panelboard

Step 1: Finding FLA of generator
445.13
FLA = 240.7 A

Step 2: Calculating FLA for conductors
445.13
240.7 A x 115% = 276.8 A

Step 3: Selecting conductors
310.10(2); Table 310.16
276.8 A requires 300 KCMIL cu.

Solution: **The size THWN conductors are required to be 300 KCMIL copper because there is no OCPD at the generator.**

SIZING CONDUCTORS FOR A GENERATOR AT 115%
NEC 445.13

Figure 3-7. The above illustrates the procedure for sizing conductors from a generator to a load using 115 percent multiplier.

PROTECTION OF LIVE PARTS
445.14

Live parts of generators operated at more than 50 volts-to-ground must not be exposed to accidental contact where accessible to unqualified persons. The basic rule is that live parts of generators must not be exposed to accidental contact. Such live parts are as follows:

(1) Brushes,
(2) Collector rings, and
(3) Other live parts.

See Figure 3-9 for a detailed illustration pertaining to this rule.

GUARDS FOR ATTENDANTS
445.15

If generators operate at more than 150 volts-to-ground, no live parts are allowed to be exposed to contact by unqualified personnel. Section **430.233** requires insulating mats or platforms around motors. Note that these protective items are also required for generators when the generator voltage is greater than 150 volts-to-ground. (**See Figure 3-10**)

Generator Tip: Generators and controllers must be guarded against accidental contact only by location as specified in **430.232 and 430.233**, and because adjustments or other maintenance may be necessary during the operation of the apparatus, suitable insulating mats or platforms are to be provided so that a qualified person cannot readily touch live parts unless standing on the mats or platforms.

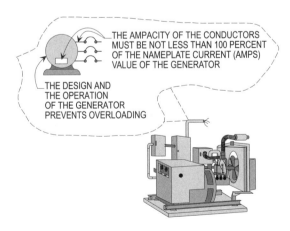

SIZING CONDUCTORS FOR A GENERATOR AT 100%
NEC 445.13, Ex.

Figure 3-8. The above illustrates the procedure for sizing the conductors from a generator to a load using the 100 percent multiplier.

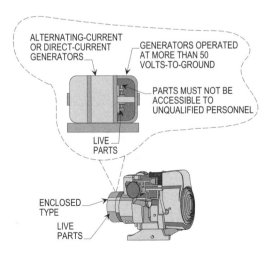

PROTECTION OF LIVE PARTS
NEC 445.15

Figure 3-9. The above illustrates the requirements for generators with exposed parts.

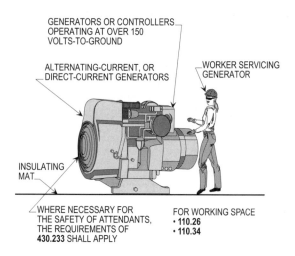

GUARDS FOR ATTENDANTS
NEC 445.15

Figure 3-10. Insulating mats or platforms must be provided for attendants servicing a generator when the voltage is greater than 150 volts-to-ground.

BUSHINGS
445.16

Where wires pass though an opening in an enclosure, conduit box, or barrier, a bushing must be used to protect the conductors from the edges of an opening having sharp edges. The bushings are required to have smooth, well-rounded surfaces where it may be in contact with the conductors. If used where oils, grease, or other contaminants may be present, the bushing must be made of a material that will not be deleteriously affected. **(See Figure 3-11)**

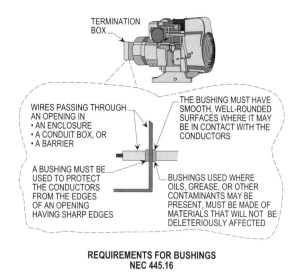

REQUIREMENTS FOR BUSHINGS
NEC 445.16

Figure 3-11. The above illustrates the rules for installing brushes to protect conductors passing through the openings of enclosures.

GENERATOR TERMINAL HOUSINGS
445.17

Generator terminal housings must comply with **430.12**. Where a HP rating is required to determine the required minimum size of the generator terminal housing, the FLC of the generator must be compared with comparable motors in **Table 430.247** through **Table 430.250**. The higher HP rating must be used whenever the generator selection is between two ratings. For example, a 460 volt, three-phase generator with an output of 345 amps must be rated at 361 amps with a HP of 300 per **Table 430.250** of the NEC.

DISCONNECTING MEANS REQUIRED FOR GENERATORS
445.18

A disconnecting means is required for generators to disconnect the generator and all protective devices and control apparatus. **(See Figure 3-12)**

A disconnecting means for the generator is not required if the driving means for the generator can be readily shut down and there is no other generator or other source of voltage in parallel with it.

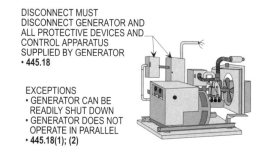

DISCONNECT REQUIRED FOR GENERATORS
NEC 445.18

Figure 3-12. A disconnecting means is required to disconnect all protective devices and control apparatus from generator.

Chapter 3: Generators and the NEC

Section   **Answer**

_____   T   F   **1.** Standard-type generators are designed to operate in dry places.

_____   T   F   **2.** DC generators must be provided with overcurrent protection.

_____   T   F   **3.** Balancer sets consist of one smaller DC generator used with a larger two-wire generator.

_____   T   F   **4.** If generators operate at more than 150 volts-to-ground, no live parts are allowed to be exposed to contact by qualified personnel.

_____   T   F   **5.** Live parts to generators are not to be exposed to accidental contact.

_____ _____   **6.** To aid designers and electrical personnel every generator must have information on the _____.

_____ _____   **7.** Ungrounded phase conductors from a generator must be rated at no less than _____ percent of the nameplate current rating.

_____ _____   **8.** Conductors can be protected at _____ percent of the rated generated current if overload conditions are to be prevented.

_____ _____   **9.** Live parts of generators operated at more than _____ volts-to-ground must not be exposed to accidental contact where accessible to unqualified personnel.

_____ _____   **10.** Generators and controllers must be _____ against accidental contact.

Emergency System Generators

This chapter covers systems that are legally required to be installed and that supply loads essential to safety and life, such as emergency lighting, essential refrigeration and ventilation, and signaling systems. Emergency systems are also installed in places of assembly, such as theaters, schools, stadiums, or locations where large numbers of people may gather. Such systems must be designed to assure safe evacuation by providing electric power for adequate emergency lighting, proper fire detection, reliable operation of fire pumps, dependable alarm signals, communications, etc.

GENERATOR SET
700.12(B)(1) THRU (B)(6)

A generator is supplied by a prime mover that must be accepted by the authority having jurisdiction and sized as covered in **700.5**. It is required to have automatic starting of the prime mover when the normal source of power fails, and is required to have a transfer switch for all electrical equipment supplied by the emergency circuit. To prevent immediate retransfer in cases of short-time restoring of the normal source of power, a time-delay feature allowing for a 15 minute setting must be provided.

Internal combustion engines used for prime movers are required to have an on-site fuel supply that will function at full demand for not less than 2 hours of operation. Fuel transfer pumps must be connected to the emergency power system where power is needed for the operation of the fuel transfer pumps to deliver fuel to a generator set day tank.

Prime movers are not to rely solely upon public utility gas systems for the fuel supply. Automatic transferring means must be provided for transferring from one fuel supply to another when a dual fuel supply is used.

> **Generator Tip:** When acceptable to the authority having jurisdiction, other than on-site fuels can be used when there is a low probability of the failure of the on-site fuel delivery system and the power from the outside electrical utility company occurring at the same time.

If a storage battery is used for control or signal power or as a means of starting prime movers, it must be suitable for that type of service and equipped with an automatic charging means independent of the generator set. A battery charger must be connected to the emergency system where the battery charger is required for the operation of the generator set. Dampers must be connected to the emergency system where power is required for the operation of dampers used to ventilate the generator set. Note that monthly check of batteries should be done.

When an emergency generator requires more than 10 seconds to develop power, an auxiliary power supply is acceptable to energize the emergency system until the regular generator is capable of picking up the load. **(See Figure 4-1)**

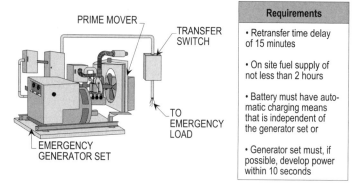

	Requirements
	• Retransfer time delay of 15 minutes
	• On site fuel supply of not less than 2 hours
	• Battery must have automatic charging means that is independent of the generator set or
	• Generator set must, if possible, develop power within 10 seconds

BASIC REQUIREMENTS FOR A GENERATOR SET
NEC 700.12(B)(1) thru (B)(6)

Figure 4-1. The above illustration lists the rules for a generator set under certain conditions of use.

PURPOSE OF EMERGENCY GENERATORS
700.1

The same types of emergency systems might not be suitable for all applications. The conditions must be evaluated as to whether the emergency system will be needed for a long period of time or a short period of time, and how much capacity the emergency system must have to supply the emergency demands.

In the case of interruption of service to a hospital, whether from within or without, the emergency system might be required to provide a large amount of power for a long period of time. Such a situation requires a complete evaluation of the possible needs, the type of system, and its capacity to serve loads and comply with the requirements of the NEC.

> **Generator Tip:** Emergency systems are usually installed in places of assembly to provide illumination in the event of a normal power outage so that there will be a means of safe exit and panic control in those buildings that may be occupied by a large number of people. Such places of assembly are hotels, theaters, sports arenas, health care facilities, and similar institutions.

Emergency systems may supply power for ventilation that may be essential for sustaining life, for fire protection and alarm systems, elevators, fire pumps, safety communications, or industrial processes where interruption of current can cause serious life, safety, or health hazard problems. **(See Figure 4-2)**

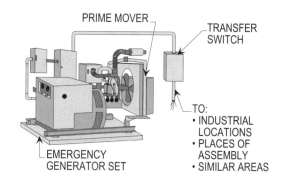

LOADS SERVED BY A GENERATOR SET
NEC 700.1, FPN 3

Figure 4-2. A generator set can be used to supply emergency loads such as places of assembly, special types of equipment, industrial related loads, and similar loads and equipment.

SIZING GENERATORS
700.5(A)

The capacity of the emergency system must be sized adequately to handle the requirements of all the equipment to be operated simultaneously without overloading the generator. The equipment must be designed and fully capable of handling the available fault-current at its terminals. **(See Figure 4-3)**

Sizing Generator
Step 1: Calculating Running Amps

Step 1: Calculating Running Amps
MT loads   = 124 A (100 HP)
                  = 96 A (75 HP)
                  = 77 A (60 HP)
Ltg. load   = 62 A
Rec. load   = 25 A
Other loads   = 122 A
Running Amps   = 506 A

Step 2: Calculating Starting Amps
(Applying 125% multiplier to all motors)
Motor load
124 A x 125%   = 155   A
96 A x 125%   = 120   A
77 A x 125%   = 96.25 A
Lighting load
62 A x 100%   = 62   A
Rec. load
25 A x 100%   = 25   A
Other loads
122 A x 100%   = 122   A
Starting Amps = 580.25 A

Step 3: Selecting kVA of generator
Running kVA
kVA = 506 A x 480 V x 1.732 ÷ 1000
kVA = 421
Starting kVA
kVA = 580.25 A x 480 V x 1.732 ÷ 1000
kVA = 482

Solution: **The generator must have a starting capacity of 482 kVA and a running capacity of 421 kVA respectively.**

TRANSFER SWITCH AND EQUIPMENT 700.6(A) THRU (D)

The transfer switch and equipment is required to be automatically operated and to be identified for emergency service, or to be approved by the authority having jurisdiction. When installing the transfer equipment, it must be so designed and installed so that accidental interconnection of the normal and the emergency source will not occur with the operation of the transfer equipment. Automatic transfer switches must be electrically operated and mechanically held. Transfer equipment may supply only emergency loads. **(See Figure 4-4)**

Generator Tip: A means for isolating the transfer switch is permitted. If isolation switches are installed, accidental parallel operations are to be avoided.

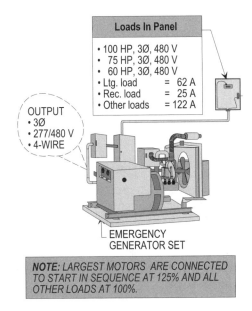

Loads In Panel
• 100 HP, 3Ø, 480 V
• 75 HP, 3Ø, 480 V
• 60 HP, 3Ø, 480 V
• Ltg. load   = 62 A
• Rec. load   = 25 A
• Other loads = 122 A

OUTPUT
• 3Ø
• 277/480 V
• 4-WIRE

EMERGENCY GENERATOR SET

NOTE: *LARGEST MOTORS ARE CONNECTED TO START IN SEQUENCE AT 125% AND ALL OTHER LOADS AT 100%.*

SIZING A GENERATOR
NEC 700.5(A)

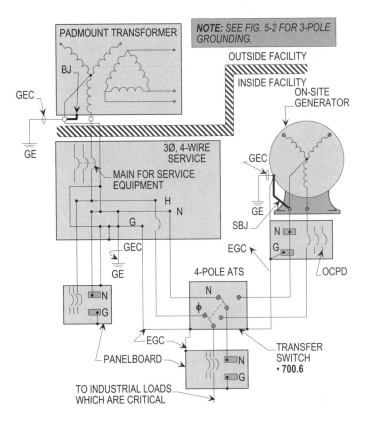

USING FOUR-POLE TRANSFER SWITCHES
NEC 700.6

Figure 4-3. Sizing an emergency generator set to supply loads at an industrial process area.

Figure 4-4. A four-pole transfer switch can be used to supply critical loads in different facilities.

WIRING IDENTIFICATION
700.9(A)

All boxes and enclosures that contain emergency circuits must be marked so that they will be readily identified as being a part of the emergency circuit or system.

WIRING SYSTEMS
700.9(B)

Emergency source wiring, including its source of disconnecting overcurrent protection devices supplying the emergency load, is to be kept entirely separate from all other wiring and equipment, raceways, cables, and cabinets that contain other than emergency wiring. Wiring of two or more emergency circuits supplied from the same source are permitted in the same raceway, cable, box or cabinet. These systems must be located and designed to avoid damage due to vandalism, flooding, icing, and other adverse conditions. **(See Figure 4-5)**

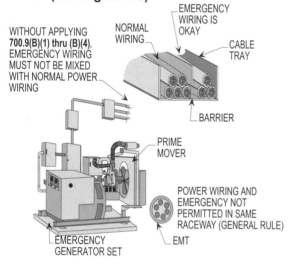

ROUTING EMERGENCY WIRING
NEC 700.9(B)

Figure 4-5. Emergency wiring and normal power are not allowed to occupy the same raceway, cable tray, cable, etc.

WIRING SYSTEMS
700.9(B)(1) THRU (4)

The emergency wiring system can be mixed with other power under any of the following conditions.

(B)(1)
In transfer equipment enclosures, the transfer equipment must supply only emergency loads.

(B)(2)
In exit or emergency luminaires (lighting fixtures), a supply from two sources is permitted.

(B)(3)
In a common junction box attached to exit or emergency luminaires (lighting fixtures) a supply from two sources is permitted.

(B)(4)
In a common junction box attached to unit equipment, that contains only the branch-circuit supplying the unit equipment and the emergency circuit supplied by the unit equipment is permitted.

GROUNDING
700.8(B)

In cases where the grounded circuit conductor from an emergency source is connected to a grounding electrode conductor anywhere except at that source, a sign must be posted at the location of the remote grounding connection identifying all emergency and normal sources connected at that location. **(See Figure 4-6)**

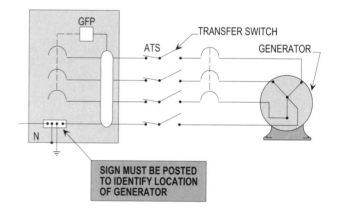

REQUIREMENTS FOR GENERATOR GROUNDING
NEC 700.8(B)

Figure 4-6. If a generator is grounded at another location, such remote location must identify the whereabouts of the generator.

Note: For the different methods used to ground a generator See Pages 6-5 through 6-9 in Chapter 6 of this book.

Generator Tip: A sign must be placed at the service-entrance equipment indicating the type and location of all on-site emergency power sources. However, a sign is not required for individual unit equipment such as battery packs specified in **700.12(F)**. Review **700.12(F) and Ex.** for detailed rules and regulations pertaining to these requirements.

Chapter 4: Emergency System Generators

Section	Answer	
	T F	**1.** A time-delay feature must be provided with a 10 minute setting to prevent an outage in case of short-time restoring of the normal source of power.
	T F	**2.** When generator sets require more than 10 seconds to develop power, the emergency system must be provided with an auxiliary power supply.
	T F	**3.** Emergency systems may be installed to supply power for ventilation that may be essential for sustaining life.
	T F	**4.** Emergency circuits and general wiring circuits cannot occupy the same raceway.
	T F	**5.** Wiring of two or more emergency circuits supplied from the same source are permitted to be installed in the same raceway.
		6. Internal combustion engines used for prime movers are required to have an on-site fuel supply that will function at full demand for not less than _____ hours of operation.
		7. The capacity of the emergency system must be sized adequately to handle the requirements of all the equipment to be operated _____ without overloading the generator.
		8. Emergency systems are usually installed in places of assembly to provide _____ in the event of a normal power outage.
		9. The transfer switch and equipment is required to be _____ and to be identified for emergency service.
		10. If a generator is grounded at another location, such remote location must _____ the whereabouts of the generator and normal source.

5

Legally Required and Optional Standby Systems

This chapter covers legally required standby systems, which are those systems required and classified as legally required standby systems by municipal, state, federal, and/or other codes, or by any governmental agency having jurisdiction.

In the event that there is a failure of the normal power source, these systems are intended to supply power automatically to special selected loads that are not classified as emergency systems.

Legally required standby power systems also supply such loads as heating and refrigeration systems, communication systems, ventilation and smoke removal systems, sewage disposal, lighting, and industrial processes that, when stopped during a power outage, could create hazards or hamper rescue or fire fighting operations.

This chapter also covers optional standby systems that are intended to protect private business or property where life safety does not depend on the performance of the system. Optional standby systems are not those systems that are classified as emergency or legally required standby systems. These systems serve as an alternate power source for industrial and commercial buildings, farms, and residences by supplying such loads as heating and refrigeration systems, data processing and communications systems, and industrial processes that, when stopped during any power outage, could cause discomfort, serious interruption of the process, or damage to the product or process.

PART III - LEGALLY REQUIRED STANDBY SYSTEMS GENERATOR REQUIREMENTS
701.11

In selecting a legally required standby generator, consideration must be given to the type of service to be rendered, whether of short-time duration or long-time duration.

Consideration should also be given to the location or design, or both, of all equipment to minimize the hazards that might cause complete failure due to floods, fires, icing, and vandalism.

Generator Tip: The assignment of the degree of reliability of a recognized legally required standby supply system depends on the careful evaluation of the variables at each particular installation.

GENERATOR SET
701.11(B)(1) THRU (B)(5)

A generator set driven by a prime mover must be sized by **701.6** and be acceptable to the authority having jurisdiction. A means must be provided for automatically starting the prime mover upon failure of the normal service and to automatically transfer all electrical circuits. A time-delay feature permitting a 15 minute setting must be provided to avoid retransfer in case of a short-time reestablishment of the normal source.

Where internal combustion engines are used as the prime mover, an on-site fuel supply must be provided with a fuel supply of not less than 2 hours full-demand operation. Prime movers are not to solely depend upon a public utility gas system for their fuel supply or municipal water supply for their cooling systems. Automatically transferring of one fuel supply to another where dual fuel supplies are used must be provided.

Generator Tip: Where acceptable to the AHJ, the use of other than on-site fuels is allowed where there is a low probability of a simultaneous failure of both the off-site fuel delivery system and power from the outside electrical utility company.

Where a storage battery is used for control or signal power, or as the means of starting the prime mover, it must be suitable for the purpose and be equipped with an automatic charging means independent of the generator set. **(See Figure 4-1)**

PURPOSE OF LEGALLY REQUIRED GENERATOR SYSTEM
701.2

Legally required standby systems are those required by municipal, state, federal, or by other codes, or any governmental agency having jurisdiction, the intent of which is to supply power to selected loads other than those classified as emergency systems. In the event that the normal power source fails, this system must provide the necessary power.

WHERE USED
701.2, FPN

Typical installations of legally required standby systems are intended for operation to serve loads such as heating, refrigerator systems, communication systems, ventilation and smoke removal systems, sewage disposal, and industrial processes, which, if the normal operation of the normal power supply fails, could create hazards or hinder rescue or fire fighting operations. **(See Figure 5-1)**

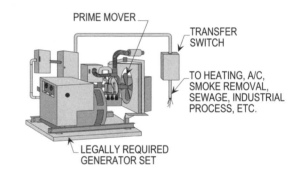

LOADS SERVICED BY A GENERATOR SET
NEC 701.2, FPN

Figure 5-1. The above illustrates the loads that are served by a legally required standby system.

SIZING GENERATORS
701.6

It is mandatory that a legally required standby system has adequate capacity and rating for supplying all equipment intended to be operated at the same time.

The alternate power source may be permitted to supply legally required standby and optional standby loads, when these loads are automatically picked up for load shedding so as to ensure power to the legally required standby circuits. **Note:** See **Figure 4-3**.

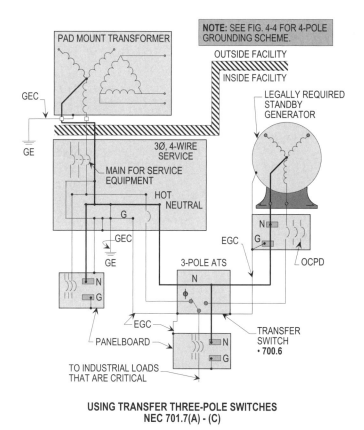

USING TRANSFER THREE-POLE SWITCHES
NEC 701.7(A) - (C)

Figure 5-2. A three-pole transfer switch can be used to supply critical loads in different facilities.

TRANSFER SWITCHES AND EQUIPMENT 701.7(A) THRU (C)

Automatic transfer equipment must be identified or approved for standby use by the AHJ. The automatic transfer equipment must be so designed so that no accidental connection of the normal and alternate sources of supply occurs at the same time in the operation of any transfer equipment. Again, the AHJ must approve the transfer equipment. Isolation equipment may be used to isolate the transfer equipment. However, if isolation equipment is used, any inadvertent parallel operation must be avoided. Automatic transfer switches (ATS) must be electrically operated and mechanically held. (**See Figure 5-2**)

WIRING LEGALLY REQUIRED SYSTEMS 701.10

Legally required standby system wiring can occupy the same raceways, cables, boxes, and cabinets with other general wiring systems. This is a big advantage over emergency wiring systems which, are not allowed this privilege. (**See Figure 5-3**)

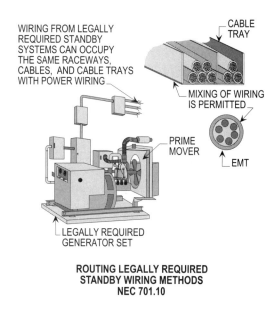

ROUTING LEGALLY REQUIRED
STANDBY WIRING METHODS
NEC 701.10

Figure 5-3. Legally required standby wiring can occupy the same raceway, cable tray, and cable as power wiring.

OVERCURRENT PROTECTION DEVICES - ACCESSIBILITY 701.15

Only authorized persons are allowed to have access to overcurrent protection devices of legally required standby circuits. This will prevent tampering or interference with the operation of the OCPD's and circuits.

Generator Tip: The alternate power source that supplies legally required standby systems is not required to have ground-fault protection of equipment per **701.17**. For selectively coordinated supply side OCPD's, see **700.18**.

GROUNDING 701.9(B)

Where the grounded circuit conductor connected to the emergency source is connected to a grounding electrode conductor at a location remote from the emergency source, there must be a sign at the grounding location that identifies all emergency and normal sources connected at that location. In addition, a mandated sign must be placed at the service-entrance indicating type and location of all on-site legally required standby power sources. However, a sign is not required for individual unit equipment such as battery packs per **701.11(G), Ex**. (**See Figure 4-6**)

PART I - OPTIONAL STANDBY SYSTEMS AND THE PURPOSE OF OPTIONAL STANDBY GENERATOR SYSTEMS
702.2

Optional standby systems are intended only for the protection of business or property and do not include places where life safety is dependent on the performance of the system. They may be operated manually or automatically.

With brownouts and blackouts over the country, many individuals and especially ranches, farms, and dairy operations, may have a standby source of power to eliminate losses and unscheduled outages from the loss of power.

WHERE USED
702.2, FPN

Optional standby systems are typically installed to provide an alternate source of electrical power for facilities such as industrial and commercial buildings, farms, ranches, and residences, heating or refrigeration systems, data processing and communication systems, and industrial processes, which, if stopped during a power outage, could cause interruption to the process or damage to the product, etc. **(See Figure 5-4)**

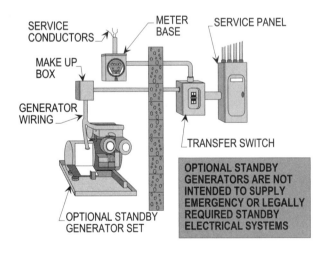

OPTIONAL STANDBY GENERATOR
NEC 702.2, FPN

Figure 5-4. Optional standby generators can be used to supply any loads except those supplied by emergency and legally required standby generator sets.

SIZING GENERATORS
702.5

Optional standby systems must be sized to have the capacity and rating adequate to supply all the equipment intended to be operated at the same time. However, such generators can be sized to supply only the loads that are selected by the user. **(See Figure 5-5)**

TRANSFER SWITCHES AND EQUIPMENT
702.6

Transfer equipment needs to be suitable for its intended use, and designed and installed so as to prevent an accidental connection with the normal or alternate sources of power. Once again, the authority having jurisdiction must give approval of such transfer equipment. **(See Figure 5-6)**

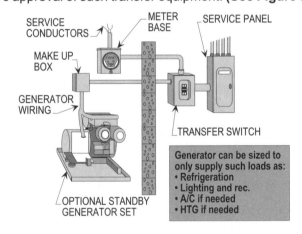

SIZING OPTIONAL STANDBY GENERATORS
NEC 702.5

Figure 5-5. Optional standby generators can be used to supply specific loads and not the entire load supplied by the electrical system.

WIRING OPTIONAL STANDBY SYSTEMS
702.9

The wiring from the optional standby equipment is permitted to be in the same raceways, cables, boxes, and cabinets as other general wiring. This is a big advantage over emergency systems, which do not allow such privileges. **(See Figure 5-7)**

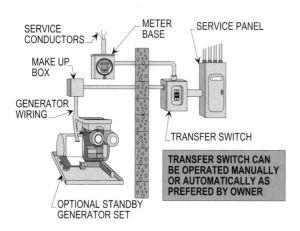

**RULES FOR TRANSFER SWITCHES
NEC 702.6**

Figure 5-6. Transfer switches used to transfer the generator power to the premise can be accomplished by a manual or automatic means.

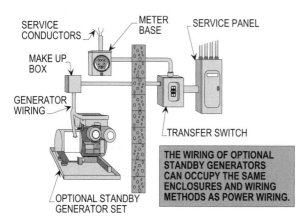

**ROUTING OPTIONAL STANDBY WIRING METHODS
NEC 702.9**

Figure 5-7. The wiring of optional standby systems can be run with the power wiring of the facility.

GROUNDING
702.8(B)

Where the grounded circuit conductor that is connected to the emergency source is connected to a grounding electrode conductor at a location remote from the emergency source, there must be a sign at the grounding location that identifies all emergency and normal sources connected at that location. In addition, a mandated sign must be placed at the service-entrance indicating the type and location of all on-site optional standby power sources. However, a sign is not required for individual unit equipment such as battery packs per **702.8(A)**. **(See Figure 4-6)**

Note: For the different methods used to ground a generator, see pages 6-5 through 6-9 in Chapter 6.

Separately Derived Sysytem 720.10(A). Where a portable optional standby source is used as a separately derived system, it shall be grounded to a grounding electrode inaccordance with **250.30**.

Nonseparately Derived Sysytem 720.10(B). Where a portable optional standby source is used as a non separately derived system, the equipment grounding conductor shall be bonded to the system grounding electrode.

For detail illustrations describing these grounding requirements, see **Figures 6-16 and 6-17** in Chapter 6 of this book.

Chapter 5: Legally Required and Optional Standby Systems

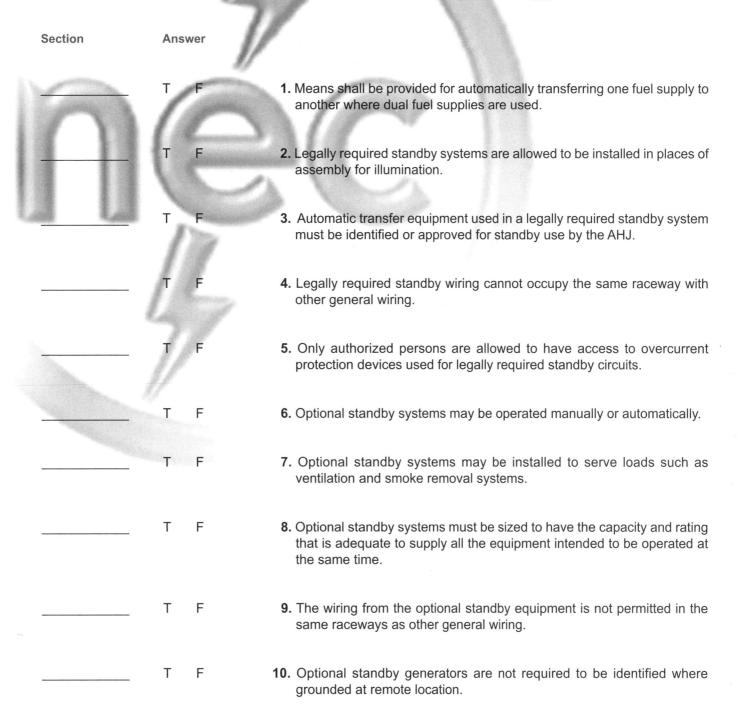

Section	Answer	
_____	T F	**1.** Means shall be provided for automatically transferring one fuel supply to another where dual fuel supplies are used.
_____	T F	**2.** Legally required standby systems are allowed to be installed in places of assembly for illumination.
_____	T F	**3.** Automatic transfer equipment used in a legally required standby system must be identified or approved for standby use by the AHJ.
_____	T F	**4.** Legally required standby wiring cannot occupy the same raceway with other general wiring.
_____	T F	**5.** Only authorized persons are allowed to have access to overcurrent protection devices used for legally required standby circuits.
_____	T F	**6.** Optional standby systems may be operated manually or automatically.
_____	T F	**7.** Optional standby systems may be installed to serve loads such as ventilation and smoke removal systems.
_____	T F	**8.** Optional standby systems must be sized to have the capacity and rating that is adequate to supply all the equipment intended to be operated at the same time.
_____	T F	**9.** The wiring from the optional standby equipment is not permitted in the same raceways as other general wiring.
_____	T F	**10.** Optional standby generators are not required to be identified where grounded at remote location.

6

Generators Supplying Essential Loads for Hospitals

This chapter covers essential electrical loads that are designed to be supplied by all types of alternate power sources, all distribution systems, and ancillary equipment that have been designed to ensure electrical power continuity to designated areas and functions of a health care facility when the normal power source is disrupted. In addition, it must also be designed to minimize the disruption of power in the internal wiring system.

Note: The essential electrical systems used in hospitals are the emergency system and the equipment system, respectively.

EMERGENCY SYSTEM
517.30(B)(2); 517.31

Feeders or branch-circuits that conform to **Article 700** and are intended to supply power from an alternate source to a limited number of designated functions that are vital for the protection of life and for the patient's safety must operate within 10 seconds of the interruption of the normal power source per **517.31**.

LIFE SAFETY BRANCH
517.2

The life safety branch is a subsystem of the emergency system consisting of feeders and branch-circuits that meet the requirements of **Article 700**. The intent is to provide enough power to adequately secure the power source to the patients and other personnel. These circuits must automatically be connected to the alternate source of power upon interruption of the normal supply of power.

CRITICAL BRANCH
517.2

The critical branch is a subsystem of an emergency system consisting of feeders and branch-circuits supplying energy for illumination, any special power-circuits, and receptacles that are selected to serve those areas where proper functioning is essential to patient care. Such circuits must be connected to an alternate power source by means of one or more transfer switches that are energized from the temporary power source when the normal power source is interrupted.

EQUIPMENT SYSTEM
517.30(B)(3)

The equipment system consists of feeders or branch-circuits arranged for delayed automatic or manual connection to the power source. This ordinarily serves the three-phase power loads as defined in **517.2**.

This equipment system is used to supply power to major pieces of electrical equipment that are essential for either hospital operations or patient care. **(See Figure 6-1)**

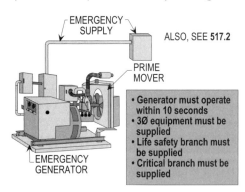

ALSO, SEE **517.2**

- Generator must operate within 10 seconds
- 3Ø equipment must be supplied
- Life safety branch must be supplied
- Critical branch must be supplied

LOADS SERVED BY AN EQUIPMENT SYSTEM NEC 517.30(B)(3)

Figure 6-1. The equipment system must supply certain types of critical loads in hospitals.

TRANSFER SWITCHES
517.30(B)(4)

Each branch of the essential electrical system must be served by at least one transfer switch. This is illustrated in Figures 517.30, No. 1 and 517.30, No. 2 of the NEC, which should be carefully reviewed. One transfer switch is allowed to serve more than one branch of the essential system, up to 150 kVA as illustrated in Figure 517.30, No. 2. The exact number of transfer switches for a facility must be based on a good engineering design that considers load, switch design, switch reliability and dependability. (**See Figure 6-2**)

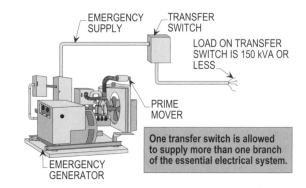

One transfer switch is allowed to supply more than one branch of the essential electrical system.

REQUIREMENTS FOR TRANSFER SWITCHES NEC 517.30(B)(4)

Figure 6-2. The number of transfer switches required are based upon the types, number, and amount of loads served.

OTHER LOADS
517.30(B)(5)(1);(2)

Loads supplied by the generating equipment that are not specifically listed in **517.32, 517.33** and **517.34** must be served by their own transfer switch. Such loads must not be:

(1) Transferred if the transfer will overload the generating equipment, and

(2) Be automatically shed upon generating equipment overloading conditions.

See Figure 6-3 for details of such rules.

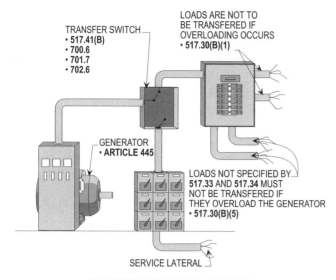

TRANSFER SWITCH
• 517.41(B)
• 700.6
• 701.7
• 702.6

LOADS ARE NOT TO BE TRANSFERRED IF OVERLOADING OCCURS
• 517.30(B)(1)

GENERATOR
• ARTICLE 445

LOADS NOT SPECIFIED BY 517.33 AND 517.34 MUST NOT BE TRANSFERRED IF THEY OVERLOAD THE GENERATOR
• 517.30(B)(5)

SERVICE LATERAL

TRANSFER SWITCH SUPPLIES OTHER LOADS
NEC 517.30(B)(5)

Figure 6-3. Transfer switches must not be overloaded when transferring from the normal power to the emergency power.

WIRING REQUIREMENTS
517.30(C)(1);(2)

Certain rules must be applied to wiring that is routed to locations where the life safety and critical branches of the emergency system are used. Such wiring must be separated from other wiring. However, in some cases, junction boxes, fixture enclosures, and transfer switch wiring can be mixed. (**See Figure 6-4**)

This emergency system must be limited to circuits which are essential to maintaining life and safety. There are two parts of this emergency system:

(1) Life safety branch, and
(2) Critical branch.

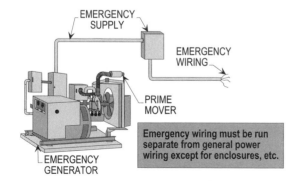

EMERGENCY SUPPLY

EMERGENCY WIRING

PRIME MOVER

Emergency wiring must be run separate from general power wiring except for enclosures, etc.

EMERGENCY GENERATOR

EMERGENCY WIRING MUST BE RUN SEPARATELY
NEC 517.30(C)

Figure 6-4. The above illustrates that general wiring must not be run in raceways, etc. with emergency wiring.

SEPARATION FROM OTHER CIRCUITS
517.30(C)(1) through (C)(4)

The life safety and critical branches of the emergency system are not allowed to be installed in a common raceway, enclosure, or box with any other wiring systems, except in one of the following conditions:

(1) In transfer switches,

(2) In the exit or emergency lighting fixtures which are supplied by two sources,

(3) In a junction box feeding a fixture such as in "(2)" above, and

(4) Wiring of two or more emergency circuits supplied from the same branch.

Note: The wiring of the equipment system is permitted to share raceways, etc. with other wiring systems that are not part of the emergency system.

See Figure 6-5 for the rules of installing the wiring (separately) of the life safety and critical branches per **517.30(C)(1)**.

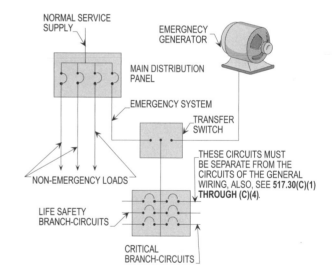

NORMAL SERVICE SUPPLY

EMERGENCY GENERATOR

MAIN DISTRIBUTION PANEL

EMERGENCY SYSTEM

TRANSFER SWITCH

THESE CIRCUITS MUST BE SEPARATE FROM THE CIRCUITS OF THE GENERAL WIRING, ALSO, SEE 517.30(C)(1) THROUGH (C)(4).

NON-EMERGENCY LOADS

LIFE SAFETY BRANCH-CIRCUITS

CRITICAL BRANCH-CIRCUITS

WIRE SEPARATION OF THE LIFE SAFETY AND CRITICAL BRANCHES
NEC 517.30(C)(1)

Figure 6-5. The above illustrates the requirements for installing the wiring of the life safety and critical branches.

ISOLATION AND PROTECTION
517.30(C)(2) AND (C)(3)

Isolated power systems that are installed in anesthetizing locations or in special environments must be supplied by an individual, dedicated circuit supplying no other loads. For mechanical protection, the wiring of an emergency system must be installed in nonflexible metal raceways, or Type MI cable or Schedule 80 RNC.

517.30(C)(3)(1) THROUGH (5)

Cords of appliances and other pieces of equipment connected to the emergency system are exempted from such rule. The secondary circuits of communication or signaling systems that are supplied by transformers do not have to be enclosed in metal raceways, except as required by Chapters 7 and 8 of the NEC.

Schedule 80 rigid nonmetallic conduit and Schedule 40 PVC encased in at least 2 in. of concrete or EMT is permitted, except for branch-circuits that serve patient care areas.

Flexible metal raceways and cable assemblies can also be used as a wiring method if they are installed in listed prefabricated medical headwalls, listed office furnishings or where necessary for flexible connections to equipment.

Note: Review (1) through (5) to **517.30(C)(3)** very carefully before installing the above wiring methods in such areas.

CAPACITY OF SYSTEMS
517.30(D)

Essential electrical systems must be sufficient to supply any demand placed on them. In the past, this normally meant that they were overengineered and provided more than enough capacity under any and all conditions of use.

These feeders, in the past, were usually computed by the rules of **Articles 215 and 220** respectively. However, the 1996 edition of the NEC allows the use of demand calculations to be used for sizing the generator set or sets if based upon the following criteria:

(1) Prudent demand factors and historical data, or

(2) Connected load, or

(3) Feeder calculation procedures described in Article 220, or

(4) Any combination of the above.

See Figure 6-6 for a detailed illustration of the rules for calculating such loads.

SOURCES OF POWER
517.35; 517.44(C), FPN

Basically, essential systems are required to have a minimum of two sources of power available. One may be the normal source and the other may be the alternate source(s) for use when the normal power is interrupted, or may be a generator set(s) driven by a prime mover and located on the facility. Where the normal power consists of a generating unit(s) on the premises, the alternate source may be another generating set(s) or an external utility source. Extreme care must be exercised in the location of equipment to protect it from damage, floods, etc. **(See Figure 6-7)**

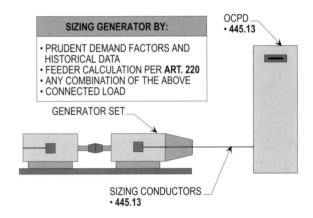

PROCEDURES FOR SIZING GENERATORS
NEC 517.30(D)

Figure 6-6. Emergency generators can be sized by demand factors, historical data, or calculations per **Article 220** of the NEC.

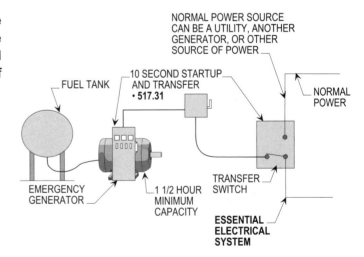

TYPES OF POWER SOURCES FOR
NORMAL POWER SUPPLY
NEC 517.35; 517.44(C), FPN

Figure 6-7. The above illustrates the power sources permitted to be used for the normal power supply.

Generator Tip: Facilities whose normal source of power is obtained from two or more central stations experience electrical service reliability that is greater than that of facilities whose normal supply of power are served from only a single source. Such a source of electrical power consists of power supplied from two or more electrical generators or two or more electrical services supplied from separate utility distribution networks that have local power in the input sources and that are arranged so as to provide mechanical and electrical separation. This is so that a fault between the facilities and the generating source will not be likely to cause the interruption of more than one of the service feeder facilities.

GENERATOR GROUNDING FOR 480 V TO 1000 V SYSTEMS
250.36 (A) THROUGH (G)

When a high-impedance grounded neutral system is utilized for a 480 volt to 1000 volt system in compliance with **250.36(A) thru (G)**, the (grounding connections) must be made by the rules and regulations of this section, as follows:

(High-Impedance Grounded Neutral Systems)

(A) The grounding impedance (usually a resistor) must be connected between the system neutral point and the grounding electrode conductor. The neutral point may be that of a wye transformer connection, or a neutral point may be derived from a 480 volt delta system by the use of a zig-zag grounding autotransformer.

(B) The neutral conductor from the neutral point to the grounding impedance must be fully insulated for it to operate at a substantial voltage above ground.

(C) The system neutral must not be connected to ground except through the impedance unit.

(D) The neutral conductor from the neutral point to the grounding impedance may be installed in a separate raceway.

(E) The equipment bonding jumper (the connection between the system equipment grounding conductors and the grounded end of the grounding impedance) must be an unspliced conductor run from the first disconnect or overcurrent device of the system to the ground end of the impedance device.

(F) The grounding electrode conductor must not be connected anywhere from the ground end of the impedance to the equipment ground bus or terminal in the service equipment or the first disconnect means.

(G) Where the grounding electrode conductor connection is made at the grounding impedance, the equipment bonding jumper must be sized per **250.66**, based on the size of the service entrance conductors for a service or the derived phase conductors for a separately derived system. If the grounding electrode conductor is connected at the first system disconnecting means or overcurrent device, the equipment bonding jumper must be sized the same as the neutral conductor in **250.36(B)**.

See Figure 6-8 for the grounding methods mostly used to ground industrial electrical power sources.

GENERATOR GROUNDING OF 1000 VOLTS OR MORE
250.186 (A) THROUGH (D)

All high-voltage electrical systems utilizing an impedance grounded neutral technique must comply with the following:

(A) The impedance must be inserted into the grounding circuit between the grounding electrode and the neutral point of the transformer or generator.

(B) The neutral of this type of system must be marked and insulated with the same quality of insulation as the phase conductors.

(C) The system can only be connected to the neutral through the impedance.

(D) In this kind of system, equipment grounding conductors can be connected to the ground bus and grounding electrode conductor, and brought to the system ground. It may be bare.

See Figure 6-8 for a detailed illustration of high-impedance grounding.

METHODS OF GROUNDING
250.130(A); (B)

The decision to ground or not to ground a generator is a choice that designers will have to make at one time or another during their careers. By definition, an ungrounded system is a system that has no intentional connection to ground. A grounded system is a system that has an intentional connection to ground.

A generator that has no intentional connection to ground is known in the industry as an ungrounded system. However, it is connected to ground through the stray capacitance of the ungrounded phase conductors. If a ground-fault does not occur, the neutral of an ungrounded system operates close to ground potential. The neutral voltage is held at such potential by the balanced stray capacitance between each ungrounded phase conductor and ground. (**See Figure 6-9**)

PURPOSE OF GENERATOR GROUNDING
250.4(A)(1) THRU (5)

One of the most important, but usually the most misunderstood and controversial elements of an industrial electrical power system design, is the subject of grounding. The term *grounding* is often used to describe circuit and system and equipment grounding, although each have different objectives to accomplish.

Electrical systems and circuit conductors are grounded to limit voltage due to lightning, line surges, or unintentional contact with other "higher" voltage lines. System grounding ensures longer insulation life for electrical equipment such as motors, generators, and transformers by suppressing overvoltages associated with different types of faults. System grounding also stabilizes the voltage-to-ground under normal operation and improves protection of the electrical system by providing fast and selective operation of protective devices in the event of ground-faults.

Equipment grounding consists of a network of grounding conductors used to ground nonelectrical conductive material that encloses or is adjacent to energized conductors. Similar to circuit and system grounding, equipment grounding also limits the voltage-to-ground and provides fast and selective operation of overcurrent protective devices in the event of ground-faults. The two major objectives of generator grounding are as follows:

(1) To improve personnel safety, and
(2) To improve protection of equipment.

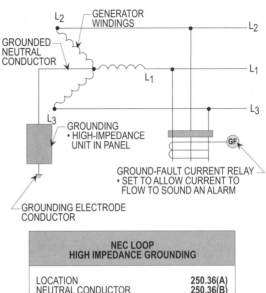

NEC LOOP HIGH IMPEDANCE GROUNDING	
LOCATION	250.36(A)
NEUTRAL CONDUCTOR	250.36(B)
SYSTEM NEUTRAL	250.36(C)
NEUTRAL ROUTING	250.36(D)
EQUIPMENT BONDING JUMPER	250.36(E)
ELECTRODE LOCATION	250.36(F)
EQUIPMENT BONDING JUMPER SIZE	250.36(G)

HIGH IMPEDANCE GROUNDING
OF 480 V TO 1000 V SYSTEMS
NEC 250.36

Figure 6-8. The above illustrates the rules for high-impedance grounding of generators rated 1000 volts or less. **Note:** The same rules apply per 250.186 for grounding generators of 1000 volts or greater.

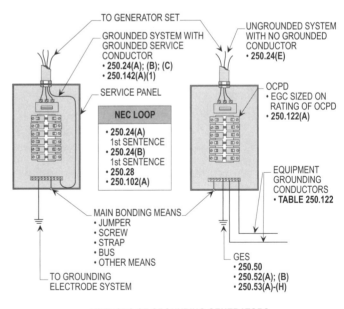

METHODS OF GROUNDING GENERATORS
SUPPLYING ELECTRICAL EQUIPMENT
NEC 250.130(A); (B)

Figure 6-9. The above illustrates the grounding of electrical equipment supplied from a grounded or ungrounded generator set.

PERSONNEL SAFETY
250.4(A)(1) THRU (5)

An equipment grounding system improves personnel safety and protects personnel from electrical shock and other hazards as follows:

(1) Reducing electric shock hazards,

(2) Providing adequate current-carrying capability to carry the high currents produced by a ground-fault without creating a fire or explosive hazard to the electrical equipment or its elements,

(3) Providing a low-impedance return path for ground-fault current necessary for the operation of the overcurrent protection devices, and

(4) Limiting voltage on the system to line-to-ground magnitudes.

See Figure 6-10 for the rules pertaining to personnel safety grounding and bonding.

EQUIPMENT PROTECTION
210.20(C); 240.3; 250.4

Proper system grounding improves the protection of equipment by:

(1) Providing a low-impedance return path for ground-fault current necessary for the operation of the overcurrent protection devices,

(2) Improving differential relay protection of motors, generators, and transformers,

(3) Limiting voltage on the system to line-to-ground magnitudes,

(4) Minimizing transient overvoltages to acceptable levels,

(5) Allowing the use of grounded neutral type arresters, and

(6) Reducing electrical arc/flashes or blast hazards.

See Figure 6-11 for the benefits of grounding and bonding equipment for safety.

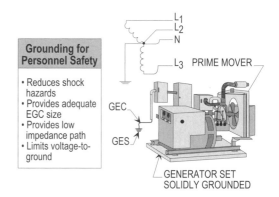

BENEFITS OF PERSONNEL SAFETY GROUNDING
NEC 250.4(A)(1) thru (5)

Figure 6-10. The above illustrates the importance of grounding electrical systems for personnel safety.

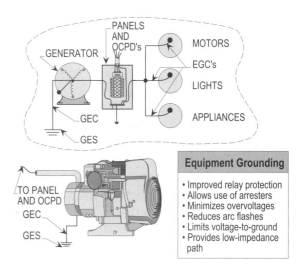

GROUNDING EQUIPMENT FOR SAFETY
NEC 240.4(A)(1) thru (A)(5); 250.134(B)

Figure 6-11. The above illustrates the benefits of grounding the non-current-carrying metal parts of equipment.

METHODS OF HIGH-IMPEDANCE GROUNDING

The following grounding methods are normally used for grounding generators:

(1) Solidly-grounded generators,
(2) Resistance-grounded generators, and
(3) Reactance-grounded generators.

SOLIDLY-GROUNDED GENERATORS

A solidly-grounded system has an intentional and direct connection to ground normally through the middle wire or neutral point of a generator's winding. Note that there is no intentional impedance added in the path from the neutral-to-ground.

In solidly grounded systems, line-to-ground fault-currents can be very high and they may exceed three-phase fault-currents. Solidly grounding generators are sometimes used in industrial facilities, however, it is not always the preferred grounding scheme for generators by most designers. (**See Figure 6-12**)

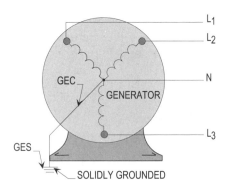

SOLIDLY GROUNDED GENERATORS
NEC 250.20(D), FPN's 1 AND 2

Figure 6-12. A solidly-grounded generator allows the maximum current available to flow and trip the OCPD's.

RESISTANCE-GROUNDED GENERATORS

In resistance-grounded systems the neutral is connected to ground through a resistor. There are two types of resistance grounded systems and they are as follows:

(**1**) Low-resistance grounding, and
(**2**) High-resistance grounding.

LOW-RESISTANCE GROUNDING

Low-resistance grounding is accomplished by inserting a resistance between a generator neutral and ground. When a line-to-ground fault occurs, the voltage across the resistor equals the normal line-to-neutral voltage of the system and the ground-fault current equals the line-to-neutral voltage divided by the size of the grounding resistor. (**See Figure 6-13**)

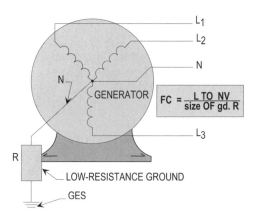

$$FC = \frac{L\ TO\ NV}{size\ OF\ gd.\ R}$$

LOW-RESISTANCE GROUNDED GENERATOR
NEC 250.36; 250.186

Figure 6-13. Low resistance-grounded generators allows the designer to regulate the amount of fault-to-ground current flowing in the system.

> **Note:** Finding fault-current
> FC = fault-current
> L to NV = line-to-neutral voltage
> V of gd.R = Size of the grounding resistor
> Formula:
> $$FC = \frac{L\ to\ NV}{V\ of\ gd.\ R}$$

HIGH-RESISTANCE GROUNDING

This grounding scheme is accomplished by sizing a resistor to provide a resistive fault-current slightly greater than or equal to three times the normal current flowing in the stray line-to-ground capacitance per ungrounded phase. (**See Figure 6-14**)

REACTANCE-GROUNDED GENERATORS

This grounding scheme is one in which a reactor is connected between the system neutral and ground. Reactance grounding of generators is only used to limit ground-fault current to a value no greater than the generator three-phase fault-current level. Therefore, it is used in very few applications. When selecting the grounding technique, verify the size of the generator and its use. For example, Is the generator small or large and fed from a utility transformer or a separately derived system? Is such generator used as the sole supply? The size of the generator and the way it is supplied or used usually determines its grounding scheme. (**See Figure 6-15**)

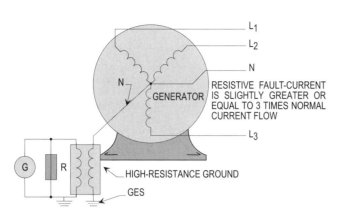

HIGH-RESISTANCE GROUNDED GENERATOR
NEC 250.36; 250.186

Figure 6-14. The above illustrates the characteristics of high-resistance grounded generator systems.

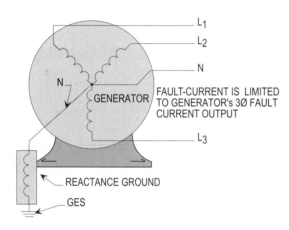

REACTANCE-GROUNDED GENERATOR
NEC 250.36; 250.186

Figure 6-15. The above illustrates the use of reactance-grounded systems to ground generators.

PORTABLE GENERATOR GROUNDING FOR SEPARATELY DERIVED SYSTEMS 702.10(A); 250.20(D), FPN 1

At installations where the generator operates as a separately derived system and the transfer switch interrupts all conductors, including the grounded circuit condcutor, the generator must comply with the normal grounding electrode requirements outlined in **250.30 of the NEC**. Note that this rule consummates the intent that an independent connection to the premises grounding electrode system in the standby panelboard be provided. (See 250.20(D), FPN 1 of the NEC.) **(See Figure 6-16)**

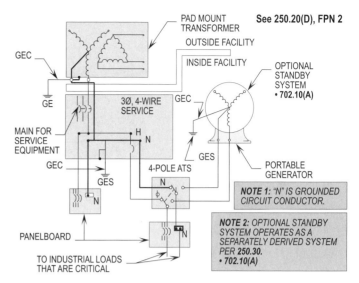

**OPTIONAL STANDBY SYSTEM -
SEPARATELY DERIVED SYSTEM**
702.10(A); 250.20(D), FPN 1

Figure 6-16. The above illustrates the grounding of a portable generator used as a separately derived system.

PORTABLE GENERATOR GROUNDING FOR SEPARATELY DERIVED SYSTEMS 702.10(B); 250.20(D), FPN 2

The grounding requirements for portable, optional type generators that are not separately derived and where the transfer switch only interupts the ungrounded conductors are covered in this section of the NEC. In this installation, the equipment grounding (bonding) conductor must be bonded to the grounding electrode system. **(See 250.20(D); FPN 2 and Figure 6-17)**

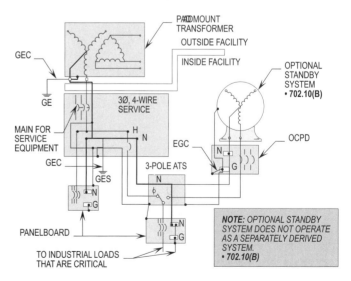

**OPTIONAL STANDBY SYSTEM -
NONSEPARATELY DERIVED SYSTEM**
702.10(B); 250.20(D), FPN 2

Figure 6-17. The above illustrates the grounding of a portable generator used as a nonseparately derived system.

Chapter 6: Generators Supplying Essential Loads for Hospitals

Section   **Answer**

_____   T   F   **1.** Emergency systems must operate within 15 seconds of the interruption of the normal power source.

_____   T   F   **2.** The equipment system is to supply power to major pieces of electrical equipment necessary for either hospital operation or patient care.

_____   T   F   **3.** Each branch of the essential electrical system must be served by at least one transfer switch.

_____   T   F   **4.** The life safety and critical branches of the emergency system are allowed to be installed in a common raceway.

_____   T   F   **5.** Schedule 40 PVC encased in at least 2 in. of concrete is permitted to be installed for branch-circuits that serve patient care areas.

_____   T   F   **6.** The neutral conductor from the neutral point of the generator to the grounding impedance may be installed in a separate raceway when the voltage range is 480 volts to 1000 volts.

_____   T   F   **7.** A solidly-grounded generator system has an intentional and direct connection to ground normally through the middle wire or neutral point of a generator's winding.

_____   T   F   **8.** In resistance-grounded generator systems the neutral is connected to ground through a resistor.

_____   T   F   **9.** Low-resistance grounding is accomplished by using a resistor to provide a resistive fault-current slightly greater than or equal to three times the normal current flowing in the stray line-to-ground capacitance per ungrounded phase.

_____   T   F   **10.** In reactance-grounded generator systems, the reactor is connected between the system neutral and ground.

_____ _____   **11.** Emergency systems are limited to _____ branch and _____ branch-circuits, which are essential to maintaining life and safety.

_____ _____   **12.** The equipment system consists of feeders or branch-circuits arranged for either delayed _____ or _____ connection to the power source.

_____ _____   **13.** One transfer switch is allowed to serve more than one branch of the essential system, up to _____ kVA.

_____ _____   **14.** For mechanical protection of emergency systems, the wiring of an emergency system must be installed in nonflexible metal raceways or Type _____ cable.

_____ _____   **15.** Essential electrical systems shall have a minimum of _____ independent sources of power.

_____ _____   **16.** The impedance for generator grounding of 1000 volts or more must be inserted into the grounding circuit between the _____ electrode and the neutral point of the transformer or generator.

_____ _____   **17.** A system that has no intentional connection to ground is called an _____ system.

_____ _____   **18.** A system that has an intentional connection to ground is called an _____ system.

_____ _____   **19.** Electrical systems and circuit conductors are grounded to _____ voltage due to lightning, line surges, or unintentional contact with other "higher" voltage lines.

_____ _____   **20.** By inserting a resistance between a generator neutral and ground a _____ - resistance grounding scheme is accomplished.

Transformers

Part Two

A vital part in maintaining an uninterrupted electrical service is the **transformer**. In a period of great industrial activity there are likely to be unusual power demands. New types of industrial machinery and high-energy-efficient motors and equipment creating new uses for electricity are likely to impose greater loads, thereby making transformers even more vital than they are now in meeting customer needs. Dependable operation of transformers is therefore becoming increasingly necessary.

Transformers are used for the transmission of electrical power from the generating plant and ultimately to the consumer. A step-down transformer is used for electrical energy in the form of alternating current (AC) at a high-voltage and stepped down to a lower voltage. A step-up transformer is used for electrical energy in the form of alternating current (AC) at a low-voltage and stepped up to a higher voltage. The *National Electrical Safety Code* (NESC) provides the requirements for utility owned transformers, while privately owned transformers not only adhere to the NESC, but in some cases to the NEC also applies.

Part Two reviews the basic theory, operation, construction and troubleshooting procedures that are necessary for a full understanding of the performance of a transformer.

Transformer Theory

Transformer windings are connected in either series or parallel to obtain the different voltages required for supplying various loads. Basic voltages are 120 volt, single-phase; 120/240 volt, single-phase; 120/208 volt, three-phase; and 277/480 volt three-phase. Higher voltages are available for other specific applications. These voltages are usually 2400/4160 volts, three-phase; 12,470 volt, three-phase; and 13,800 volts, respectively.

To obtain the different voltage levels for a transformer and to supply the various loads, the windings are connected in either series or parallel. These windings or loads must be balanced between phases and from each phase-to-neutral to prevent possible overloads. Transformer windings are generally connected in a wye or an open or closed delta connected system. The primary and secondary sides of the transformer may have combination connections in order to obtain different voltage configurations.

TRANSFORMER PRINCIPLES

The amperage of a single-phase transformer is found by dividing the kVA rating of the transformer by the primary or secondary voltage. This calculation determines the amount of current that a transformer will deliver, under normal operating conditions, when supplying various loads.

For example: What is the amperage for a 20 kVA transformer with a 240 volt, single-phase secondary output?

Step 1: Finding amperage of Sec.
A = 20 kVA x 1000 ÷ 240 V
A = 83 amps

Solution: The transformer amperage is 83 amps for the secondary output.

For example: What is the amperage for a 20 kVA transformer with a 480 volt, single-phase secondary output?

Step 1: Finding amperage of Sec.
A = 20 kVA x 1000 ÷ 480 V
A = 42 amps

Solution: The transformer amperage is 42 amps for the secondary output.

The voltage (V), current (A), and impedance (Z) is determined by the number of turns on the primary and secondary windings of a transformer. Based upon the number of turns on the primary and secondary windings of a transformer, the following characteristics of the voltage, current, and impedance shall apply:

Number of turns are the same

(1) Input voltage and output voltage are the same
(2) Impedance remains constant
(3) Input current and output current are the same

Fewer turns on the primary than the secondary

(1) Voltage is stepped up
(2) Current is stepped down

Fewer turns on the secondary than the primary

(1) Voltage is stepped down
(2) Current is stepped up

See Figure 7-1 for applying the number of turns on the primary and secondary of a transformer.

The transformer voltage, amperage, and turns ratio are determined by the ratio of the number of turns on the primary windings to the number of turns on the secondary windings. The transformer kVA or volt-amp rating is the same value for the primary and secondary outputs. **(See Figure 7-2)**

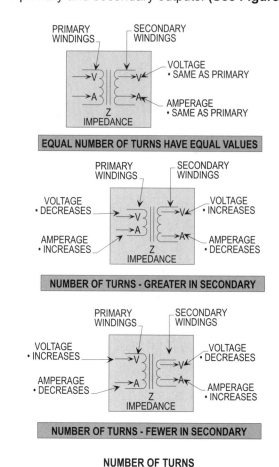

Figure 7-1. The output voltage and amperage is determined by the number of turns on the primary and the secondary of a transformer.

WINDINGS

Depending on the job to be performed, a transformer winding may be connected in a number of ways. Depending on the desired voltage, two or more transformer windings may also be connected together in a number of ways. The most commonly used transformer connections are delta and wye. When applying voltage configurations for these connections, high-voltage systems are connected in series and low-voltage systems are connected in parallel.

SINGLE-PHASE OUTPUT

When the secondary of a transformer supplies 120/240 volt, single-phase loads, there will be 120 volts between either one of the phase lines and the neutral. When the secondary of a transformer supplies 120/240 volt, single-phase loads, there will be 240 volts between both the phase lines. Lighting, receptacle, and appliance loads are supplied from the 120 volt lines. Water heaters, air-conditioning, and electrical heating are supplied from the 240 volt lines. Note that these transformers may be connected for 120 volt or 240 volt, single-phase systems.

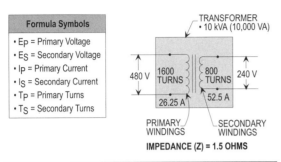

Formula Symbols
• E_P = Primary Voltage
• E_S = Secondary Voltage
• I_P = Primary Current
• I_S = Secondary Current
• T_P = Primary Turns
• T_S = Secondary Turns

Applying Formula To Find Voltage, Amperage, And Turns		
Finding Voltage	**Finding Amperage**	**Finding # of Turns**
Primary Voltage	**Primary Amperage**	**Primary Turns**
$E_P = \dfrac{E_S \times I_S}{I_P}$	$I_P = \dfrac{E_S \times I_S}{E_P}$	$T_P = \dfrac{E_P \times T_S}{E_S}$
$E_P = \dfrac{240\,V \times 52.5\,A}{26.25}$	$I_P = \dfrac{240\,V \times 52.5\,A}{480\,V}$	$T_P = \dfrac{480\,V \times 800\,T}{240\,V}$
$E_P = \dfrac{12,600\,VA}{26.25}$	$I_P = \dfrac{12,600\,VA}{480\,V}$	$T_P = \dfrac{384,00}{240\,V}$
$E_P = 480\,V$	$I_P = 26.25\,A$	$T_P = 1600$
Secondary Voltage	**Secondary Amperage**	**Secondary Turns**
$E_S = \dfrac{E_P \times I_P}{I_S}$	$I_S = \dfrac{E_P \times I_P}{E_S}$	$T_S = \dfrac{E_S \times T_P}{E_P}$
$E_S = \dfrac{480\,V \times 26.25\,A}{52.5\,A}$	$I_S = \dfrac{480\,V \times 26.25\,A}{240\,V}$	$T_S = \dfrac{240\,V \times 1600\,T}{480\,V}$
$E_S = \dfrac{12,600\,VA}{52.5\,A}$	$I_S = \dfrac{12,600\,VA}{240\,V}$	$T_S = \dfrac{384,000\,VA}{480\,V}$
$E_S = 240\,V$	$I_S = 52.5\,A$	$T_S = 800\,T$

FINDING VOLTAGE, AMPERAGE, AND NUMBER OF TURNS

Figure 7-2. The voltage, amperage, and number of turns in transformer windings are determined by applying the proper formula.

WYE-CONNECTED TRANSFORMERS

The voltage between phase-to-phase and phase-to-neutral will always be the same on a wye-connected, three-phase, four-wire transformer. The voltage of the power conductors (phase-to-phase) will always be more than the voltage between any one of the phase conductors and the neutral. For example, if the voltage between the power conductors of any two phases for a 120/208 volt, three-phase, four-wire transformer is 208 volts, the voltage from any phase power conductor to ground will be 120 volts. The voltage between any two phase conductors for a wye-connected transformer is derived by multiplying the voltage-to-ground by the square root of 3 (1.732). The voltage from any phase conductor to ground for a wye-connected transformer is derived by dividing the phase-to-phase voltage by the square root of 3 (1.732). **(See Figure 7-3)**

When all three ungrounded power conductors and neutrals are connected to form a wye-connected secondary, the transformer's output produces a three-phase voltage. When connecting only two ungrounded power conductors plus a neutral to a wye-connected supply, the voltage obtained will be a single-phase system.

The ungrounded phase-to-phase voltage for a wye-connected transformer is the same, but the coil voltage is equal to the square root of 3 (1.732) divided into the phase-to-phase voltage. Each phase leg is connected through the winding to a common connection where they all meet to form a wye-connected secondary. **(See Figure 7-4)**

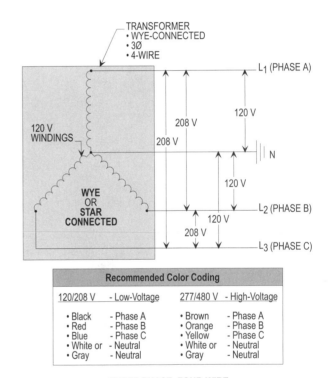

Recommended Color Coding			
120/208 V - Low-Voltage		**277/480 V** - High-Voltage	
• Black	- Phase A	• Brown	- Phase A
• Red	- Phase B	• Orange	- Phase B
• Blue	- Phase C	• Yellow	- Phase C
• White or	- Neutral	• White or	- Neutral
• Gray	- Neutral	• Gray	- Neutral

**THREE-PHASE, FOUR-WIRE,
WYE-CONNECTED TRANSFORMER**

Figure 7-3. Phase-to-phase voltage for a wye-connected transformer is found by multiplying the phase-to-neutral voltage by square root of 3 (1.732). Phase-to-neutral voltage is found by dividing the phase-to-phase voltage by the square root of 3 (1.732).

BALANCED CURRENT FLOW

The winding voltage in a wye-connected system is not the same as the phase-to-phase voltage. The winding voltage is multiplied by the square root of 3 (1.732) to find the phase-to-phase voltage or by dividing the phase-to-phase voltage by 1.732 to find the winding's voltage. The flow of current in the windings of a wye system is the same as the line current. **(See Figure 7-5)**

The windings in a wye system will develop more heat than delta-connected windings because they are pulling the same current as the line. The windings of a delta connected system only pulls 58 percent of the line current.

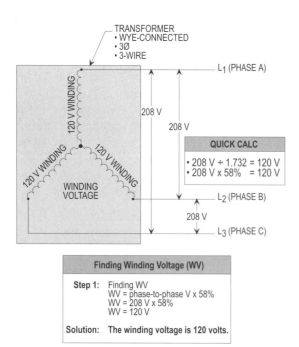

Finding Winding Voltage (WV)

Step 1:	Finding WV WV = phase-to-phase V x 58% WV = 208 V x 58% WV = 120 V
Solution:	The winding voltage is 120 volts.

THREE-PHASE, THREE-WIRE, WYE-CONNECTED TRANSFORMER

Figure 7-4. The winding voltage in a wye-connected transformer system can be found by dividing the phase-to-phase voltage by the square root of 3 (1.732).

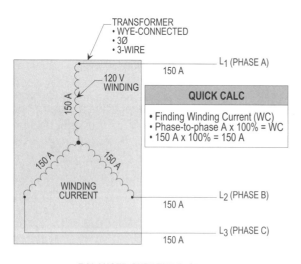

BALANCED CURRENT FLOW IN WYE-CONNECTED TRANSFORMER

Figure 7-5. The flow of current in a wye system is the same in the windings as the line current.

UNBALANCED CURRENT FLOW

The grounded neutral conductor of a three-wire, 120/208 volt feeder-circuit is required to be the same size as the ungrounded phase conductors for a feeder-circuit derived from a four-wire, 120/208 volt system. The reason is, the grounded neutral of a three-wire circuit consisting of two phase conductors and the neutral to a four-wire, three-phase system carries approximately the same amount of current as the ungrounded phase conductors. Therefore, per 220.22, a reduction in ampacity is not allowed.

NEUTRAL CURRENT FLOW

A 120/208 volt or 277/480 volt wye-connected system is different than a 120/240 volt, single-phase or 120/240 volt delta-connected system when determining the flow of current in the neutral conductor. To find the amount of current flow in the neutral of a 120/208 volt or 277/480 volt wye-connected system, use the formula in **Figure 7-6** and replace the values as necessary to calculate another neutral value.

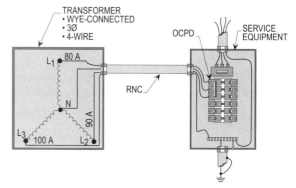

Finding Current Flow In Neutral

Step 1:	Finding A $A = \sqrt{I2L1 + I2L2 + I2L3 - (IL1 \times IL2) - (IL2 \times IL3) - (IL3 \times IL1)}$ $A = \sqrt{80^2 + 90^2 + 100^2 - (80 \times 90) - (90 \times 100) - (100 \times 80)}$ $A = \sqrt{6,400 + 8,100 + 10,000\ A - 7,200 - 9,000 - 8,000}$ $A = \sqrt{24,500 - 24,200}$ $A = \sqrt{300}$ $A = 17.3$
Solution:	The current flow in the neutral is 17.3 amps.

NEUTRAL CURRENT FLOW IN WYE-CONNECTED SYSTEM

Figure 7-6. The above illustrates the procedure for finding the current in amps for the neutral of a three-phase, four-wire wye-connected system.

DELTA-CONNECTED TRANSFORMERS

A delta-connected system is a good installation when used for short-distance distribution systems. This type of system is most commonly used for neighborhood and small commercial loads close to the supplying substation. In a delta-connected system only one voltage is available between any two lines. The coil voltage in a delta-connected system is the same as the phase-to-phase voltage. The windings of a delta-connected transformer may be connected by one of the following:

(1) Open connected delta systems, or
(2) Closed connected delta systems.

A triangle is used to show a delta-connected system. A wire from each connection point of the triangle represents a three-phase, three-wire delta system. Between any two wires the voltage is the same. **(See Figure 7-7)**

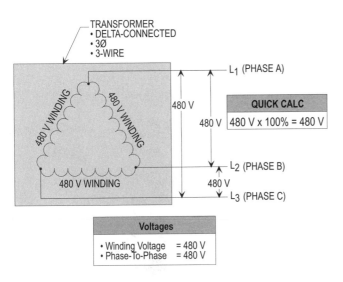

Figure 7-7. Between any two conductors the voltage is 480 volts. The winding voltage is also 480 volts.

OPEN CONNECTED WINDINGS

Only two transformers are used when connecting the windings of an open delta-connected system. One transformer is always larger than the other due to 120 volt loading.

CLOSED CONNECTED WINDINGS

Three transformers are used when connecting the windings of a closed delta-connected system. Depending on the three-phase and single-phase loads served when using a closed delta-connected system, one transformer may be larger than the other two.

BALANCED CURRENT FLOW

The winding voltage and phase-to-phase voltage are the same in a delta-connected system. The winding current and line current are not the same in a delta-connected system. In a delta-connected system the flow of current has two paths to follow at each closed end where the phase conductors terminate. The amount of current in a delta-connected winding is 58 percent of the line current measured on each phase. The multiplier (58 percent) is found by dividing 1 by the square root of 3 ($1 \div \sqrt{3}$ (1.732) = 58%). For example, if the current of each phase is 150 amps, the coil current would be 87 amps. **(See Figure 7-8)**

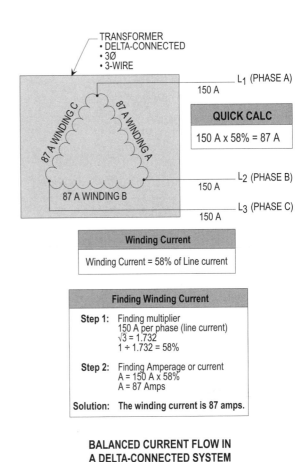

Figure 7-8. The winding current of a balanced delta system is found by multiplying the phase (line) amperage by 58 percent.

UNBALANCED CURRENT FLOW

When the current flow in a delta-connected system is unbalanced, the current flow in L_1 is found by the square root of the other winding currents ($B^2 + C^2 + BC$). The current flow in L_2 and L_3 are found by substituting the appropriate winding current values. **(See Figure 7-9)**

NEUTRAL CURRENT FLOW

The unbalanced current in a four-wire, three-phase delta-connected system is carried by the neutral between phases A and C. This portion of a delta-connected system has only 120 volt ungrounded phase conductors. Note that 120 volts is derived from tapping one of the 240 volt windings. From the tap to each outside phase conductor (phases A and C) 120 volts is derived. The current flow in phase B must travel through one 240 volt and one 120 volt winding to reach the tap, which is connected to ground. Note that 208 volts (phase B to ground) is derived by measuring the voltage-to-ground (120 V + 240 V = 360 V) and dividing by $\sqrt{3}$ (360 V ÷ 1.732 = 208 V). **(See Figure 7-10)**

SINGLE-PHASE LOAD BALANCING

Single-phase, three-wire transformer systems have a secondary voltage of 120/240 volts. One of the windings can be overloaded if loads are not distributed as evenly as possible on each 120 volt winding. Where a 240 volt winding is center-tapped and connected to ground, there will be two 120 volt windings for a transformer. The 30 kVA, single-phase transformer is balanced for each 120 volt winding. Dividing by 2 (30 kVA ÷ 2 = 15 kVA) will derive a capacity of each 120 volt winding.. Each balanced 120 volt winding can be loaded to 15 kVA or less. Proper balancing of the 30 kVA transformer prevents the winding from overheating. **(See Figure 7-11)**

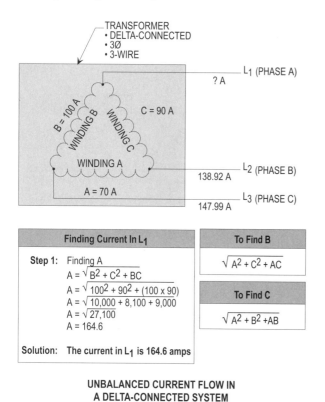

UNBALANCED CURRENT FLOW IN
A DELTA-CONNECTED SYSTEM

Figure 7-9. As illustrated, the current flow of an unbalanced delta-connected system in L_1 is found by the square root of the other winding currents ($B_2 + C_2 + BC$). The current flow in L_2 and L_3 are found by substituting the appropriate winding current values.

BALANCING LOADS ON TRANSFORMER WINDINGS

The loads connected to single-phase and three-phase transformers must be balanced as evenly as possible. Branch-circuit loads in panelboards shall be divided as evenly as possible on phases A, B, and C to assure that transformer windings are not overloaded.

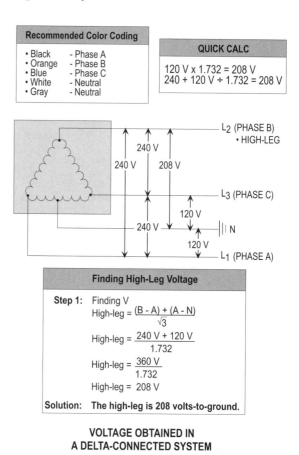

VOLTAGE OBTAINED IN
A DELTA-CONNECTED SYSTEM

Figure 7-10. The above illustrates that only two 120 volt circuits and one 208 volt circuit to ground is produced in a four-wire, three-phase, delta-connected system.

For example, the 30 kVA transformer has 120/240 volt secondary to serve loads of 24 kVA at 240 volts and two 3 kVA loads at 120 volts. Each 120 volt winding must be divided as evenly as possible to balance the single-phase load. The 240 volt loads must be balanced using the same procedure. To prevent overheating of windings, the loads must be properly balanced. The load is unbalanced if the two 120 volt, 3 kVA loads are connected to one 120 volt winding instead of one load to each winding.

THREE-PHASE LOAD BALANCING

Three-phase, four-wire transformer systems usually have a secondary voltage of 120/208 volts or 277/480 volts. When balancing the load, each phase of a three-phase transformer must be considered as a single-phase transformer.

For example, a 40 kVA transformer has a 120/208 volt secondary to serve five loads of 12 kVA, 8 kVA, 6 kVA, 5 kVA, and 3 kVA at 120 volts, single-phase. Each 120 volt phase of the 40 kVA transformer can be loaded up to 13.3 kVA (40 kVA ÷ 3 = 13.3 kVA). **(See Figure 7-12)**

DERATING FOR HIGH ALTITUDE

High-altitude operation of distribution transformers can also be a problem. Dry-type transformers are air-cooled, and require a flow of fresh air in and around the transformer windings to maintain normal operating temperatures. At very high altitudes the air becomes thinner and transformer cooling is not as efficient as at lower latitudes. The NEMA standard is based on normal operation at an altitude of 3300 feet above sea level. So, for every additional 330 feet, derate the transformer load capacity by $^3/_{10}$ of one percent (.3%) for safe and reliable loading. **(See Figure 7-13)**

For derating the load capacity of motors installed in high-altitudes, see title head "TEMPATURE RISE" on page 19-2.

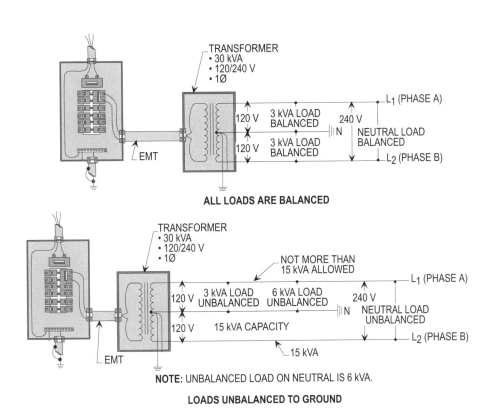

Figure 7-11. To prevent overheating, single-phase transformer loads must be balanced (as close as possible) phase-to-ground and phase-to-phase.

Figures 7-12 & 7-13 on next page.

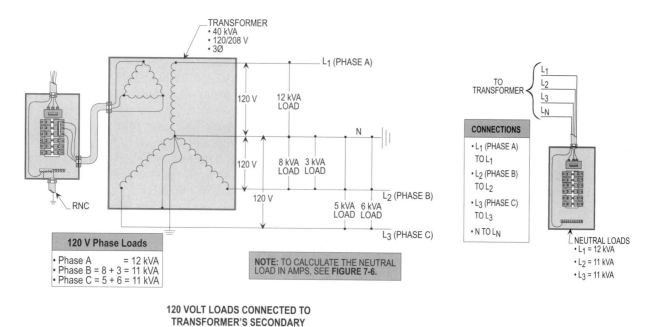

120 VOLT LOADS CONNECTED TO
TRANSFORMER'S SECONDARY

Figure 7-12. Each 120 volt phase of the 40 kVA transformer must not exceed 13.3 kVA (40 ÷ 3 = 13.3 kVA). A neutral load of 13.3 kVA or less is permitted, which will not overload windings.

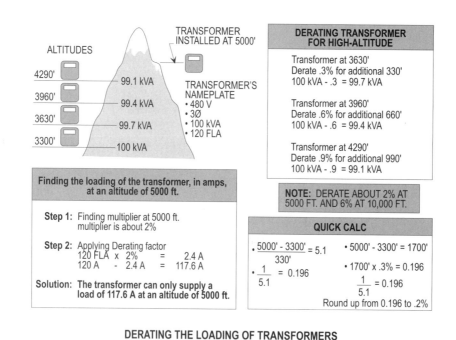

DERATING THE LOADING OF TRANSFORMERS
DUE TO HIGH-ALTITUDE

Figure 7-13. The above illustration shows the rule-of-thumb method for determining the FLC in amps for transformers installed in high altitudes.

Chapter 7: Transformer Theory

Section   **Answer**

_____   T F   **1.** The primary or secondary amperage of a transformer is found by the kVA rating of the transformer multiplied by the primary or secondary voltage.

_____   T F   **2.** The voltage, current, and impedance of a transformer is determined by the number of turns on the primary and secondary windings of a transformer.

_____   T F   **3.** The voltage between any two ungrounded phase conductors will always be the same on a wye-connected, three-phase, four-wire transformer system.

_____   T F   **4.** The winding voltage in a wye-connected system will always be the same as the phase-to-phase voltage.

_____   T F   **5.** Only one voltage level is available between any two lines for a 3-conductor delta-connected systems.

_____   T F   **6.** The winding voltage in a delta-connected system will always be the same as the phase-to-phase voltage.

_____   T F   **7.** The unbalanced neutral current in a four-wire, three-phase delta-connected system is carried by the neutral between phases A and B.

_____   T F   **8.** The loads connected to single-phase transformers are not required to balanced as evenly as possible.

_____   T F   **9.** A three-phase, four-wire wye transformer usually has a secondary voltage of 120/208 volts or 277/480 volts.

_____   T F   **10.** When balancing the single-phase load, each phase of a three-phase transformer must be considered as a single-phase transformer.

_____ _____   **11.** When the number of primary and secondary turns for the windings are the same, the _____ remains constant.

_____ _____   **12.** When there are fewer turns for the windings on the secondary than the primary, the _____ is stepped up.

_____ _____   **13.** When the secondary of a transformer supplies 120/240 volt, single-phase loads, there will be _____ volts between either one of the phase lines and the neutral.

_____ _____   **14.** When connecting two ungrounded power conductors to a wye-connected transformer, the voltage obtained is _____ -phase.

15. The grounded neutral conductor of a three-wire, 120/208 volt feeder-circuit is required to be the same size as the _____ phase conductors for a feeder-circuit that is derived from a four-wire, 120/208 volt system.

16. The windings of a delta-connected transformer may be connected for either a _____ or _____ delta operation.

17. When connecting the windings of an open delta-connected system only _____ transformers are used.

18. The amount of current in the phase windings of a delta-connected system is _____ percent of the line current measured on each ungrounded phase conductor.

19. Branch-circuit loads in panelboards must be _____ as evenly as possible on phases A, B, and C.

20. Where a single-phase _____ volt winding is center-tapped and connected to ground, there will be two 120 volt windings available.

Installing Transformers

Transformers and transformer vaults must be designed, installed and protected per Article 450 in the NEC. Based upon their design and type, a transformer installation can be located either inside of a building or on the outside, sometimes exposed to adverse weather conditions.

Transformers are installed not only to provide a level of safety for nonqualified personnel as well as for qualified personnel, note that a provision of accessibility must also be designed into the installation. Other safety factors include ventilation of transformer vaults and the compliance of minimum fire-resistant standards for the walls, doors, and roof that are associated with a transformer installation.

MARKINGS
450.11

Transformers must be provided with a marking on the nameplate giving the following information:

(1) Name of manufacturer,
(2) Rated kVA,
(3) Frequency,
(4) Primary and secondary voltage,
(5) Impedance for transformers rated 25 kVA and larger,
(6) Clearances for transformers with ventilating openings, and
(7) Amount and kind of insulating liquid.

GUARDING
450.8

Transformers should be isolated in a room or accessible only to qualified personnel to prevent accidental contact with live parts. To safeguard live parts from possible damage, the transformer may be elevated. The following are acceptable means of safeguarding live parts as required in **110.27(A)** and **110.34(E)**.

(1) Transformers should be isolated in a room or accessible only to qualified personnel.

(2) Permanent partitions or screens may be installed.

(3) Transformers should be elevated at least 8 ft. above the floor to prevent unauthorized personnel from contact.

> **Transformer Tip:** Signs indicating the voltage of live exposed parts of transformers, or other suitable markings, must be used in areas where transformers are located.

VENTILATION OF TRANSFORMERS
450.9

Transformers must be located and installed in rooms or areas that are not subjected to exceedingly high temperatures in order to prevent overheating and possible damage to windings. Transformers with ventilation openings must be installed so that the openings are not blocked by walls or other obstructions that could obstruct air flow.

ACCESSIBILITY OF TRANSFORMERS
450.13

Transformers must be located where readily accessible to qualified personnel for inspection and maintenance. Where it is necessary to use a ladder, lift, or bucket truck to get to a transformer, it is not considered readily accessible. See definition of accessible and readily accessible in **Article 100** of the NEC. **(See Figure 8-1)**

OPEN INSTALLATIONS
450.13(A)

Dry-type transformers, rated at not over 600 volts and located on open walls or steel columns, do not have to be readily accessible. It is permissible to gain access to this type of installation using a portable ladder, bucket lift, etc. **(See Figure 8-2)**

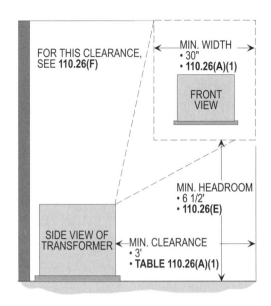

TRANSFORMERS MUST BE READILY ACCESSIBLE
NEC 450.13

Figure 8-1. The general rule of **450.13** requires transformers to be readily accessible for maintenance, repair, and service. See AHJ for this requirement.

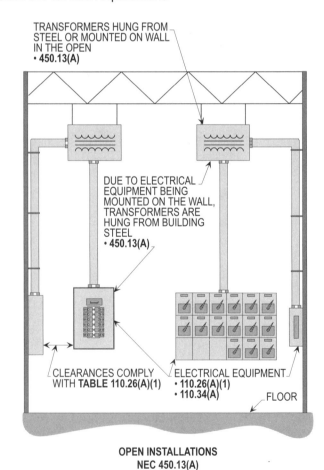

OPEN INSTALLATIONS
NEC 450.13(A)

Figure 8-2. Transformers hung from a wall or ceiling do not have to be readily accessible.

HOLLOW SPACE INSTALLATIONS
450.13(B)

Dry-type transformers, rated at not over 600 volts and 50 kVA or less, may be installed in hollow spaces of buildings. These transformers cannot be permanently closed in and there must be access to the transformers, but they do not have to be readily accessible per **Article 100** of the NEC. It was not clear in the 1993 or in previous editions of the NEC if dry-type transformers, not exceeding 600 volts, nominal, and rated 50 kVA or less, were permitted to be installed in the space above suspended ceilings (with removable panels) even if the transformer is accessible and provided with proper working clearances. Note that the space, where the transformer is installed, must comply with the ventilation requirements of **450.9**. If such ceiling space is used as a return air space for air-conditioning, **300.22(C)** must be reviewed and the provisions of this section must also be complied with. **(See Figure 8-3)**

> **Transformer Tip:** The two Subsections (**450.13(A) and (B)**) are for dry-type transformers. These subsections do not apply for oil or askarel-filled transformers due to possible oil damage or fire because of a rupture occurring in the case.

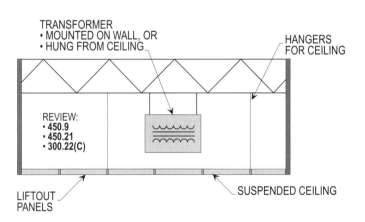

HOLLOW SPACE INSTALLATIONS
NEC 450.13(B)

Figure 8-3. Transformers mounted in a suspended ceiling (hollow space) do not have to be readily accessible.

CLEARANCES OF DRY-TYPE TRANSFORMERS INSTALLED INDOORS
450.21(A) AND (B)

The rules for installing dry-type transformers indoors can be summed up as follows:

NEC 450.21(A)

(1) Dry-type transformers (not totally enclosed) rated 112 1/2 kVA or less and 600 volts or less, must have a fire-resistant heat insulating barrier between transformers and combustible material, or without a barrier, must be separated at least 12 in. from the combustible material where the voltage is 600 volts or less. **(See Figure 8-4)**

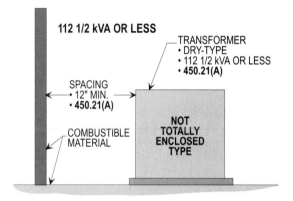

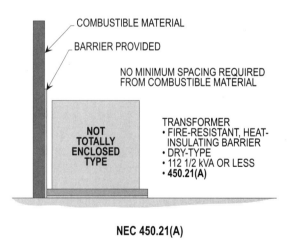

NEC 450.21(A)

Figure 8-4. Dry-type transformers rated 112 1/2 kVA or less and 600 volts or less must have a fire-resistant heat insulating barrier between transformers and combustible material, or, without a barrier, separation must be at least 12 in. from the combustible material where the voltage is 600 volts or less.

(2) Dry-type transformers rated 112 1/2 kVA or less and 600 volts or less, are not required to have a 12 in. separation or barrier, if they are completely enclosed except for vent openings. **(See Figure 8-5)**

(3) All indoor dry-type transformers of over 35,000 volts shall be installed in a vault. Vault requirements must fully comply with Part III to **Article 450** of the NEC. **(See Figure 8-6)**

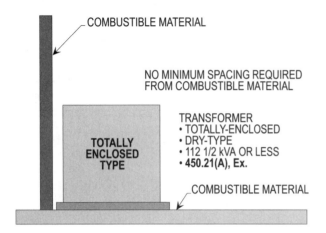

NEC 450.21(A), Ex.

Figure 8-5. Dry-type transformers rated 112 1/2 kVA or less and 600 volts or less are not required to have a 12 in. separation or barrier if they are completely enclosed except for vent openings.

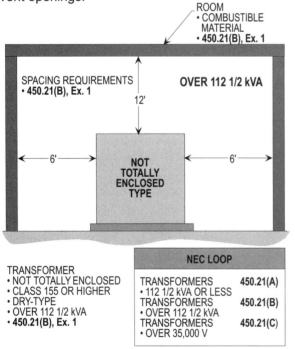

NEC 450.21(B), Ex. 1

Figure 8-6. All indoor dry-type transformers of over 35,000 volts shall be installed in a vault.

NEC 450.21(B), Ex. 1

(4) Dry-type transformers greater than 112 1/2 kVA, having Class 155 or higher insulation systems, must be installed in a fire-resistant transformer room. **(See Figure 8-7)**

NEC 450.21(B), Ex. 2

(5) Dry-type transformers greater than 112 1/2 kVA, having Class 155 or higher insulation systems, must have a fire-resistant heat-insulating barrier placed between transformers and combustible material, or, if no barrier, must be separated by at least 6 ft. horizontally and 12 ft. vertically from the combustible material. **(See Figure 8-8)**

Note that XFMR's ventilate from the bottom to protect from fire hazards. In case of a short-circuit a protective barrier should be installed beneath the XFMR opening.

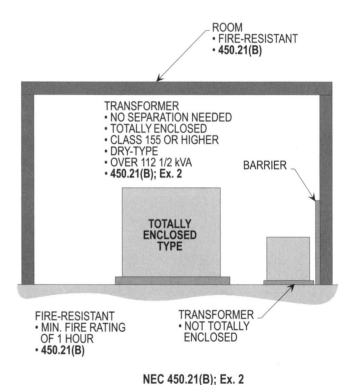

NEC 450.21(B); Ex. 2

Figure 8-7. Dry-type transformers greater than 112 1/2 kVA, having Class 155 or higher insulation systems, must have a fire-resistant heat-insulating barrier placed between transformers and combustible material, or, if no barrier, must be separated at least 6 ft. horizontally and 12 ft. vertically from the combustible material.

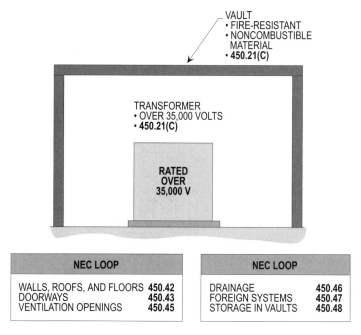

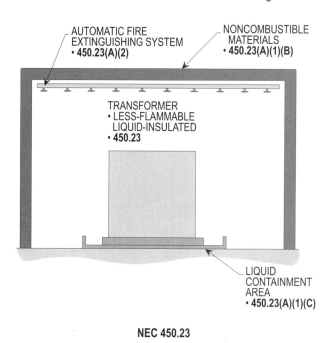

NEC LOOP	
WALLS, ROOFS, AND FLOORS	**450.42**
DOORWAYS	**450.43**
VENTILATION OPENINGS	**450.45**

NEC LOOP	
DRAINAGE	**450.46**
FOREIGN SYSTEMS	**450.47**
STORAGE IN VAULTS	**450.48**

NEC 450.21(C)

NEC 450.23

Figure 8-8. Dry-type transformers greater than 112 1/2 kVA, having Class 155 or higher insulation systems, must be installed in a fire-resistant transformer room or be totally enclosed.

Figure 8-9. Transformers using a "listed" high fire point liquid may be installed indoors, but only in "noncombustible" areas of "noncombustible" buildings. The NEC sets the minimum fire-point at 300°C (572°F).

DRY-TYPE TRANSFORMERS INSTALLED OUTDOORS 450.22

Dry-type transformers installed outdoors must have weatherproof enclosures. See the definition in **Article 100** of the NEC for the difference between "weatherproof" and "watertight."

LESS-FLAMMABLE LIQUID-INSULATED TRANSFORMERS 450.23

Transformers using a "listed" high fire point liquid may be installed indoors, but only in "noncombustible" areas of "noncombustible" buildings. The NEC sets the minimum fire-point at 300°C (572°F). This is the minimum temperature at which the liquid ignites. Such transformers may be installed indoors for voltages up to 35,000. Higher voltages require a vault when installed indoors due to the safety requirements for the higher voltage and associated equipment. **(See Figure 8-9)**

NONFLAMMABLE FLUID-INSULATED TRANSFORMERS 450.24

Transformers using a "dielectric" nonflammable liquid may be installed indoors in any location, for voltages up to 35,000. Higher voltages require a vault when installed indoors due to the safety required for the higher voltage and associated equipment.

For the purpose of this section, a nonflammable dielectric fluid is one that does not have a flash point or fire point that is flammable in the air.

ASKAREL-INSULATED TRANSFORMERS INSTALLED INDOORS 450.25

Askarel is a liquid that does not burn; therefore, it is safer than oil to be used as a transformer liquid. However, even though arcing in askarel-insulated transformers does produce large amounts of gas, it is nonexplosive.

Askarel-insulated transformers of over 25 kVA shall be furnished with a relief vent such as a chimney to relieve the pressure built up by gases that may be generated within the transformer by arcing.

In rooms that are well ventilated, the vent may be discharged directly into the room. In rooms that are poorly ventilated the vent must be piped to a flue or chimney that is capable of carrying the gases out of the room. Or, as an alternative for such ventilation, the transformer can be fitted with a gas absorber placed inside the case. When there is a gas absorber, the vent may also be discharged into the room.

Askarel-insulated transformers of more than 35,000 volts must be installed in a vault because of the oil and higher voltage being hazardous to personnel. **(See Figure 8-10)**

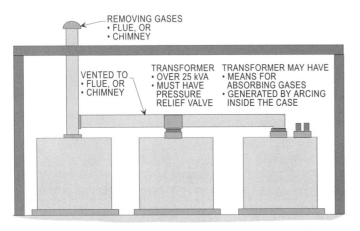

TRANSFORMERS SHALL BE INSTALLED IN A
VAULT WHEN OVER 35,000 VOLTS
NEC 450.25

Figure 8-10. Askarel-insulated transformers of over 25 kVA must be furnished with a relief vent such as a chimney to relieve the pressure built up by gases that may be generated within the transformer.

OIL-INSULATED TRANSFORMERS INSTALLED INDOORS
450.26

The rules for installing oil-insulated transformers installed indoors can be summed up as follows:

(1) Indoor, oil-filled transformers, greater than 600 volts, must be installed in a vault, with the following exceptions, where, regardless of voltage, a vault is not required:

(a) Electric furnace transformers, with a total rating of 75 kVA or less, may be located in a fire-resistant room.

(b) Oil-filled transformers may be installed in a building without a vault, provided the building is accessible to qualified personnel only and used solely for the provision of electric service to other buildings.

(2) A transformer of 75 kVA is permitted to supply a voltage of 600 volts or less that is an integral part of charged particle accelerating equipment. Such transformer can be installed without a vault, if located in a building or room of noncombustible of fire-resistant construction and provided with suitable arrangements to prevent a transformer oil fire from spreading to other combustible material.

(3) If suitable provisions are provided to prevent a possible oil fire from igniting other materials, oil-filled transformers of 600 volts or less may be installed without a vault. When installed without a vault, the total kVA ratings of all transformers allowed in a room or section of a building is limited to 10 kVA for non-fire-resistant buildings, and 75 kVA for fire-resistant buildings. **Note:** Before installing such transformers, review Ex.'s (1) through (6) very carefully.

See **Figure 8-11** for installation rules when applying **Ex. 1 to 450.26.**

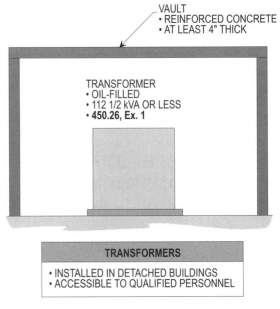

NEC 450.26, Ex. 1

Figure 8-11. Oil-insulated transformers rated at 112 1/2 kVA or less can be installed in a vault constructed of reinforced concrete not less than 4 in thick.

OIL-INSULATED TRANSFORMERS INSTALLED OUTDOORS
450.27

When oil-filled transformers are installed on or adjacent to combustible buildings or material, the building or material must be safeguarded from possible fire originating in a transformer. Fire-resistant barriers and water-spray systems are enclosures that are permitted to be safeguards, if installed by the NEC. **(See Figures 8-12(a) and (b))**

MODIFICATION OF TRANSFORMERS
450.28

When modifications are applied to a transformer in an existing installation, the following rules and regulations must be adhered to:

(1) The type of insulating liquid installed must be marked on the transformer.

(2) Any modifications must comply with applicable requirements for the modified transformer.

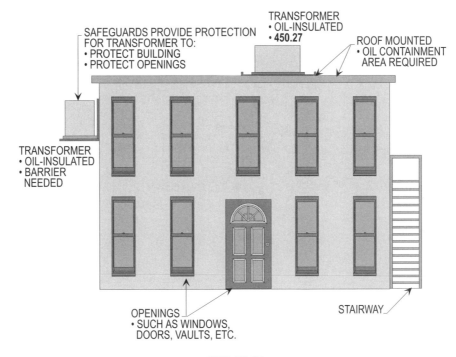

NEC 450.27

Figure 8-12(a). This illustration represents an oil containment area as a permitted safeguard from possible fire originating in a transformer. **Note:** Review **450.27(1) through (4)** very carefully.

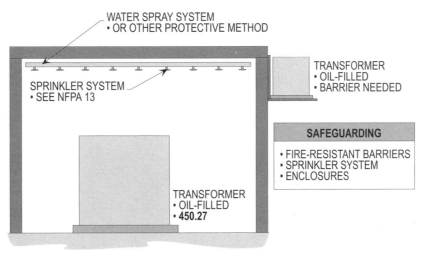

NEC 450.27

Figure 8-12(b). This illustration represent a water-spray system that is used as a permitted safeguard.

Chapter 8: Installing Transformers

Section	Answer		
_____	T F	**1.**	Transformers must be provided with markings on the nameplate.
_____	T F	**2.**	Transformers must be elevated at least 10 ft. above the floor to prevent unauthorized personnel from contact.
_____	T F	**3.**	Transformers hung from wall or ceiling have to be readily accessible.
_____	T F	**4.**	Transformers mounted in ceiling do not have to be readily accessible.
_____	T F	**5.**	All indoor dry-type transformers over 25,000 volts must be installed in a vault.
_____	T F	**6.**	Transformers using a "dielectric" nonflammable liquid may be installed indoors in any location, for voltages up to 35,000 volts.
_____	T F	**7.**	Askarel-insulated transformers of over 35 kVA must be furnished with a relief vent.
_____	T F	**8.**	Askarel transformers of more than 35,000 volts must be installed in a vault.
_____	T F	**9.**	Electric furnace transformers with a total rating of 75 kVA or less may be located in a fire-resistant room.
_____	T F	**10.**	The type of insulating liquid installed must be marked on the transformer when modifications are applied to a transformer.
_____ _____		**11.**	Transformers must be isolated in a room or accessible only to _____ personnel to prevent accidental contact with live parts.
_____ _____		**12.**	Transformers must be located where _____ accessible to qualified personnel for inspection and maintenance.
_____ _____		**13.**	Dry-type transformers greater than 112 1/2 kVA, having Class 155 or higher insulation systems, without a barrier must be separated at least _____ ft. horizontally and _____ ft. vertically from the combustible material.
_____ _____		**14.**	Dry-type transformers less than 112 1/2 kVA, without a barrier must be separated at least _____ in. from the combustible material where the voltage is 600 volts or less.
_____ _____		**15.**	Dry-type transformers rated 112 1/2 kVA or less and 600 volts or less, must have a _____ in. separation, if they are completely enclosed.
_____ _____		**16.**	Dry-type transformers installed outdoors must have _____ enclosures.
_____ _____		**17.**	Less-flammable liquid-insulated transformers may be installed indoors for voltages up to _____ volts.

Transformer Vaults

Vaults can be used to house dry-type transformers that are rated over 35 kV or transformers filled with combustible material that is used as an aid in cooling their windings.

Vaults must be designed and built with specific rules and regulations.

Wherever possible, transformer vaults must be located at an outside wall of the building. These rules are intended to allow direct ventilation to the outside of the building without the use of ducts, flues, etc. per **450.45**.

WALLS, ROOFS, AND FLOORS
450.42

Requirements for the construction of transformer vaults are set forth in this Section. The floor of a vault, its walls, and roof must be made of fire-resistant material, such as concrete, and capable of withstanding heat from a fire within, for at least three hours. A 6 in. thickness is recommended for the walls and roof per **450.42, FPN 2**. The floor, when laid and in contact with the earth, must be at least 4 in. thick. **(See Figure 9-1)**

DOORWAYS
450.43

The door to a transformer vault must be built according to the standards of the National Fire Protection Association that requires a three hour fire rating. The door sill must be at least 4 in. high. This is to prevent any oil that may accumulate on the floor from running out of the transformer room and moving to other areas. Doors must be kept locked at all times to prevent access by unqualified persons to the vault.

> **Transformer Tip:** Personnel doors must swing out and be equipped with panic bars, pressure plates, or other devices that open under simple pressure.

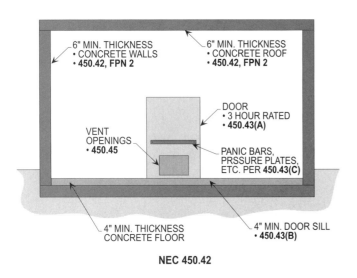

NEC 450.42

Figure 9-1. Floors, walls, and roofs must be of fire-resistant material, such as concrete, and capable of withstanding heat from a fire within for at least three hours. A 6 in. thickness such as reinforced concrete is specified for the walls and roof. The floor, when laid and in contact with the earth, must be at least 4 in. thick.

VENTILATION OPENINGS
450.45

Where ventilation is directed to the outside, without the use of ducts or flues, the vent opening must have an area of at least three square inches for each kVA of transformer capacity, but never less than one square foot in area. The vent opening must be fitted with a screen or grating, and also with an automatic closing damper. If ducts are used in the vent system, the ducts must have sufficient capacity to maintain a suitable vault temperature. **(See Figure 9-2)**

DRAINAGE
450.46

Drains should be provided in a transformer vault in order to drain off oil that might accumulate on the floor due to a leak in a transformer caused by an accident. This rule is designed to prevent a fire hazard from occurring.

WATER PIPES AND ACCESSORIES
450.47

Piping for the protection of the equipment or for water-cooled transformers may be present in a vault. However, no other piping or duct system can enter or pass through the vault. Neither are valves or other fittings of a foreign piping or duct system ever permitted in a vault containing transformers. **(See Figure 9-3)**

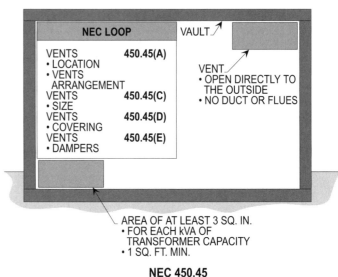

NEC 450.45

Figure 9-2. Where ventilation is directed to the outside, without the use of ducts or flues, the vent opening must have an area of at least three square inches for each kVA of transformer capacity, but never less than one square foot in area.

STORAGE IN VAULTS
450.48

No storage of any kind may be put in a transformer vault other than the transformers and equipment necessary for their operation. This typically means that transformer vaults are not to be used for warehouses or storage areas but to contain transformers and accessories only. The reason that the vault is to be kept clear is due to the high-voltage and safety measures that are needed for personnel servicing the equipment. Also, consideration must be given to materials that can be considered a fire threat under certain conditions. **(See Figure 9-4)**

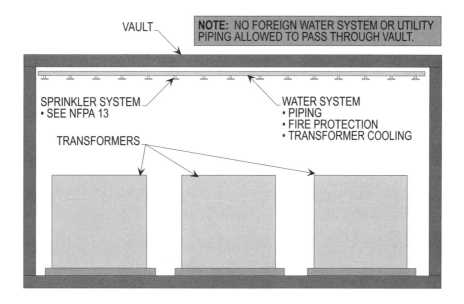

NEC 450.47

Figure 9-3. Only piping for the protection of equipment within the vault or piping to water-cooled transformers may be present in a transformer vault.

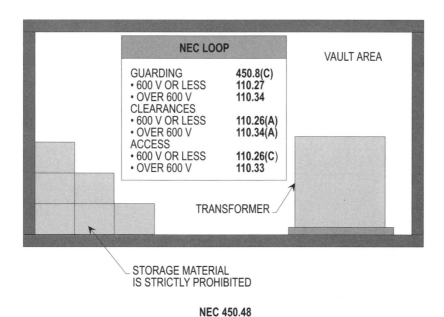

NEC 450.48

Figure 9-4. No storage of any kind can be put in a vault other than the transformers and the equipment necessary for their operation.

Chapter 9: Transformer Vaults

Section	Answer			
_____	T	F	1.	The floor, walls, and roof of a transformer vault must be of fire-resistant material.
_____	T	F	2.	Doors to transformer vaults must not be required to be kept locked at all times.
_____	T	F	3.	Transformer vaults must be provided with floor drains to drain off oil that might accumulate.
_____	T	F	4.	Other piping or duct systems are allowed to enter or pass through a transformer vault.
_____	T	F	5.	Storage is permitted within a transformer vault.

6. The floor, when laid and in contact with the earth, must be at least _____ in. thick.

7. The walls and roof must have a _____ in. thickness.

8. The door to a transformer vault must have at least a _____ hour fire rating.

9. The door sill must be at least _____ in. high.

10. A ventilation opening must have an area of at least _____ square inches for each kVA of transformer capacity.

Sizing Transformers and Connections

The total volt-amps of all loads in a building must be used to size a transformer. Depending on the load requirements of a building, single-phase and three-phase voltage or a combination of such may be used to supply the building. The windings can be connected in configurations necessary to provide the proper voltage needed to supply the load requirements of the facility.

SIZING WYE-CONNECTED SECONDARIES

The size transformers required to supply a wye-connected secondary system can be found by applying one of the following two procedures:

(1) Adding the total single-phase and three-phase loads together for a single three-phase transformer, or

(2) Multiplying the load in VA by 1/3 (.33) to derive three single-phase transformers.

The kVA rating of three single-phase transformers, if they are separately connected together, will add up to a single three-phase transformer. A single three-phase transformer rating can be sized and selected from the total volt-amps. One transformer with three windings is sized by adding the total VA of all the loads together and selecting the transformer's kVA rating based upon this value per Table 12 from the Troubleshooting Tables in the back of this book. **(See Figure 10-1)**

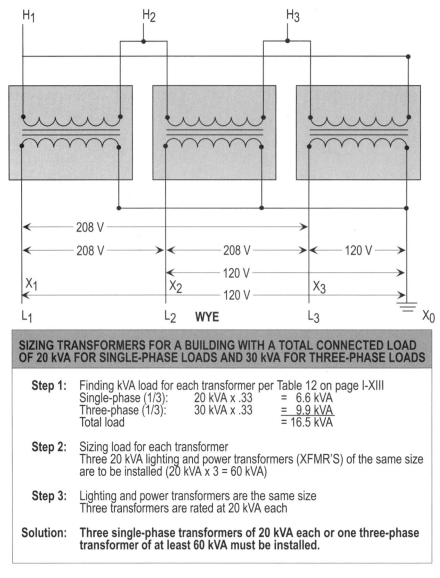

Figure 10-1. Sizing wye-connected transformers with single-phase and three-phase loads.

SIZING CLOSED DELTA-CONNECTED SECONDARIES

The size transformer required to supply a closed delta-connected secondary system can be found by using the following methods:

(1).

(a) Multiply the single-phase load in VA by 67 percent.

(b) Multiply the three-phase load in VA by 33 percent.

Now add the kVA load of (a) and (b) together to derive one lighting and power transformer.

(2).

(a) Multiply the single-phase and three-phase loads in VA by 33 percent.

Use this total in kVA to derive two power transformers.

See Figure 10-2 for the rules and regulations for sizing the transformer used in a closed delta-connected system.

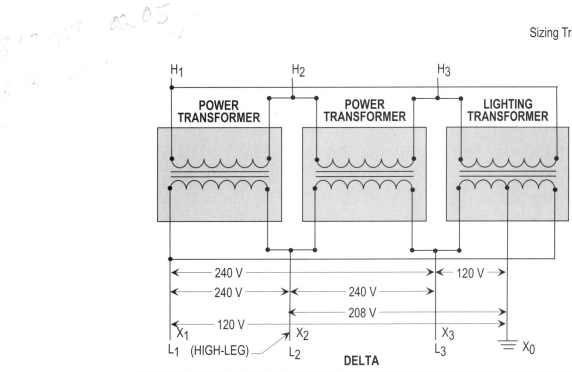

Figure 10-2. Sizing closed delta-connected transformers with single-phase and three-phase loads.

SIZING TRANSFORMERS FOR A BUILDING WITH A TOTAL CONNECTED LOAD OF 20 kVA FOR SINGLE-PHASE LOADS AND 30 kVA FOR THREE-PHASE LOADS

Step 1: Finding kVA load for lighting transformers per Table 12 on page I-XIII
Single-phase (2/3): 20 kVA x 67% = 13.4 kVA
Three-phase (1/3): 30 kVA x 33% = 9.9 kVA
Total load = 23.3 kVA

Solution: **Lighting transformer is 25 kVA.**

Step 1: Finding kVA load for power transformers
Single-phase(1/3): 20 kVA x 33% = 6.6 kVA
Three-phase (1/3): 30 kVA x 33% = 9.9 kVA
Total load = 16.5 kVA

Solution: **Two power transformers of 20 kVA each are required.**

Note: Lighting transformer is 25 kVA and the two power transformers are 20 kVA each.

CLOSED DELTA-CONNECTED SECONDARIES

SIZING OPEN DELTA-CONNECTED SECONDARIES

Open delta-connected secondary systems can be determined by calculating the single-phase load at 100 percent and the three-phase load at 58 percent and using the total value to size the transformer. By adding these two loads together, the size of a lighting transformer can be determined. A power transformer can be sized by calculating the three-phase load at 58 percent which is the reciprocal of the square root of 3 (1 ÷ 1.732 = 58%). This reduced total is then used to size the power transformer which will be smaller in rating than the lighting transformer. **(See Figure 10-3)**

SIZING AUTOTRANSFORMERS

Autotransformers are used to boost or buck voltage by multiplying the nameplate kVA by 1000 and then dividing by the secondary voltage. Autotransformers must be equipped with a kVA, amperage, and secondary voltage rating having enough capacity to supply the load served.

For more information on autotransformers, see **210.9, 215.11, 450.4**, and **450.5** of the NEC.

For example: What are the secondary amps for an autotransformer rated 2.5 kVA with a secondary voltage of 24 volts? The supply voltage is boosted from a 208 to 230 volt, single-phase system?

Step 1: Finding sec. A
Sec. A = kVA x 1000 ÷ Sec. V
Sec. A = 2.5 kVA x 1000 ÷ 24 V
Sec. A = 104

Solution: The secondary amperage is 104 amps.

For example: By multiplying the output volts by the secondary amps, then dividing by 1000, the kVA of the autotransformer can be sized?

Step 1: Finding secondary kVA
kVA = output V x Sec. A ÷ 1000
kVA = 230 V x 104 A ÷ 1000
kVA = 23.9

Solution: The kVA of the load is 23.9 kVA.

Note: The size of the autotransformer must be capable of handling a load of 23.9 kVA.

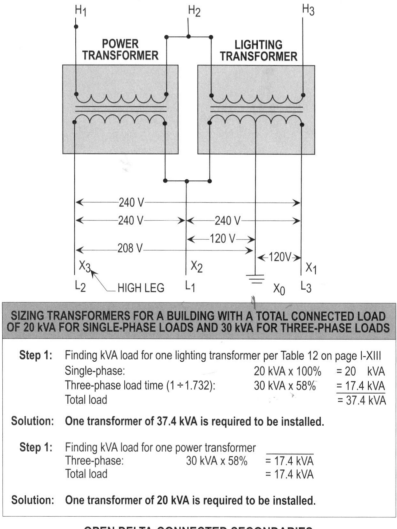

OPEN DELTA-CONNECTED SECONDARIES

Figure 10-3. Sizing open delta-connected transformers with single-phase and three-phase loads.

For example: What is the required rating for an autotransformer with a 24 volt secondary voltage serving a 230 volt motor with a connected load of 12,000 volt-amps?

Step 1: Finding A
A = load served ÷ supply V
A = 12,000 VA ÷ 230 V
A = 52

Step 2: Sizing VA (Round Up kVA)
AXFMR = A x sec. V
AXFMR = 52 A x 24 V
AXFMR = 1248 VA

Solution: The autotransformer must supply a load of 1.25 kVA. (1248 ÷ 1000 = 1.25 kVA)

Transformer Tip: The autotransformer must have a rating of at least 1248 VA with a transformation voltage of 24 volts and supply a load of 12,000 VA.

See Figure 10-4 for a detailed procedure for sizing an autotransformer to supply a motor circuit in a commercial or industrial application. Note that the supply voltage used is too low for the motor to operate properly. The autotransformer is to be used to boost the voltage from 185 volts to 208 volts.

See Figure 10-5 for a detailed procedure for sizing an autotransformer to supply a motor circuit in a commercial or industrial application. Note that the supply voltage is too high for the motor to operate properly. The voltage can be reduced by a buck-type autotransformer if the supply voltage is too high. The autotransformer is used to reduce the voltage from 269 volts to 240 volts.

SIZING CONNECTIONS FROM THE SECONDARY OF TRANSFORMERS 240.21(B); (C)

Overcurrent protection devices of circuits must be located at the point where the service to those circuits originates. However, it is permitted to make connections from the secondary side of transformers. Sizing connections not over 10 ft. long must be designed and installed per **240.21(C)(2)**. Connections not over 25 ft. long must be designed and

installed per **240.21(B)(3), (C)(5) and (C)(6)**. Transformer secondary conductors of separately derived systems for industrial locations are sized per **240.21(C)(3)**. Outside transformer connections are sized per **240.21(C)(4)**. Overcurrent protection must be provided by the rules found in **450.3(B)** and **Table 450.3(B)**. **(See Figure 10-6)**

SIZING CONNECTIONS NOT OVER 10 FT. LONG 240.21(C)(2)

Conductors not over 10 ft. long are not required to have overcurrent protection at the connection to a transformer secondary, if all of the following conditions are met:

(1) Connected conductors have a current rating not less than the combined calculated load of the circuits supplied by the connected conductors and their ampacity is not less than the rating of the overcurrent protection device at the termination of the connected conductors.

(2) The connected conductors shall not extend beyond the switchboard, panelboard, disconnecting means, or control devices they supply.

(3) Connected conductors are enclosed in a raceway that extends from the connection to the enclosure of any enclosed switchboard, panelboard, or control devices, or to the back of an open switchboard.

(4) The rating of the overcurrent device on the secondary side of the connecting conductors protect the conductors and the output of the transformer from overload. If more than 10 percent of the single-pole circuit breakers of a panelboard, rated at 30 amps or less, are used for branch-circuits to lighting and appliance loads with neutral connections, the panelboard is classified as a lighting and appliance panelboard per **408.34** and **408.35**. When the panelboard is classified as a lighting and appliance panelboard, overcurrent protection is restricted to two mains or less per **408.36(A)**.

See Figure 10-7 for the proper procedure for making a connection using the 10 ft. transformer rule.

SIZING CONNECTIONS NOT OVER 25 FT. LONG 240.21(B)(3); (C)(5)

Conductors, not over 25 ft. long, supplying a transformer plus primary and secondary may be tapped from a feeder and without overcurrent protection at the tap but only when all of the following conditions are met:

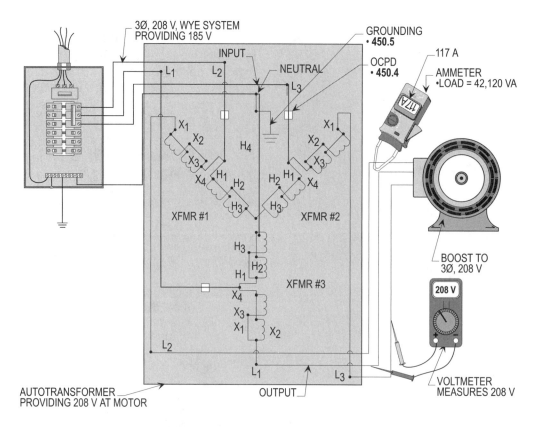

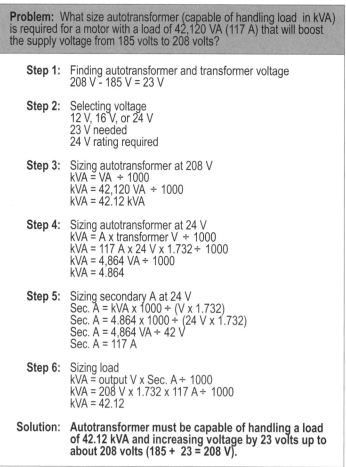

Problem: What size autotransformer (capable of handling load in kVA) is required for a motor with a load of 42,120 VA (117 A) that will boost the supply voltage from 185 volts to 208 volts?

Step 1: Finding autotransformer and transformer voltage
208 V - 185 V = 23 V

Step 2: Selecting voltage
12 V, 16 V, or 24 V
23 V needed
24 V rating required

Step 3: Sizing autotransformer at 208 V
kVA = VA ÷ 1000
kVA = 42,120 VA ÷ 1000
kVA = 42.12 kVA

Step 4: Sizing autotransformer at 24 V
kVA = A x transformer V ÷ 1000
kVA = 117 A x 24 V x 1.732 ÷ 1000
kVA = 4,864 VA ÷ 1000
kVA = 4.864

Step 5: Sizing secondary A at 24 V
Sec. A = kVA x 1000 ÷ (V x 1.732)
Sec. A = 4.864 x 1000 ÷ (24 V x 1.732)
Sec. A = 4,864 VA ÷ 42 V
Sec. A = 117 A

Step 6: Sizing load
kVA = output V x Sec. A ÷ 1000
kVA = 208 V x 1.732 x 117 A ÷ 1000
kVA = 42.12

Solution: **Autotransformer must be capable of handling a load of 42.12 kVA and increasing voltage by 23 volts up to about 208 volts (185 + 23 = 208 V).**

SIZING AUTOTRANSFORMERS

Figure 10-4. Sizing and selecting an autotransformer to boost the supply voltage to the motor.

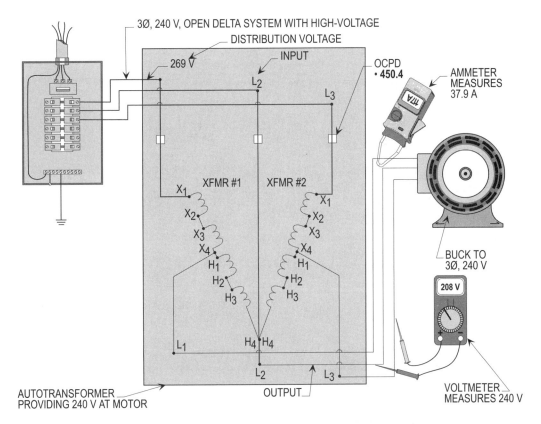

SIZING AUTOTRANSFORMERS

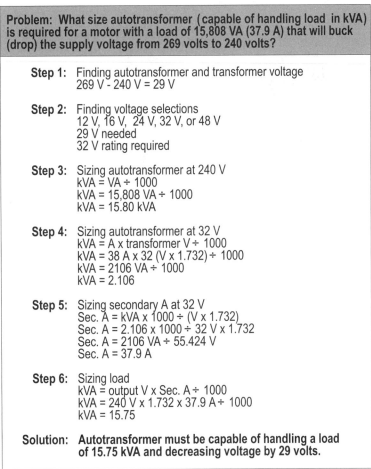

Figure 10-5. Sizing and selecting an autotransformer to buck the supply voltage to the motor.

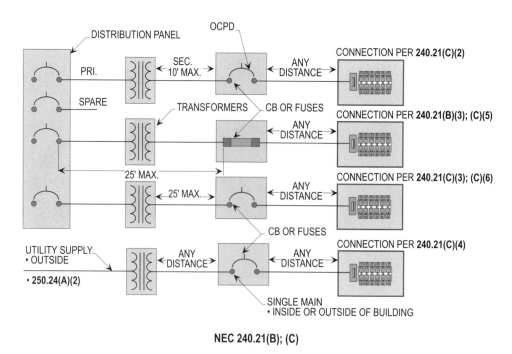

Figure 10-6. The illustration above shows the four most used transformer connections utilized to supply electrical systems.

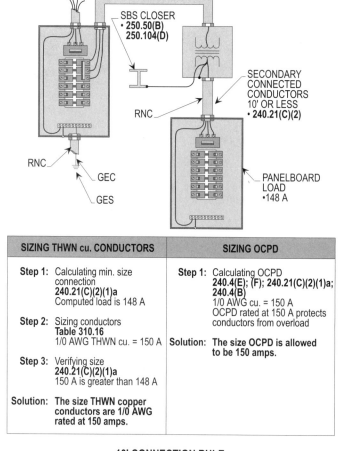

SIZING THWN cu. CONDUCTORS	SIZING OCPD
Step 1: Calculating min. size connection **240.21(C)(2)(1)a** Computed load is 148 A	**Step 1:** Calculating OCPD **240.4(E); (F); 240.21(C)(2)(1)a; 240.4(B)** 1/0 AWG cu. = 150 A OCPD rated at 150 A protects conductors from overload
Step 2: Sizing conductors **Table 310.16** 1/0 AWG THWN cu. = 150 A	**Solution:** The size OCPD is allowed to be 150 amps.
Step 3: Verifying size **240.21(C)(2)(1)a** 150 A is greater than 148 A	
Solution: The size THWN copper conductors are 1/0 AWG rated at 150 amps.	

10' CONNECTION RULE
NEC 240.21(C)(2)

Figure 10-7. The above illustration shows the procedure for sizing a 10 ft. connection from the secondary of a transformer.

(1) Tap conductors supplying the primary have an ampacity of at least 1/3 of the rating of the feeder being tapped.

(2) Connected conductors supplied by the secondary have an ampacity of at least 1/3 of the rating of the feeder being connected, based on the primary-to-secondary voltage ratio.

(3) The total length of one primary plus one secondary conductor is not be over 25 ft.

(4) The primary and secondary conductors are enclosed in an approved raceway or other approved means.

(5) Secondary conductors are terminated in a single circuit breaker or set of fuses, sized to protect the secondary.

See Figure 10-8 for the proper procedure for making a secondary connection using the 25 ft. rule.

TRANSFORMER SECONDARY CONDUCTORS OF SEPARATELY DERIVED SYSTEMS FOR INDUSTRIAL LOCATIONS 240.21(C)(3)

Conductors, not over 25 ft. long, may be connected to a transformer secondary of a separately derived system for industrial locations, without overcurrent protection at the connection, when all of the following conditions are met:

(1) Ampacity of connected conductors are equivalent to the current rating of the transformer and the overcurrent protection devices do not exceed the ampacity of the connected conductors.

(2) All overcurrent devices are grouped.

(3) Connected conductors are protected from physical damage.

Note: Also, See **240.21(C)(6)** for 25 ft. rule.

See Figures 10-9 and 10-15 for the procedure to be applied when a 25 ft. connection is installed from the secondary side of a transformer.

OUTSIDE TRANSFORMER CONNECTION 240.21(C)(4)

Outside conductors can be connected to a feeder or at the transformer secondary, without overcurrent protection at the

SIZING PRIMARY CONDUCTORS	
Step 1:	Calculating pri. tap **240.21(B)(3)(1)** 1/3 of 250 A = 83 A
Step 2:	Selecting conductors **Table 310.16** 83 A requires 4 AWG cu.
Solution:	**The size THWN copper conductors are 4 AWG.**

SIZING SECONDARY CONDUCTORS	
Step 1:	Calculating sec. connection **240.21(B)(3)(2)** 480 V ÷ 208 V x 1/3 x 250 A = 192 A
Step 2:	Selecting conductors **Table 310.16** 192 A requires 3/0 AWG
Solution:	**The size THWN copper conductors are 3/0 AWG.**

SIZING SECONDARY OCPD	
Step 1:	Selecting OCPD in sec. connection **240.4(E); (F); 240.21(B)(3)(2); 240.21(C)(5); 240.6(A)** 200 A (3/0) requires 200 A
Solution:	**The size OCPD is 200 amps.**

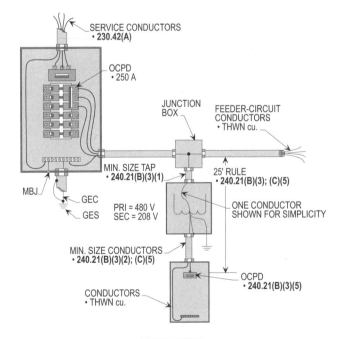

25' XFMR RULE
PRIMARY AND SECONDARY
NEC 240.21(B)(3); (C)(5)

Figure 10-8. The primary tap for this connection rule must be at least 1/3 of the OCP device protecting the larger feeder-circuit conductors. The secondary connected conductors must be at least 1/3 of the OCPD protecting the feeder-circuit conductors based on the primary-secondary transformer ratio.

connection, when all of the following conditions are complied with:

(1) The connected conductors are suitably protected from physical damage in an approved manner.

(2) The conductors terminate at a single circuit breaker or a single set of fuses that will limit the load to the ampacity of the conductors. This single overcurrent protection device can supply any number of additional overcurrent devices on its load side.

(3) The overcurrent protection device for the conductors is an integral part of a disconnecting means or is located immediately adjacent thereto.

(4) The disconnecting means for the conductors are installed at a readily accessible location, either outside of a building or structure or inside, nearest the point of entrance of the conductors.

See Figure 10-10 for the rules pertaining to outside transformer connections from the secondary side of transformers.

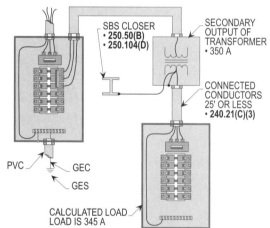

SIZING THWN cu. CONDUCTORS	SIZING OCPD
Step 1: Calculating min. size connection **240.21(C)(3)(1)** Computed load is 345 A	**Step 1:** Calculating OCPD **240.3(E); 240.21(C)(3)(1); 240.3(B)** 500 KCMIL cu. = 380 A OCPD rated at 350 A protects conductors from overload
Step 2: Sizing conductors **Table 310.16** 500 AWG THWN cu. = 380 A	
Step 3: Verifying size **240.21(C)(3)(1)** 380 A is greater than 345 A	**Solution:** The size OCPD is allowed to be 350 amps.
Solution: The size THWN copper conductors are 500 KCMIL rated at 380 amps.	

25' CONNECTION RULE
NEC 240.21(C)(3)

Figure 10-9. The above illustration shows the procedure for sizing a 25 ft. connection from the secondary of a transformer installed in an industrial plant.

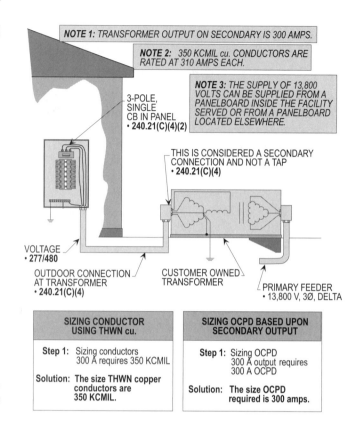

SIZING CONDUCTOR USING THWN cu.	SIZING OCPD BASED UPON SECONDARY OUTPUT
Step 1: Sizing conductors 300 A requires 350 KCMIL	**Step 1:** Sizing OCPD 300 A output requires 300 A OCPD
Solution: The size THWN copper conductors are 350 KCMIL.	**Solution:** The size OCPD required is 300 amps.

OUTSIDE CONNECTION RULE
NEC 240.21(C)(4)

Figure 10-10. The above illustration shows the rules for sizing the conductors and OCPD for a feeder connection from a transformer located outside. The 13,800 volts can be supplied from a panelboard inside the facility.

DETERMINING A LIGHTING AND APPLIANCE PANELBOARD
408.34(A); 408.35; 408.36(A)

If more than 10 percent of the single-pole circuit breakers of a panelboard are rated at 30 amps or less and used for branch-circuits to lighting and appliance loads with neutral connections, the panelboard is classified as a lighting and appliance panelboard per **408.34(A) and 408.35**. When the panelboard is classified as a lighting and appliance panelboard, overcurrent protection is restricted to two mains or less per **408.36(A)**.

Note that a main circuit breaker is not required to be installed in a panelboard supplied by a feeder-circuit from a service panel or subfeed panel supplied from a feeder-circuit.

TRANSFORMER SECONDARY CONDUCTORS IN LENGTHS OF OVER 10 FT. TO 25 FT.
240.21(C)(6)

Conductors over 10 ft. and up to 25 ft. in length can be connected to the secondary size of a transformer. When applying this section of the code, the 25 ft. (7.5 m) secondary connection shall be terminated in a single OCPD (CB or fuses) to limit the load and to also comply with the 1/3 rule when multiplied by the secondary-to-primary voltage ratio. The secondary conductors are required to be protected from physical damage and abuse. **(See Figure 10-11)**

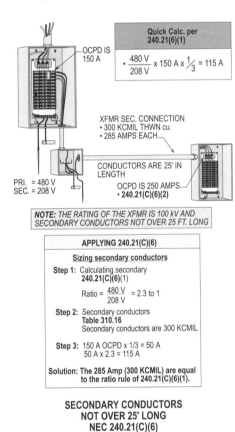

Figure 10-11. The above shows the rules for making a 25 ft. secondary conductor connection in other than industrial locations.

SUPERVISED INDUSTRIAL INSTALLATIONS FEEDER AND BRANCH-CIRCUIT CONDUCTORS
240.92(A)

Feeder and branch-circuit conductors must be protected at the point where the conductors receive their supply.

However, it permits a variation of requirements for transformer secondary conductors terminated from separately derived systems and outside feeder connections. **(See Figure 10-12)**

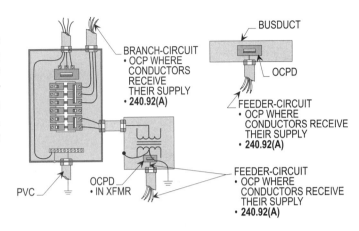

NEC 240.92(A)

Figure 10-12. The above illustration shows feeder and branch-circuit conductors protected at the point where the conductors receive their supply. (Also, see **240.21(C)(4)** for a similar rule).

SUPERVISED INDUSTRIAL INSTALLATIONS CONNECTIONS UP TO 100 FT.
240.92(B)(1)(1)

Unprotected lengths of secondary conductors are allowed up to 100 ft., if the transformer primary overcurrent device is sized at a value (reflected to the secondary by the transformer secondary to primary voltage ratio) of not more than 150 percent of the secondary conductor ampacity. **(See Figure 10-13)**

SUPERVISED INDUSTRIAL INSTALLATIONS CONNECTIONS UP TO 100 FT.
240.92(B)(1)(2)

Secondary conductors can be connected up to 100 ft. in length, if protected by a differential relay with a trip setting equal to or less than the conductor ampacity. Note that a differential relay provides superior short-circuit protection at a trip open value that is almost always well below the conductor ampacity. **(See Figure 10-13)**

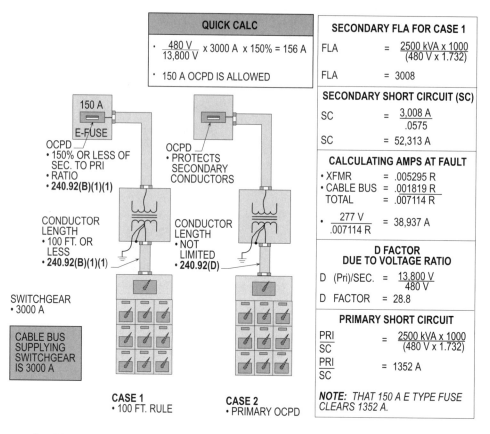

QUICK CALC
• $\dfrac{480\ V}{13,800\ V}$ x 3000 A x 150% = 156 A
• 150 A OCPD IS ALLOWED

SECONDARY FLA FOR CASE 1	
FLA	= $\dfrac{2500\ kVA \times 1000}{(480\ V \times 1.732)}$
FLA	= 3008

SECONDARY SHORT CIRCUIT (SC)	
SC	= $\dfrac{3,008\ A}{.0575}$
SC	= 52,313 A

CALCULATING AMPS AT FAULT	
• XFMR	= .005295 R
• CABLE BUS	= .001819 R
TOTAL	= .007114 R
• $\dfrac{277\ V}{.007114\ R}$	= 38,937 A

D FACTOR DUE TO VOLTAGE RATIO	
D (Pri)/SEC.	= $\dfrac{13,800\ V}{480\ V}$
D FACTOR	= 28.8

PRIMARY SHORT CIRCUIT	
$\dfrac{PRI}{SC}$	= $\dfrac{2500\ kVA \times 1000}{(480\ V \times 1.732)}$
$\dfrac{PRI}{SC}$	= 1352 A

NOTE: *THAT 150 A E TYPE FUSE CLEARS 1352 A.*

150 A
E-FUSE

OCPD
• 150% OR LESS OF SEC. TO PRI
• RATIO
• 240.92(B)(1)(1)

CONDUCTOR LENGTH
• 100 FT. OR LESS
• 240.92(B)(1)(1)

SWITCHGEAR
• 3000 A

CABLE BUS SUPPLYING SWITCHGEAR IS 3000 A

OCPD
• PROTECTS SECONDARY CONDUCTORS

CONDUCTOR LENGTH
• NOT LIMITED
• 240.92(D)

CASE 1
• 100 FT. RULE

CASE 2
• PRIMARY OCPD

Consider a 2500 kVA transformer with a 13.8 kV to 480/277 ratio and 500 mVA available short-circuit current on the primary.

For 100 circuit feet of 3000 amp cable bus, and a three-phase bolted fault at the end of the bus (worst case), about 38,937 amps will flow from the system or 1352 amps on the primary, which will clear a typical 150 E fuse (which meets the maximum 150 percent requirement) within 42 seconds.

This time vs. current value is well within the rating of the secondary conductors. (Cable bus has a resistance of .001819 and the XFMR has a resistance of .005295.)

NEC 240.92(B)(1)(1); (2); (3)

Figure 10-13. The above illustration shows methods of providing short-circuit and ground-fault protection for transformers and conductors.

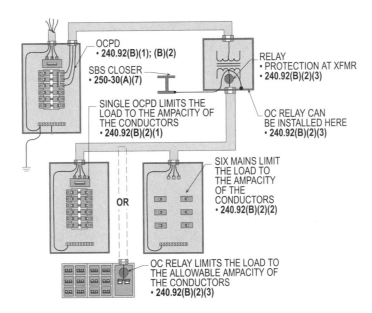

NEC 240.92(B)(2)

Figure 10-14. The above illustration shows methods of providing overload protection.

SUPERVISED INDUSTRIAL INSTALLATIONS ENGINEERING SUPERVISION 240.92(B)(1)(3)

Conductors up to 100 ft. in length are allowed, if calculations are made under engineering supervision and it is determined that the secondary conductors will be protected within recognized time versus current limits for all short-circuits and ground-fault conditions that could occur. **(See Figure 10-13)**

SUPERVISED INDUSTRIAL INSTALLATIONS OVERLOAD PROTECTION 240.92(B)(2)

When providing overload protection, the secondary conductors can be terminated in a single overcurrent protection device or in the lugs of the bus, if not more than six overcurrent protection devices with a combined rating, are installed and do not exceed the ampacity of the conductors. Another method of protection is to provide overload current relaying with the ability (designed into) to trip either the primary overcurrent protection devices or the those downstream, so that the load current doesn't exceed the conductor's ampacity. **(See Figure 10-14)**

> **Transformer Tip:** In some installations, the short-circuit and ground-fault protective arrangements may provide overload protection. Note that if engineering calculations prove this to be the case, separate overload protection isn't really needed.

SUPERVISED INDUSTRIAL INSTALLATIONS OUTSIDE FEEDER CONNECTIONS 240.92(C)

Section **240.92(C)** allows alternate means of protecting transformer secondary conductors in supervised industrial installations where the transformer is located outside. The secondary conductors must be protected against (1) overloads, with the additional stipulation that (2) they are suitably protected against physical damage. **(See Figure 10-15)**

Note: For connections of unlimited lengths, see **240.92(D)** of the NEC and case 2 in **Figure 10-13**.

> **Transformer Tip:** Such protection can be provided by six or less OCPD's, where the total rating does not exceed the ampacity of the conductors routed per **240.92(C)** and **240.21(C)(4)**. Note that up to six OCPD's can be used instead of just one installed at the feeder termination.

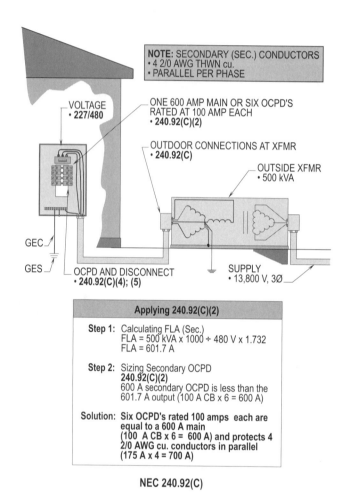

NOTE: SECONDARY (SEC.) CONDUCTORS
• 4 2/0 AWG THWN cu.
• PARALLEL PER PHASE

VOLTAGE
• 227/480

ONE 600 AMP MAIN OR SIX OCPD'S
RATED AT 100 AMP EACH
• 240.92(C)(2)

OUTDOOR CONNECTIONS AT XFMR
• 240.92(C)

OUTSIDE XFMR
• 500 kVA

GEC

GES

OCPD AND DISCONNECT
• 240.92(C)(4); (5)

SUPPLY
• 13,800 V, 3Ø

Applying 240.92(C)(2)
Step 1: Calculating FLA (Sec.) FLA = 500 kVA x 1000 ÷ 480 V x 1.732 FLA = 601.7 A
Step 2: Sizing Secondary OCPD **240.92(C)(2)** 600 A secondary OCPD is less than the 601.7 A output (100 A CB x 6 = 600 A)
Solution: Six OCPD's rated 100 amps each are equal to a 600 A main (100 A CB x 6 = 600 A) and protects 4 2/0 AWG cu. conductors in parallel (175 A x 4 = 700 A)

NEC 240.92(C)

Figure 10-15. The above illustration shows an alternate means allowed for protecting conductors connected to a transformer located outside.

Chapter 10: Sizing Transformers and Connections

Section   Answer

_____   T   F   1. The ampacity of a closed delta-connected secondary system can be found by multiplying the three-phase load by 67 percent.

_____   T   F   2. The capacity of an open delta-connected secondary systems can be found by multiplying the three-phase load by 58 percent.

_____   T   F   3. The secondary current of an autotransformer can be found by multiplying the name-plate kVA by 1000, then dividing by the secondary voltage.

_____   T   F   4. Conductors that are 10 ft. or less in length, are permitted to be connected to the transformer's secondary but must be installed with overcurrent protection at their termination point.

_____   T   F   5. In a supervised industrial plant with proper engineering, a transformer secondary connection of 100 ft. or less is permitted

_____   T   F   6. Transformer secondary conductors of separately derived systems for industrial locations must be permitted without overcurrent protection at the connection where not exeeding 100 ft. in length.

_____   T   F   7. Outside conductors must be protected from physical damage when connected and classified as a feeder-circuit.

_____   T   F   8. Lighting and appliance panelboards shall have more than 10 percent of the single-pole circuit breakers of a panelboard rated at 20 amps or less.

_____ _____ 9. What size wye-connected transformers are required for a building with a total connected load of 25 kVA for single-phase loads and 40 kVA for three-phase loads?

_____ _____ 10. What size closed delta-connected transformers are required for a building with a total connected load of 25 kVA for single-phase loads and 40 kVA for three-phase loads?

_____ _____ 11. What size open delta-connected transformer (lighting and power) is required for a building with a total connected load of 25 kVA for single-phase loads and 40 kVA for three-phase loads?

_____ _____ 12. What are the secondary amps for an autotransformer rated 2 kVA with a secondary voltage of 24 volts?

_____ _____ **13.** What is the size and rating of an autotransformer with a 24 volt secondary voltage serving a 230 volt motor with a connected load of 10,000 volt-amps?

_____ _____ **14.** What size THWN copper conductors and OCPD are required for a 10 ft. transformer secondary connection with a computed load of 168 amps?

_____ _____ **15.** What size primary and secondary THWN copper conductors and secondary OCPD is required for a 25 ft. connection from a feeder with a 200 amp OCPD having a 480 volt primary and 208 volt secondary?

_____ _____ **16.** What size connected THWN copper conductors and OCPD is required for a separately derived system installed in an industrial location with a computed load of 312 amps? Note that the load on the secondary side never exceeds 248 amps.

_____ _____ **17.** What size connected THWN copper conductors and OCPD are required for an outside transformer with an output on the secondary of 280 amps? Note that the load on the secondary side never exceeds 224 amps.

Protecting Transformers

Transformers must be sized to supply power to the loads served and allow high inertia loads to start and run. In addition, they must be protected by properly sized OCPD's and be equipped with conductors having enough capacity to supply the loads. OCPD's and conductors must be designed and installed in such a manner to safely protect the windings of such power sources from dangerous overloads, short-circuits and ground-faults.

The OCPD's and conductors are sometimes adjusted in size in order to allow loads with high inrush currents on the transformer's secondary side to start and operate. Note that, for simplicity, the size of the OCPD's in this chapter are used for the protection of transformers rated over 600 volts and less than 600 volts are selected from **240.6(A)** of the NEC. The size of OCPD's used on electrical systems over 600 volts are selected from ANSI 37.06 for CB's and ANSI 37.46 for buses.

CALCULATING PRIMARY AND SECONDARY CURRENT

The transformer's primary amp rating must be equivalent to the amps of the connected load when installing a feeder-circuit to supply the primary of a transformer to step-up or step-down the voltage. To determine the FLA of a transformer, the kVA times 1000 must be divided by the voltage multiplied by 1.732, if the voltage of the power supply is three-phase. **(See Figures 11-1(a) and (b))**

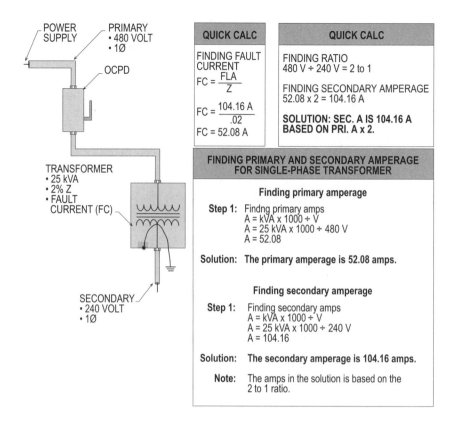

FINDING AMPERAGE OF SINGLE-PHASE TRANSFORMER

Figure 11-1(a). Finding the amperage for a single-phase transformer.

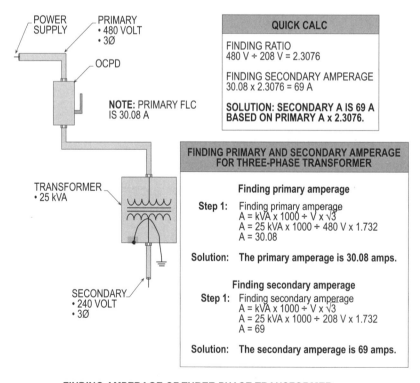

FINDING AMPERAGE OF THREE-PHASE TRANSFORMER

Figure 11-1(b). Finding the amperage for a three-phase transformer.

FINDING AMPERAGE

The kVA or amp rating for the primary or secondary of a transformer can be determined for a single-phase system by applying one of the following formulas:

kVA = volts x amps ÷ 1000
amps = kVA x 1000 ÷ volts

The following formula must be applied to determine the ratio of a transformer having a 480 volt primary and 240 volt secondary:

primary ÷ secondary
480 V ÷ 240 V
2:1 ratio

The FLC amp rating for the primary or secondary can be determined for a three-phase system by applying the following formula:

kVA = volts x 1.732 x amps ÷ 1000
amps = kVA x 1000 ÷ volts x 1732

CALCULATING FAULT-CURRENTS

When determining the fault-current (FC) of a transformer, the interrupting current (IC) rating of the OCPD must be determined for the amount of fault-current to be delivered at the terminals of the transformer when a short-circuit develops at that point. The current-limiting characteristic of a transformer at its terminals is called impedance. The impedance of a transformer (always expressed as a percentage) is used for sizing the interrupting capacity (IC) rating of fuses and circuit breakers used to protect the primary of a transformer.

> **For example:** What is the interrupting capacity (IC) or fault-current (FC) rating of a 25 kVA transformer with a 1.5 percent impedance supplied by a 120/240 volt, single-phase secondary?
>
> **Step 1:** Finding FLA
> FLA = kVA x 1000 ÷ V
> FLA = 25 kVA x 1000 ÷ 240 V
> FLA = 104 A
>
> **Step 2:** Finding IC
> FC = FLA ÷ impedance
> FC = 104 A ÷ .015
> FC = 6933 A
>
> **Solution: The FC is 6933 amps.**

Note that the greater the impedance rating is, the lower the fault-current will be, in amps.

OVERCURRENT PROTECTION
450.3(A); (B)

There are two sets of rules for providing overcurrent protection for transformers that are rated over 600 volts (**450.3)(A)**) and transformers rated 600 volts or less (**450.3(B)**). The OCPD may be placed in the primary only or in the primary and secondary, whichever provides the best protection and allows proper operation.

TRANSFORMERS RATED OVER 600 VOLTS - PRIMARY ONLY
450.3(A); TABLE 450.3(A)

The term *primary* is often inferred in the field as being the high side, and the term *secondary* as the low side of the transformer. This is really not the proper terminology. The primary is the input side of the transformer, and the secondary is the output side. Thus, voltage has nothing to do with the terms whether it is high or low.

Each transformer has to be protected by an overcurrent device in the primary side, if supervised and of any impedance. If the overcurrent protection device is fuses, they must be rated not greater than 250 percent (2.5 times) of the rated primary current of the transformer. When circuit breakers are used they must be set at not greater than 300 percent (3 times) of the rated primary current in amps. **(See Figure 11-2)**

This overcurrent protection device may be mounted in the vault or at the transformer, if approved for such purpose. It may also be mounted in the panelboard and designed to protect the windings and circuit conductors supplying the transformer from short-circuits and overloads.

In the case of a vault, the OCPD can be installed outdoors on a pole, with a disconnecting means installed in the vault to disconnect the supply conductors.

APPLYING NOTE 1
TABLE 450.3(A), NOTE 1

Where 250 percent (2.5 times) of the rated primary current of the transformer does not correspond to a standard rating of a fuse, the next higher standard rating (**240.6(A)**) is permitted per **Table 450.3(A), Note 1**. **Note:** See supervised locations in the NEC.

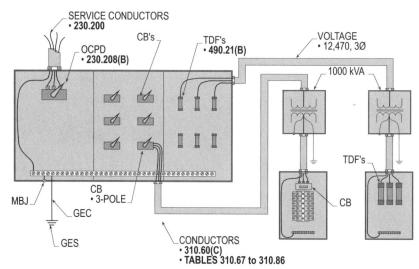

FINDING INDIVIDUAL OCPD FOR THE PRIMARY SIDE OF THE TRANSFORMER

Sizing OCPD using CB's

Step 1: Finding FLA of primary
450.3(A)
FLA = kVA x 1000 ÷ V x √3
FLA = 1,000 x 1000 ÷ 12,470 V x 1.732
FLA = 46.3 A

Step 2: Calculating FLA for OCPD
450.3(A); Table 450.3(A)
46.3 A x 300% = 138.9 A

Step 3: Selecting OCPD
Table 450.3(A), Note 1; 240.6(A)
138.9 A requires 150 A

Solution: The size circuit breaker is 150 amps.

Sizing OCPD using TDF's

Step 1: Calculating FLA for TDF's
450.3(A); Table 450.3(A)
46.3 A x 250% = 115.8 A

Step 2: Selecting OCPD
Table 450.3(A), Note 1; 240.6(A)
115.8 A allows 125 A

Solution: The size time delay fuses are 125 amps.

Note 1: Supervised location with any impedance per **Table 450.3(A)**.

Note 2: For simplicity, OCPD's are selected from **240.6(A)** of the NEC.

INDIVIDUAL PROTECTION
NEC 450.3(A); TABLE 450.3(A)

Figure 11-2. If the overcurrent protection devices are fuses, they shall be rated not greater than 250 percent (2.5 times) of the rated primary current of the transformer. When circuit breakers are used they must be set not greater than 300 percent (3 times) of the rated primary current.

PRIMARY AND SECONDARY OF TRANSFORMERS - OVER 600 VOLTS 450.3(A); TABLE 450.3(A)

A transformer over 600 volts, nominal, having an overcurrent device on the primary side rated to open no greater than the values listed in **Table 450.3(A)** or a transformer equipped with a coordinated thermal overload protection by the manufacturers, is not required to have individual protection in the secondary. However, a feeder overcurrent protection device rated or set to open at not greater than the values listed in **Table 450.3(A)** must be provided at the feeder.

TRANSFORMERS LOCATED IN SUPERVISED LOCATIONS 450.3(A); TABLE 450.3(A)

Overcurrent protection may be placed in the primary and secondary side of high-voltage transformers if the OCPD's are designed and installed according to the provisions listed in **Table 450.3(A)**.

Where the facility has trained engineers and maintenance personnel in supervised locations, the OCPD for the secondary can be sized at not more than 250 percent of the FLA for voltages of 600 volts or less with a rated impedance of 0-10%. With higher voltage on the secondary side of the transformer, the percentages for sizing the OCPD's are selected from **Table 450.3(A),** based upon the particular voltage and impedance level. **(See Figures 11-3(a) and (b))**

Note: See Figure 11-4 for certain design conditions that allow the primary OCPD to be used to protect the primary and secondary sides of two-wire to two-wire connected transformers and three-wire to three-wire delta connected transformers per **240.4(F)**.

TRANSFORMERS LOCATED IN NONSUPERVISED LOCATIONS 450.3(A); TABLE 450.3(A)

Overcurrent protection for a nonsupervised location may be placed in the primary and secondary side of high-voltage transformers, if the OCPD's are designed and installed according to the provisions listed in **Table 450.3(A)**.

If the secondary voltage is 600 volts or less, the OCPD and conductors on the secondary side must be sized at 125 percent of the transformer's FLC rating in amps. OCPD's sized at 125 percent of the FLC in amps protects the conductors and windings of the transformer from dangerous overload conditions. With higher voltage (over 600 volts) on the secondary side of the transformer, the percentages for sizing the OCPD's are selected from **Table 450.3(A)** based upon the particular voltage and impedance level. **(See Figure 11-5)** Note that **ANSIC 37.06** (CB's) and **ANSIC 37.46** (Fuses) are normally used to protect systems rated over 600 volts.

SIZING OCPD FOR PRIMARY SIDE	SIZING OCPD FOR SECONDARY SIDE
Step 1: Finding FLA of transformer FLA = kVA x 1000 ÷ V x √3 FLA = 500 x 1000 ÷ 4,160 V x 1.732 FLA = 69.4 A	**Step 1** Finding FLA of transformer FLA = kVA x 1000 ÷ V x √3 FLA = 500 x 1000 ÷ 480 V x 1.732 FLA = 601.7 A
Step 2: Calculating FLA for OCPD **450.3(A); Table 450.3(A)** FLA = 69.4 A x 600% FLA = 416.4 A	**Step 2:** Calculating FLA for OCPD **450.3(A); Table 450.3(A)** FLA = 601.7 A x 250% FLA = 1504.3 A
Step 3: Selecting OCPD **Table 450.3(A), Note 3, 240.6(A)** 416.4 A allows 400 A	**Step 3** Selecting OCPD **Table 450.3(A), Note 3; 240.6(A)** 1504.3 A allows 1500 A
Solution: **The size OCPD for the primary side is 400 amps.**	**Solution:** **The size OCPD for the secondary side is 1500 amps.**
Note 1: Transformer's impedance is less than 6%.	**Note 3:** If the secondary voltage is 4160, the OCPD using a CB can be sized at 300 percent and a fuse can be sized at 250 percent of transformer's FLC.
Note 2: For simplicity, the size of the OCPD's are selected from **240.6(A)**.	

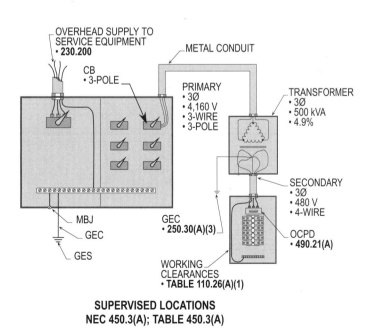

SUPERVISED LOCATIONS
NEC 450.3(A); TABLE 450.3(A)

Figure 11-3(a). Sizing the primary and secondary side of a transformer in a supervised location.

FINDING OCPD FOR THE PRIMARY AND SECONDARY SIDE OF THE TRANSFORMER

Sizing OCPD for primary side

Step 1: Finding FLA of transformer
FLA = kVA x 1000 ÷ V x √3
FLA = 450 x 1000 ÷ 13,800 V x 1.732
FLA = 18.83 A

Step 2: Calculating FLA for OCPD
450.3(A); Table 450.3(A)
FLA = 18.83 A x 600%
FLA = 112.9 A

Step 3: Selecting OCPD
Table 450.3(A), Note 3; 240.6(A)
112.9 A allows 110 A

Solution: The size OCPD for the primary side is a 110 amp circuit breaker.

Sizing OCPD for secondary side

Step 1: Finding FLA of transformer
FLA = 450 x 1000 ÷ 4160 x 1.732
FLA = 62.5 A

Step 2: Calculating FLA for OCPD
450.3(A); Table 450.3(A)
FLA = 62.5 A x 250%
FLA = 156.3 A

Step 3: Selecting OCPD
Table 450.3(A), Note 3; 240.6(A)
156.3 A allows 150 A

Solution: The size OCPD for the secondary side is 150 amp fuses.

Note: For simplicity, the size of the OCPD's are selected from 240.6(A) of the NEC.

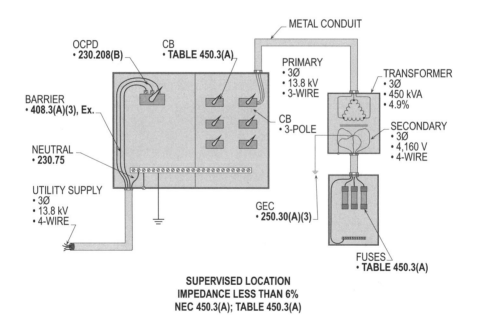

Figure 11-3(b). Sizing the primary and secondary side of a transformer in a supervised location.

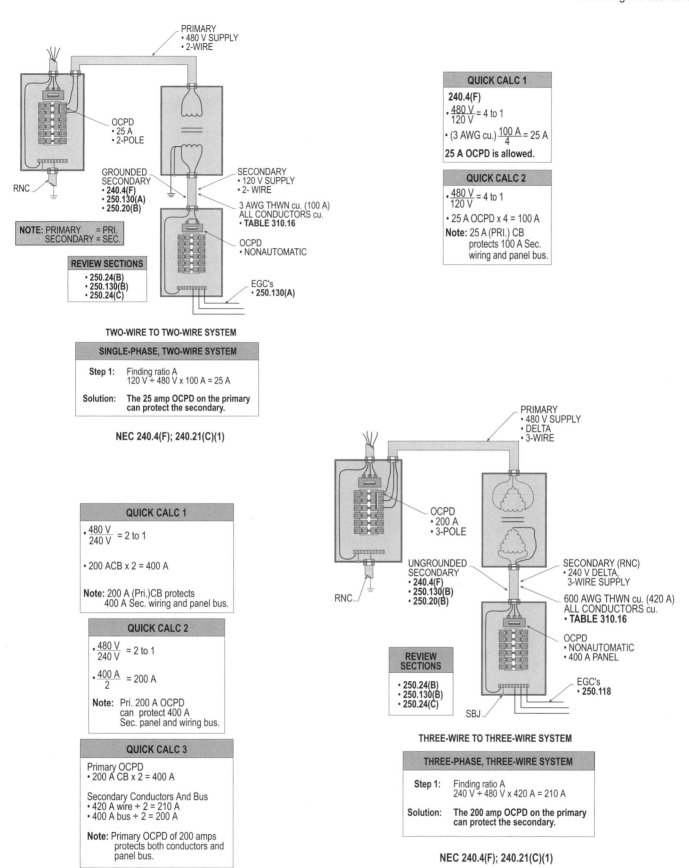

Figure 11-4. Sizing the OCPD for a single-phase, two-wire system and an ungrounded three-phase, three-wire delta system.

SIZING OCPD FOR PRIMARY SIDE	SIZING OCPD FOR SECONDARY SIDE
Step 1: Finding FLA of transformer FLA = kVA x 1000 ÷ V x √3 FLA = 500 x 1000 ÷ 4160 V x 1.732 FLA = 69.4 A	**Step 1:** Finding FLA of transformer FLA = kVA x 1000 ÷ V x √3 FLA = 500 x 1000 ÷ 480 V x 1.732 FLA = 601.7 A
Step 2: Calculating FLA for OCPD **450.3(A); Table 450.3(A)** FLA = 69.4 A x 600% FLA = 416.4 A	**Step 2** Calculating FLA for OCPD **450.3(A); Table 450.3(A)** FLA = 601.7 A x 125% FLA = 752 A
Step 3: Selecting OCPD **Table 450.3(A), Note 1; 240.6(A)** 416.4 A allows 450 A	**Step 3:** Selecting OCPD **Table 450.3(A), Note 1; 240.6(A)** 752 A allows 800 A
Solution: **The size OCPD for the primary side is 450 amps.**	**Solution:** **The size OCPD for side secondary is 800 amps.**
Note 1: Higher percentages than 125 percent per **Table 450.3(A)** may be used to size the OCPD in the secondary side where the secondary voltage is greater than 600 volts.	**Note 2:** If the secondary voltage is 4160, the OCPD using a CB can be sized at 300 percent and a fuse can be sized at 250 percent of the transformer FLC.

NOTE: NONSUPERVISED LOCATIONS, IN **TABLE 450.3(A)**, ARE CONSIDERED AS TRANSFORMERS IN ANY LOCATION HAVING IMPEDANCE RATINGS OF 0 TO 10%.

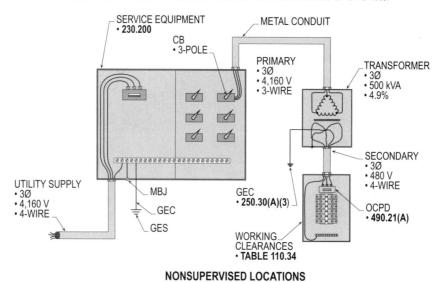

NONSUPERVISED LOCATIONS
ANY LOCATIONS
NEC 450.3(A); TABLE 450.3(A)

Figure 11-5. Sizing the primary and secondary side of a transformer in a nonsupervised location.

TRANSFORMERS RATED 600 VOLTS OR LESS - PRIMARY ONLY
450.3(B); TABLE 450.3(B)

A transformer rated 600 volts or less, nominal, having an individual overcurrent protection device on the primary side must be sized at not more than 125 percent of the transformer's full-load current rating. Note, with the OCPD and conductors sized at 125 percent or less of the transformer's FLC in amps, the supply conductors and transformer windings are considered to be protected from overload conditions. Primary protection is permitted by **450.3(B)** and as outlined in **Table 450.3(B)**. **(See Figure 11-6)**

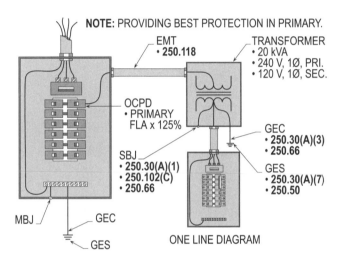

ONE LINE DIAGRAM

FINDING OCPD FOR PRIMARY SIDE OF TRANSFORMER

Step 1: Finding FLA of primary
FLA = kVA x 1000 ÷ V
FLA = 20 kVA x 1000 ÷ 240 V
FLA = 83.3 A

Step 2: Calculating OCPD (round the size down)
450.3(B); Table 450.3(B)
83.3 A x 125% = 104 A

Step 3: Selecting OCPD
Table 450.3(B); 240.6(A)
104 A allows 100 A (rounding down)

Solution: **The size OCPD in the primary side is 100 amps.**

Note: OCPD can be rounded up to 110 A per Note 1 to **Table 450.3(B)**

NEC 450.3(B); TABLE 450.3(B)

Figure 11-6. A transformer rated 600 volts or less, nominal, having an overcurrent protection device on the primary side, only, must be sized at not more than 125 percent of the transformer's full-load current rating in amps.

TRANSFORMERS RATED 9 AMPS OR MORE
450.3(B); TABLE 450.3(B)

Where the rated primary current of a transformer is 9 amps or more and 125 percent of this current does not correspond to a standard rating of a fuse or circuit breaker, the next size may be used per **240.6(A)**. **(See Figure 11-7)**

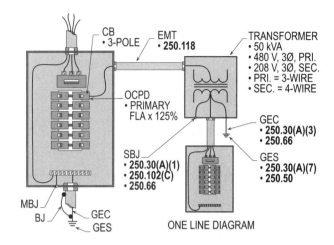

ONE LINE DIAGRAM

FINDING OCPD FOR PRIMARY SIDE OF TRANSFORMER

Sizing primary OCPD (First Level) per Table 450.3(B)

Step 1: Finding FLA of primary
FLA = kVA x 1000 ÷ V x √3
FLA = 50 kVA x 1000 ÷ 480 V x 1.732
FLA = 60.2 A

Step 2: Calculating OCPD
450.3(B); Table 450.3(B)
60.2 A x 125% = 75.3 A

Step 3: Selecting OCPD
Table 450.3(B), Note 1; 240.6(A)
75.3 A allows 80 A (rounding up)

Solution: **The size OCPD in the primary side is 80 amps.**

PRIMARY 9 AMPS OR MORE
NEC 450.3(B); TABLE 450.3(B)

Figure 11-7. Where the rated primary current of a transformer is 9 amps or more and 125 percent of the transformer's FLC in amps does not correspond to a standard rating of a fuse or circuit breaker, the next size may be used, as listed in **240.6(A)**.

TRANSFORMERS RATED 2 AMPS OR MORE BUT LESS THAN 9 AMPS
450.3(B); TABLE 450.3(B)

Where the rated primary current of a transformer is less than 9 amps but 2 amps or more, an overcurrent protection device rated or set at no more than 167 percent of the transformer's FLC may be used. **(See Figure 11-8)**

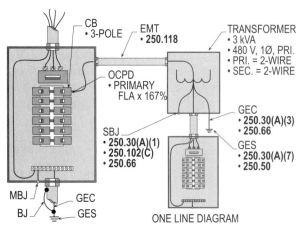

ONE LINE DIAGRAM

FINDING OCPD FOR PRIMARY SIDE OF TRANSFORMER

Sizing primary OCPD (Second Level) per Table 450.3(B)

Step 1: Finding FLA of primary
FLA = kVA x 1000 ÷ V
FLA = 3 kVA x 1000 ÷ 480 V
FLA = 6.25 A

Step 2: Calculating OCPD
450.3(B); Table 450.3(B)
6.25 A x 167% = 10.4 A

Step 3: Selecting OCPD
450.3(B); 240.6(A)
10.4 A allows 10 A

Solution: The size OCPD in the primary side is 10 amps.

PRIMARY 2 AMPS OR MORE BUT LESS THAN 9 AMPS
NEC 450.3(B); TABLE 450.3(B)

Figure 11-8. Where the rated primary current of a transformer is less than 9 amps but more than 2 amps, an overcurrent protection device rated or set at no more than 167 percent of the primary's FLC in amps may be used.

TRANSFORMERS RATED
LESS THAN 2 AMPS
450.3(B); TABLE 450.3(B)

When the rated primary current of a transformer is less than 2 amps, an overcurrent protection device rated or set at not more than 300 percent must be used. **(See Figure 11-9)**

TRANSFORMERS RATED 600 VOLTS OR LESS - PRIMARY AND SECONDARY 450.3(B); TABLE 450.3(B)

Combination protection can be provided for both the primary and secondary sides of a transformer. A current value of 250 percent of the rated primary current of the transformer may be used if 125 percent of the rated primary current of the transformer is not sufficient to allow loads with high inrush current to start and operate. However, the secondary overcurrent protection device must be sized at 125 percent of the rated secondary's full-load current of the transformer. Where the rated secondary current of a transformer is less than 9 amps, an overcurrent device rated or set at no more than 167 percent of secondary current may be used. **(See Figure 11-10)**

TRANSFORMERS RATED 9 AMPS OR MORE TABLE 450.3(B); NOTE 1

Where the rated secondary current of a transformer is 9 amps or more and 125 percent of this current does not correspond to a standard rating of a fuse or circuit breaker, the next size may be selected per **240.6(A)**.

CALCULATING OCPD'S FOR AUTOTRANSFORMERS (AXFMR'S) 450.4

Autotransformers (AXFMR's) rated at 600 volts or less are required to be protected by an OCPD installed on their primary side. The size of the OCPD is found by multiplying the full-load current rating in amps of the autotransformer times the percentage.

AUTOTRANSFORMERS RATED 9 AMPS OR MORE 450.4(A)

When sizing the OCPD for an autotransformer rated 9 amps or more, the full-load input current rating of the autotransformer in amps is multiplied by 125 percent and the next size device is permitted to be selected.

AUTOTRANSFORMERS RATED LESS THAN 9 AMPS 450.4(A), Ex.

When sizing the OCPD for an autotransformer rated less than 9 amps, the full-load input current rating of the autotransformer in amps is multiplied by 167 percent and the next lower size device must be selected.

GROUNDING AUTOTRANSFORMERS 450.5

Autotransformers can be connected to three-phase, three-wire ungrounded systems to derive a three-phase, four-wire grounded system. Three autotransformers connected in a

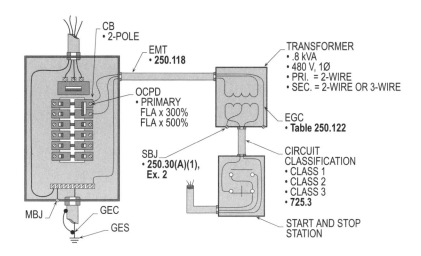

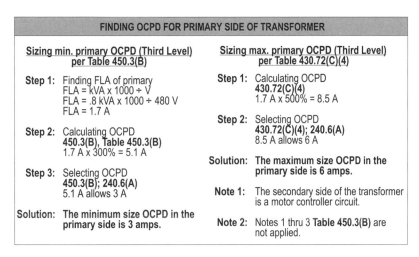

PRIMARY LESS THAN 2 AMPS
NEC 450.3(B) – (MIN.)
NEC 430.72(C)(4) – (MAX.)

Figure 11-9. When the rated primary current of a transformer is less than 2 amps, an overcurrent protection device rated or set at not more than 300 percent of the primary's FLC in amps must be used.

wye configuration to a three-phase ungrounded system converts such system to a three-phase, four-wire grounded system.

Autotransformers can be used to ground electrical systems that are not grounded. Existing ungrounded delta systems are grounded with autotransformers to derive a neutral. Three-phase zigzag transformers are generally installed and used for this purpose.

DERIVING A NEUTRAL
450.5(A)

To derive a neutral from a three-phase, three-wire ungrounded system to a three-phase, four-wire grounded

system when connecting an autotransformer, the following conditions must be applied:

(A)(1) Proper connections must be made,
(A)(2) Overcurrent protection must be provided,
(A)(3) Transformer fault sensing installed, and
(A)(4) Rating be adequately sized.

CONNECTIONS
450.5(A)(1)

Transformers must be directly connected to the ungrounded phase conductors with no switches or overcurrent protection devices installed between the connection and the autotransformer.

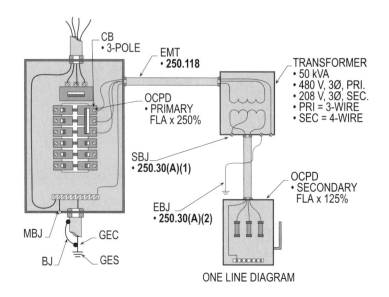

FINDING OCPD FOR PRIMARY AND SECONDARY SIDE OF TRANSFORMER	
Sizing OCPD in primary	**Sizing OCPD in secondary**
Step 1: Finding FLA of primary FLA = kVA x 1000 ÷ V x $\sqrt{3}$ FLA = 50 x 1000 ÷ 480 V x 1.732 FLA = 60.2 A	**Step 1:** Finding FLA of primary FLA = kVA x 1000 ÷ V x $\sqrt{3}$ FLA = 50 x 1000 ÷ 208 V x 1.732 FLA = 138.9 A
Step 2: Calculating OCPD **450.3(B); Table 450.3(B)** 60.2 A x 250% = 150.5 A	**Step 2:** Calculating OCPD **450.3(B); Table 450.3(B)** 138.9 A x 125% = 173.6 A
Step 3: Selecting OCPD **450.3(B); 240.6(A)** 150.5 A allows 150 A	**Step 3:** Selecting OCPD **450.3(B), Note 1; 240.6(A)** 173.6 A allows 175 A
Solution: **The size OCPD in the primary side is 150 amps.**	**Solution:** **The size OCPD in the primary side is 175 amps.**
Note: The size of the OCPD in primary must not exceed the 250% x FLA of the primary.	**Note:** **Table 450.3(B), Note 1** allows the next higher size OCPD to be used.

NEC 450.3(B); TABLE 450.3(B)

Figure 11-10. Sizing OCPD for the primary and secondary side of a transformer rated 600 volts or less.

OVERCURRENT PROTECTION
450.5(A)(2)

An overcurrent protection sensing device must be designed to trip at 125 percent of its continuous current per phase or neutral rating. The next higher standard rating may be installed where the input current is 9 amps or more and computed at 125 percent. Input current of 2 amps or less must not exceed 167 percent. **(See Figure 11-11)**

TRANSFORMER FAULT SENSING
450.5(A)(3)

A main switch or common-trip overcurrent protection device for a three-phase, four-wire system can be provided with fault sensing systems to guard against single-phasing or internal faults.

RATING
450.5(A)(4)

Autotransformers must be designed with a continuous neutral current rating sufficient to handle the maximum possible unbalanced neutral load current that could flow in the four-wire system.

GROUND REFERENCE
450.5(B)

The following conditions must be applied when autotransformers are used to detect grounds on three-phase, three-wire systems:

(B)(1) Proper rating,
(B)(2) Overcurrent protection sized adequately, and
(B)(3) Ground reference for damping transitory over-voltages.

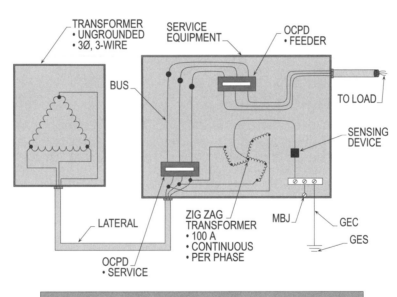

NEC 450.5(A)(2)

Figure 11-11. An overcurrent protection sensing device must be designed to trip at 125 percent of its continuous current per phase or neutral rating.

⸜)(1)

.otransformers must have a continuous neutral current ,ating sufficient for the specified ground-fault current that could develop in the system.

OVERCURRENT PROTECTION
450.5(B)(2)

The overcurrent protection device must open simultaneously with a common trip of all ungrounded conductors and be set to trip at not more than 125 percent of the rated phase current of the transformer. Note that 42 percent of the overcurrent protection device's rating may be used if connected in the autotransformer's neutral connection. **(See Figure 11-12)**

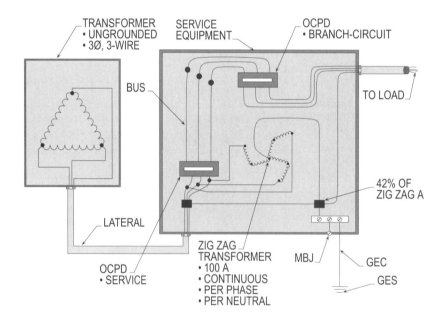

Figure 11-12. The overcurrent protection device must open simultaneously with a common trip of all ungrounded conductors and be set to trip at not more than 125 percent of the rated phase current of the transformer.

Chapter 11: Protecting Transformers

Section   **Answer**

_____   T   F   **1.** When circuit breakers are used to protect a transformer over 600 volts they must be set not greater than 250 percent of the rated primary current.

_____   T   F   **2.** Where 300 percent of the rated primary current of the transformer does not correspond to a standard rating of a fuse, the next higher standard rating is permitted.

_____   T   F   **3.** Transformers located in supervised locations can have the OCPD for the secondary sized at not more than 250 percent of the FLC for voltages, 600 volts or less.

_____   T   F   **4.** If the secondary voltage is 600 volts or less for transformers located in nonsupervised locations, the OCPD and conductors on the secondary side must be sized at 125 percent of the FLC rating.

_____   T   F   **5.** A transformer rated 600 volts or less having a primary overcurrent protection device only (rounding down) on the primary side must be sized at not more than 150 percent of the transformer's full-load current rating.

_____ _____   **6.** Where the rated primary current of a transformer is less than 9 amps but more than 2 amps, an overcurrent protection device rated or set at no more than _____ percent of the primary current is permitted.

_____ _____   **7.** When the rated primary current of a transformer is less than 2 amps, an overcurrent protection device rated or set at not more than _____ percent must be used.

_____ _____   **8.** Transformers rated 600 volts or less, may have a current value of _____ percent of the rated primary current if _____ percent of the rated primary current of the transformer is not sufficient to allow loads with high inrush current to start and operate.

_____ _____   **9.** When sizing the overcurrent protection device for an autotransformer rated 9 amps or more, the full-load input current rating of the autotransformer is multiplied by _____ percent and the next size standard device may be selected.

_____ _____   **10.** Autotransformer must be designed with a _____ neutral current rating.

_____ _____   **11.** What is the primary and secondary amperage for a 20 kVA, single-phase transformer with a 480 volt primary and 240 volt secondary?

12. What is the primary and secondary amperage for a 20 kVA, three-phase transformer with a 480 volt primary and 240 volt secondary?

13. What is the interrupting capacity (IC) rating of a 20 kVA transformer with a 1.5 percent impedance supplied by a 120/240 volt, single-phase secondary?

14. What size OCPD using a circuit breaker is required on the primary side for a 1500 kVA, 12,470 volt, three-phase transformer?

15. What size OCPD is required for the primary and secondary side of a 400 kVA, three-phase transformer with a 4160 volt primary and 480 volt secondary installed in a supervised location? (impedance (Z) is less than 6%).

16. What size OCPD is required for the primary and secondary side of 500 kVA, three-phase transformer with a 13,800 volt primary and 4160 volt secondary installed in a supervised location? (impedance (Z) is less than 6%).

17. What size OCPD is required for a two-wire to two-wire, 480 volt primary and a 240 volt secondary transformer with 3 AWG THWN copper conductors on the secondary side?

18. What size OCPD is required for a three-wire to three-wire delta, 480 volt primary and a 240 volt secondary transformer with 500 KCMIL THWN copper conductors on the secondary side?

19. What size OCPD is required for the primary and secondary side of a 400 kVA, three-phase transformer with a 4160 volt primary and 480 volt secondary installed in a nonsupervised location? (impedance (Z) is less than 6%).

20. What size OCPD is required for the primary side (only) of 25 kVA, single-phase transformer with a 240 volt primary and 120 volt secondary?

21. What size OCPD is required for the primary side (only) of .7 kVA, single-phase transformer with a 480 volt primary?

22. What size OCPD is required for the primary and secondary side of 40 kVA, three-phase transformer with a 480 volt primary and 208 volt secondary?

23. What is the amount of current needed to trip open an overcurrent protection sensing device for a zig-zag transformer with a 150 amp continuous load?

Secondary Ties

In large industrial plants and facilities, a "network" distribution system is usually utilized for supplying electrical (power) loads. Three-phase banks of transformers can be located at various points throughout the plant or facility. There are normally two high-tension primary circuits feeding these transformers. A double throw switch that is located at each transformer bank allows either primary circuit to serve any bank of transformers. The primary circuit conductors are sized with enough capacity so that either circuit is capable of carrying the entire load if a fault develops in the other circuit. Secondary voltage is usually three-phase systems rated 600 volts or less. The transformer secondaries are connected together in a network system, and these transformers feed all the loads involved, which can be used all at once or as necessary.

Note that the conductors connecting the transformer secondaries together are defined as the secondary ties of the distribution system.

TIE CIRCUITS
450.6

Secondary ties must be protected at both ends and such protection may be by fuses based upon the current-carrying capacity of the conductors per **450.6(A)** or the ties may be protected by a current limiter installed at each end per **450.6(A)(3)**. A limiter does not protect the conductors against a short-circuit, however it does provide overload protection. Usually, limiters, rather than fuses, are used for protection of the ties due to the fact they are very current limiting and will protect the circuit components from damage during short-circuit conditions.

There is normally a load center connected to the tie at the points where a transformer bank connects to the tie. The transformer is protected by a circuit breaker in the secondary leads between the transformer and load center. Circuit breaker settings may be up to 250 percent (2.5 times) of the transformer's secondary current rating, in amps, per **450.6(B)**. A reverse power relay must also be provided per **450.6(B)** that opens the circuit in case the transformer should fail for any reason. A reverse power relay is provided to prevent the current from being fed to an out-of-service transformer from the other transformers of the network. Where the secondary voltage is greater than 150 volts-to-ground, to ensure adequate protection, ties must be provided with a switch at each end per **450.6(A)(5)**.

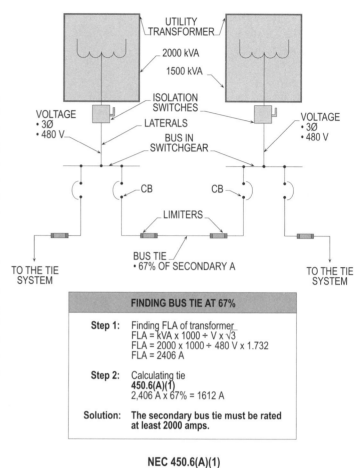

FINDING BUS TIE AT 67%	
Step 1:	Finding FLA of transformer FLA = kVA x 1000 ÷ V x √3 FLA = 2000 x 1000 ÷ 480 V x 1.732 FLA = 2406 A
Step 2:	Calculating tie **450.6(A)(1)** 2,406 A x 67% = 1612 A
Solution:	**The secondary bus tie must be rated at least 2000 amps.**

NEC 450.6(A)(1)

Figure 12-1. The ampacity of the ties connecting conductors must not be less than 67 percent of the rated secondary current, in amps, of the largest transformer in the tie circuit.

LOADS AT TRANSFORMER SUPPLY POINTS ONLY
450.6(A)(1)

Where transformers are tied together in parallel and connected by tie conductors that do not have overcurrent protection as per **Article 240** of the NEC, the ampacity of the ties connecting conductors must not be less than 67 percent of the rated secondary current, in amps, of the largest transformer in the tie circuit. **(See Figure 12-1)**

This rule applies where the loads are located at the transformer supply points per **450.6(A)(1)**.

The paralleling of transformers is common, but great care should be exercised to assure that the transformers are similar in all conditions of use. If they are not, one transformer will try to carry more of the load than the other transformer. If the transformers are of the same capacity and similar characteristics, they each, in theory, will carry 50 percent of the total load. The 67 percent allows for differences in transformer sizes and this allows for adjusting situations.

LOADS CONNECTED BETWEEN TRANSFORMER SUPPLY POINTS
450.6(A)(2)

Where the load is connected to the tie at any point between the transformer supply points and overcurrent protection is not provided by the provisions listed in **Article 240** of the NEC, the rated ampacity of the tie is required to be not less than 100 percent of the rated secondary current of the largest transformer connected to the secondary tie system, except as provided in **450.6(A)(4)**. **(See Figure 12-2)**

This rule applies, mainly, where the loads are connected between the transformer supply points per **450.6(A)(2)**.

TIE CIRCUIT PROTECTION
450.6(A)(3)

In **450.6(A)(1) and (A)(2)**, both ends of each tie connection must be provided with a protective device that opens at a certain temperature of the tie conductor. This prevents damage to the tie conductor and its insulation, and such installations may consist of:

(1) A limiter that is a fusible link cable connector. The limiter is selected and designed for the insulation, conductor material, etc., on the tie conductors.

(2) A circuit breaker actuated by devices having characteristics which are comparable can be used, if designed and sized correctly.

The above applies where the tie circuit protection is provided per **450.6(A)(3)** and the tie conductor must fully comply with all rules and regulations in such sections.

INTERCONNECTION OF PHASE CONDUCTORS BETWEEN TRANSFORMER SUPPLY POINTS 450.3(A)(4)

Where the tie consists of more than one conductor per phase, the conductor of each phase must be interconnected in order to create a load supply point. The protection required in **450.6(A)(3)** must be provided in each tie conductor at this point, except as follows:

Section **450.6(A)(4)(b)** allows the loads to be connected to the individual conductor(s) of each phase and without the protection listed in **450.6(A)(3)** above at load connection points, provided the tie conductors of each phase have a combined capacity of not less than 133 percent of the rated secondary current of the largest transformer connected to the secondary tie system. The total load of such ties must not exceed the rated secondary current of the largest transformer and the loads must be equally divided on each phase and on the individual conductors of each phase, or as close as possible. **(See Figure 12-3)**

The use of multiple conductors on each phase and the requirement that loads do not have to tie to the multiple conductors of the same phase might possibly set unbalanced current flow in the multiple conductors on the same phase.

Where the combined capacity of the multiple conductors on the same phase are rated at 133 percent of the secondary current of the largest transformer, limiters are necessary at the tie or connections to the transformers that are tied together to properly protect the components of the circuit.

The above rule applies where the interconnection of phase conductors between the transformer's supply points occur per **450.6(A)(4)(b)**.

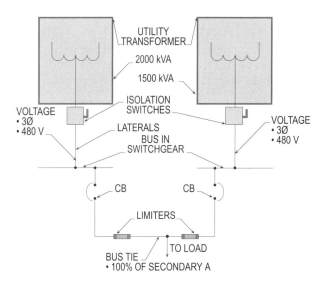

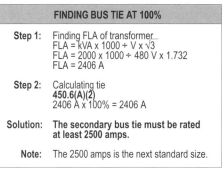

NEC 450.6(A)(2)

Figure 12-2. The rated ampacity of the tie is required to be not less than 100 percent of the rated secondary current of the largest transformer connected to the secondary tie system except as provided in **450.6(A)(4)**.

TIE CIRCUIT CONTROL 450.6(A)(5)

If the operating voltage of the secondary tie exceeds 150 volts-to-ground, there must be a switch ahead of the limiters and tie conductors that is capable of deenergizing the tie conductors and the limiters. This switch must comply with the following:

(1) The current rating of the switch must not be less than the current rating of the conductors connected to such switch.

(2) The switch must be capable of opening its rated current.

(3) The switch must not open under the magnetic forces caused from short-circuit currents.

The above applies where the tie circuit control is located as mentioned in **450.6(A)(5)**.

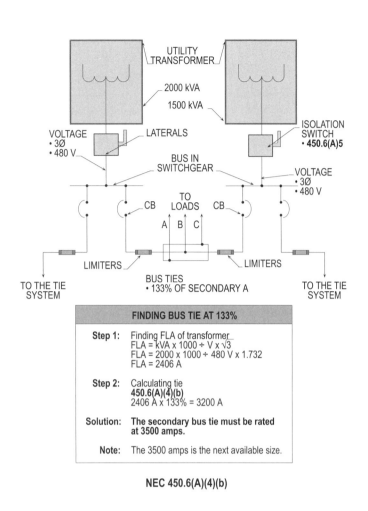

FINDING BUS TIE AT 133%

Step 1:	Finding FLA of transformer FLA = kVA x 1000 ÷ V x √3 FLA = 2000 x 1000 ÷ 480 V x 1.732 FLA = 2406 A
Step 2:	Calculating tie **450.6(A)(4)(b)** 2406 A x 133% = 3200 A
Solution:	**The secondary bus tie must be rated at 3500 amps.**
Note:	The 3500 amps is the next available size.

NEC 450.6(A)(4)(b)

Figure 12-3. The above illustrates the procedure for sizing the secondary tie at 133 percent of the secondary amps of the largest transformer.

OVERCURRENT PROTECTION (OCP) FOR SECONDARY CONNECTIONS 450.6(B)

When secondary ties from transformers are used, an overcurrent device in the secondary of each transformer that is rated or set at not greater than 250 percent (2.5 times) of the rated secondary current of the transformer must be provided. In addition, there must be a circuit breaker actuated by a reverse-current relay and the breaker must be set at not greater than the rated secondary current in amps of the transformer. Such overcurrent protection prevents overloads and short-circuit conditions. Therefore, the reverse-current relay and circuit breaker must be designed to handle any reversal of current flow into the transformer. **(See Figure 12-4)**

The above rules apply where overcurrent protection for secondary connections is installed to protect the system as required by **450.6(B)**.

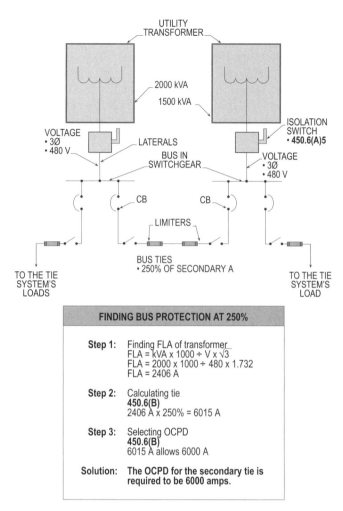

FINDING BUS PROTECTION AT 250%

Step 1:	Finding FLA of transformer FLA = kVA x 1000 ÷ V x √3 FLA = 2000 x 1000 ÷ 480 x 1.732 FLA = 2406 A
Step 2:	Calculating tie **450.6(B)** 2406 A x 250% = 6015 A
Step 3:	Selecting OCPD **450.6(B)** 6015 A allows 6000 A
Solution:	**The OCPD for the secondary tie is required to be 6000 amps.**

NEC 450.6(B)

Figure 12-4. When secondary ties from transformers are used, an overcurrent device in the secondary of each transformer that is rated or set at not greater than 250 percent (2.5 times) of the rated secondary current in amps of the transformer must be provided.

RADIAL SUPPLY SYSTEMS

Radial low-voltage and high-voltage systems are systems provided with high-voltage from the power company to the plant's transformer. Service equipment for low-voltage systems and the service equipment for high-voltage systems are installed as needed.

RADIAL LOW-VOLTAGE SYSTEMS

Radial low-voltage systems are installed with lower voltage feeders that are run to switchboards and load centers in the plant. This type of installation is costly due to larger conductors and conduits being installed and the elements and components of the load centers requiring lower voltage with higher current ratings. **(See Figure 12-5)**

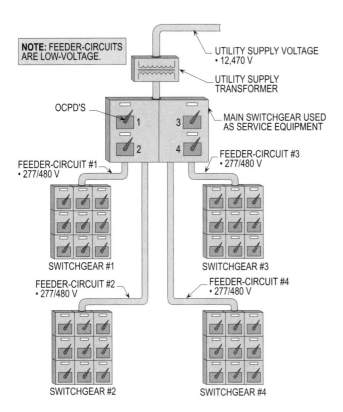

Figure 12-5. An installation of a radial low-voltage system with wiring methods and components.

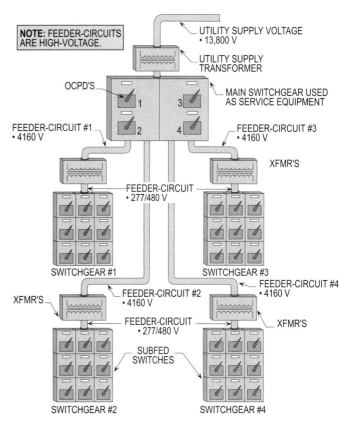

Figure 12-6. An installation of a radial high-voltage system with wiring methods and components.

RADIAL HIGH-VOLTAGE SYSTEMS

Radial high-voltage systems can be installed to transformers at each distribution point and the voltage stepped down to the desired operating level. This type of installation is less costly due to the feeders requiring smaller conduits, conductors, and equipment. This type of installation raises the voltage on a feeder-circuit and lowers the current ratings for all the elements and components for the feeder-circuit. **(See Figure 12-6)**

LOOP SUPPLY SYSTEMS

The distribution loop supply system is more superior than the radial low-voltage and high-voltage systems. Distribution loop supply systems are formed in a loop or circle by a pair of circuit breakers that are connected to the power company's supply loop. These circuit breakers are installed between each transformer and load center connected to the loop. Protection is provided by the circuit breakers for any section of the feeders connected to the loop and also provides isolation in case of a fault or disorderly shutdown.

A distribution loop supply system would alleviate problems if trouble developed between two switchgear. The circuit breakers would remove the damaged load center. For example, if trouble where to develop between switchgear 3 and switchgear 4, the circuit breakers would remove the damaged load center. All other switchgears connected to the loop would continue to be in service. If switchgear 4 were damaged in some way to create a fault condition, this fault condition would be cleared by one of the circuit breakers installed from the service or in the loop system that would disconnect switchgear 4. **(See Figure 12-7)** Note that switchgear are considered load centers.

BUS-TIE LOOPS

Bus-tie distribution loop supply systems are installed to eliminate a complete shutdown of a feeder section that has been damaged. The transformer, service equipment, or any of the elements could have this type of damage to the feeder system. The bus-tie loop feeder, in addition, has a second loop, called the bus-tie loop feeder, which is connected between the transformers forming the loop.

Circuit breakers are installed to protect these secondary connections installed at each switchgear (load center) location. A continuous circle (bus tie loop system) is formed from these loops at each switchgear from 1 through 4 and back to 1. This continuous loop provides power to the service of the section, if a fault or trouble were to develop at the transformer. **(See Figure 12-8)**

BUS-TIE CONDUCTORS

Bus-tie conductors or secondary conductors are low-voltage (600 volts or less) secondary loop connections. Bus-ties are used to connect two power sources to the secondaries of two transformers. Overcurrent devices are set at 150 percent to limit the maximum current of the capacity of the conductor where loads are connected at the supply points. If protection is not provided for conductors rated at 150 percent or less, the current of the bus-tie must be at least 67 percent of the full-load current rating of the largest transformer. **(See Figure 12-9)**

The current-carrying capacity of the bus-tie conductor must be 100 percent of the full-load current rating of the largest transformer for loads connected from the secondary bus-tie and not the transformer location. **(See Figure 12-10)**

The secondary tie can consist of a number of conductors, parallel per phase, with loads connected to individual conductors between the locations of the transformers. The combined total rating of the conductors between stations must be at least 133 percent of the full-load secondary current of the largest transformer, provided the loads are not tapped to every one of the tie conductors. **(See Figure 12-11)**

When loads are tapped, they must be equally divided on each phase and on the individual conductors of each phase, as close as possible. **(See Figure 12-12)**

BUS-TIE PROTECTION

To protect conductors from short-circuit conditions, current limiters or automatic circuit breakers must be installed at both ends of each tie. Current limiters can be used when the operating voltage is above 150 volts-to-ground. A switch must be provided at either end of the tie and equal to the conductor's ampacity. In addition, an overcurrent protection device must be installed in the secondary circuit of each transformer and set at 250 percent or less of the rated full-load current to protect the bus-tie conductors. **(See Figure 12-13)**

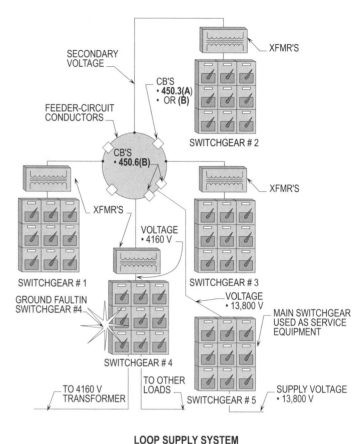

LOOP SUPPLY SYSTEM

Figure 12-7. An installation of a distribution loop supply system.

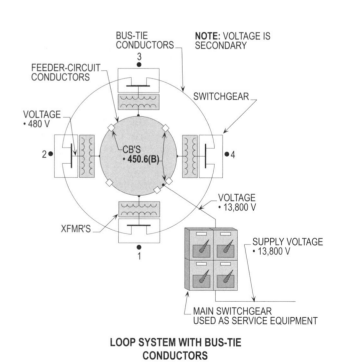

LOOP SYSTEM WITH BUS-TIE CONDUCTORS

Figure 12-8. An installation of bus-ties used in a loop system with components and equipment.

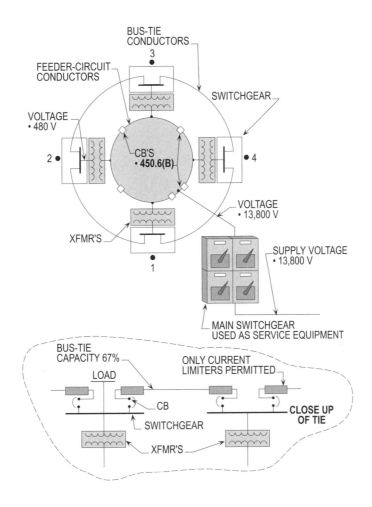

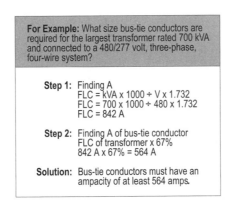

For Example: What size bus-tie conductors are required for the largest transformer rated 700 kVA and connected to a 480/277 volt, three-phase, four-wire system?

Step 1: Finding A
FLC = kVA x 1000 ÷ V x 1.732
FLC = 700 x 1000 ÷ 480 x 1.732
FLC = 842 A

Step 2: Finding A of bus-tie conductor
FLC of transformer x 67%
842 A x 67% = 564 A

Solution: Bus-tie conductors must have an ampacity of at least 564 amps.

**LOOP SYSTEM WITH 67% BUS-TIE
CONDUCTORS PER 450.6(A)(1)**

Figure 12-9. An installation of bus-tie conductors calculated at 67 percent and minimum amps selected from such.

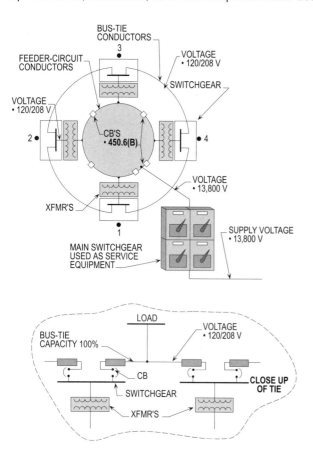

For Example: What size bus-tie conductors are required for the largest transformer rated 350 kVA and connected to a 208/120 volt, three-phase, four-wire system?

Step 1: Finding A
FLC = kVA x 1000 ÷ V x 1.732
FLC = 350 x 1000 ÷ 208 x 1.732
FLC = 972 A

Step 2: Finding A of bus-tie
FLC of transformer x 100%
972 A x 100% = 972 A

Solution: Bus-tie conductors must have an ampacity rating of at least 972 amps.

LOOP SYSTEM WITH 100% BUS-TIE CONDUCTORS PER 450.6(A)(2)

Figure 12-10. An installation of bus-ties calculated at 100 percent.

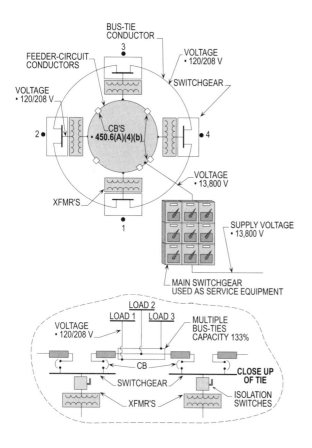

For Example: What size bus-tie conductors are required for the largest transformer rated 350 kVA and connected to a 208/120 volt, three-phase, four-wire system?

Step 1: Finding A
FLC = kVA x 1000 ÷ V x 1.732
FLC = 350 x 1000 ÷ 208 x 1.732
FLC = 972 A

Step 2: Finding A of bus-tie
FLC of transformer x 133%
972 A x 133% = 1293 A

Solution: Bus-tie conductors must have an ampacity rating of at least 1293 amps.

LOOP SYSTEM WITH 133% BUS-TIE CONDUCTORS PER 450.6(A)(4)

Figure 12-11. An installation of bus-tie conductors calculated at 133 percent.

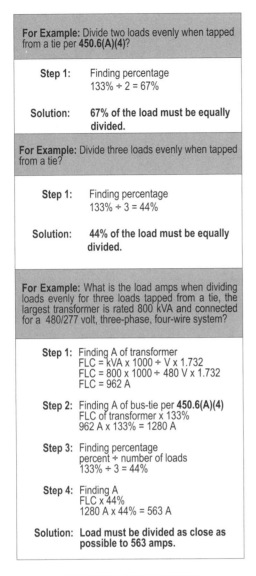

For Example: Divide two loads evenly when tapped from a tie per **450.6(A)(4)?**

Step 1: Finding percentage
133% ÷ 2 = 67%

Solution: **67% of the load must be equally divided.**

For Example: Divide three loads evenly when tapped from a tie?

Step 1: Finding percentage
133% ÷ 3 = 44%

Solution: **44% of the load must be equally divided.**

For Example: What is the load amps when dividing loads evenly for three loads tapped from a tie, the largest transformer is rated 800 kVA and connected for a 480/277 volt, three-phase, four-wire system?

Step 1: Finding A of transformer
FLC = kVA x 1000 ÷ V x 1.732
FLC = 800 x 1000 ÷ 480 V x 1.732
FLC = 962 A

Step 2: Finding A of bus-tie per **450.6(A)(4)**
FLC of transformer x 133%
962 A x 133% = 1280 A

Step 3: Finding percentage
percent ÷ number of loads
133% ÷ 3 = 44%

Step 4: Finding A
FLC x 44%
1280 A x 44% = 563 A

Solution: **Load must be divided as close as possible to 563 amps.**

DIVIDING LOADS EVENLY

Figure 12-12. An installation of loads divided evenly on ties to prevent unbalanced loading.

If reverse current exceeding the full-load current of a transformer tries to flow into the unit, a reverse-current relay must be installed to actuate a circuit breaker. The secondary windings of the transformer are disconnected from the circuit breaker actuated by the reverse-current relay due to faults or current feedback.

NETWORK POWER SYSTEMS

A simple network power system consists of two power sources fed into a loop and connected to transformers. Transformers 1 and 3 are connected to power source PS1. Transformers 2 and 4 are connected to power source PS2. In case one of the transformers develops trouble, switches are installed to provide a disconnecting means. **(See Figure 12-14)**

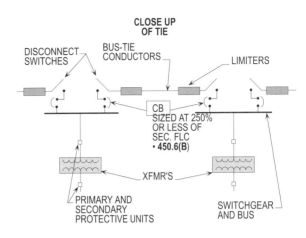

BUS-TIE PROTECTION

Figure 12-13. An installation for protection of bus-ties.

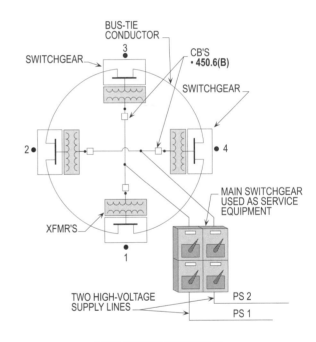

SIMPLE NETWORK SYSTEM

Figure 12-14. An installation of a simple network system.

Chapter 12: Secondary Ties

Section   Answer

T   F   **1.** Radial low-voltage systems are installed with lower voltage feeders.

T   F   **2.** Radial high-voltage systems are installed to transformers at each distribution point and the voltage is stepped up to the desired operating value.

T   F   **3.** Distribution loop supply systems are formed in a loop or circle by a pair of fuses which are connected to the power company's power to the loop.

T   F   **4.** Bus-tie loop systems are designed and installed to eliminate a complete shutdown of a feeder section that has been damaged.

T   F   **5.** The current-carrying capacity of the bus-tie must be 125 percent of the full-load current rating of the largest transformer for loads tapped from the secondary bus-tie conductors and not the transformer location.

6. Where transformers are tied in parallel and connected by tie conductors that do not have overcurrent protection, the ampacity of the ties connecting conductors must not be less than _____ percent of the rated secondary current of the largest transformer in the tie circuit.

7. Where the load is connected to the tie at any point between the transformer supply points and overcurrent protection is not provided, the rated ampacity of the tie must be not less than _____ percent of the rated secondary current in amps of the largest transformer connected to the secondary tie system.

8. The current of the bus-tie must be at least _____ percent of the full-load current rating of the largest transformer if protection is not provided for conductors rated at _____ percent or less.

Windings and Components

Transformer windings are connected for either additive or subtractive polarity and can be connected in a delta or wye configuration to supply either single-phase or three-phase voltage to service equipment or other electrical equipment. Windings must be connected for the proper polarity for the current to flow through the windings in the proper direction.

TESTING WINDINGS

When testing transformer windings for a delta or wye-connected configuration to supply single-phase or three-phase voltages to the service equipment or other electrical equipment, the following polarity checks must be made to verify if they are connected in:

(1) Additive polarity, or
(2) Subtractive polarity.

ADDITIVE POLARITY

The induced voltage in the primary and secondary windings will be in opposite directions for transformer windings connected in additive polarity. Additive connected transformers windings are wound in the same direction. However, a subtractive transformer can be used as an additive transformer by reversing the flow of current through the windings. Note that the overcurrent protection device will trip or the transformer will not operate properly if one of the windings is accidently connected in subtractive polarity. **(See Figures 13-1 through 13-4)**

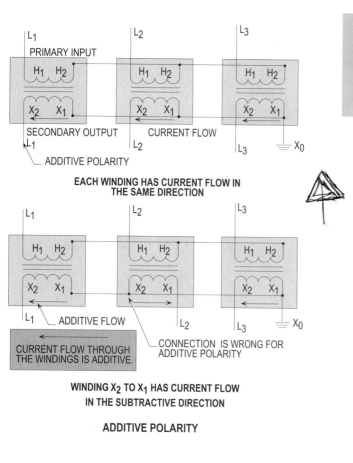

Figure 13-1. Transformer windings connected for additive polarity with one connection subtractive.

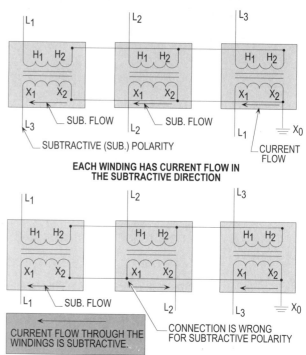

Transformer Note: This is under the assumption that the primary voltage is higher than the secondary. **Warning:** A lower voltage may have to be applied to measure the voltage safely if the voltage is high. **(See Figure 13-3)**

Figure 13-2. Transformer windings connected for subtractive polarity with one connection additive.

SUBTRACTIVE POLARITY

The induced voltage and current in the primary and secondary windings will be in the same direction for transformer windings connected in subtractive polarity. A subtractive transformer can be used as an additive transformer by reversing the flow of current through its windings. The overcurrent protection device will trip or the transformer will not operate properly if one of the windings is accidentally connected in additive polarity. **(See Figures 13-2 and 13-4)**

TESTING FOR POLARITY

Transformer windings are identified as either additive or subtractive polarity by measuring the primary and secondary voltage. The voltage would be equal to the primary and secondary added together for transformer windings that are additive connected. Note that the primary voltage is always greater than the voltage measured.

The voltage would be equal to the primary minus the secondary for transformer windings connected in subtractive polarity. The primary voltage is always more than the voltage measured between primary and secondary. **(See Figure 13-3)**

POLARITY CONNECTIONS AND IDENTIFYING TERMINALS

The letter H and accompanying numbers are used for identification of high-voltage or input terminals that are located at the left of the primary side of the transformers windings. The letter X and accompanying numbers are used for identification of low-voltage or output terminals. The primary or secondary side of a transformer can be used as input or output terminals under certain conditions.

If the windings are connected with additive polarity, the X terminal and number is located at the left of the secondary side of the transformer. If the windings are connected with subtractive polarity, H_1 on the primary side will line up with X_1 on the secondary side of the transformer. If the transformer windings are connected in additive polarity, the current flows in the opposite direction. However, if the transformer windings are connected in subtractive polarity, the current flows in the same direction. **(See Figure 13-4)**

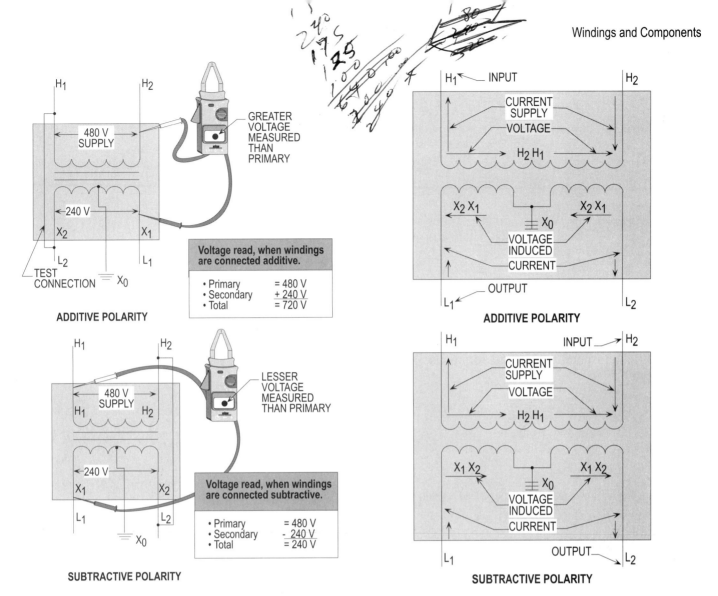

Figure 13-3. A simple test for polarity is to connect two adjacent terminals of the high- and low-voltage windings together and apply a moderate voltage to either winding. If voltage reading is greater than the primary voltage, the windings are connected additive. If voltage reading is less than the primary voltage, the connection is subtractive.

Figure 13-4. If the transformer windings are connected in additive polarity, the two windings are in the opposite direction. If the transformer windings are connected in subtractive polarity, the two windings are in the same direction.

TESTING VOLTAGE OF WINDINGS

When testing the voltage of windings for single-phase and three-phase transformers, the following two measurements are used to determine the voltage level:

(1) Phase-to-phase, and
(2) Phase-to-ground.

PHASE-TO-PHASE VOLTAGE

The voltage is measured between the phases to determine the voltage for a single-phase or three-phase transformer. For example, a 120/240 volt, single-phase transformer measured phase-to-phase is 240 volts. **(See Figure 13-5)**

PHASE-TO-GROUND VOLTAGE

The voltage is measured between the phase-to-ground to determine the voltage for a single-phase or three-phase transformer. For example, a 120/240 volt, single-phase transformer measured phase-to-ground is 120 volts. **(See Figure 13-6)**

IDENTIFYING AND CONNECTING WINDINGS

When identifying and connecting windings based upon the purpose of installation, transformers may be connected in a number of different ways. There are several ways of connecting a single-phase or three-phase transformer to the power supply and the loads to be served.

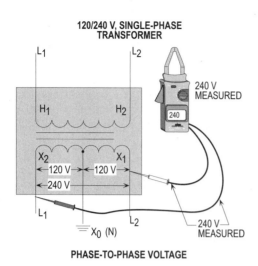

Figure 13-5. Measuring the voltage from phase-to-phase to determine the voltage.

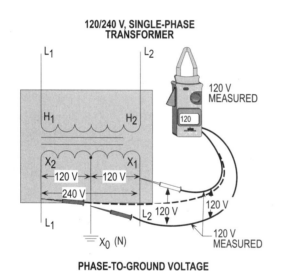

Figure 13-6. Measuring the voltage from phase-to-ground to determine the voltage level.

120/240 VOLT, SINGLE-PHASE TRANSFORMERS

When connecting the secondary terminals of a 120/240 volt single-phase transformer, the first transformer winding is connected to X_3 and the second winding connected to X_2, making the neutral connection X_0. The connection between the first winding X_3 to the second winding X_2 will series the two 120 volt windings to derive 240 volts from L_1 to L_2. To obtain the neutral, a jumper from X_3 to X_2 is tapped and connected to ground. Note that 120 volts is derived from the connections between L_1 to X_0 (N) and L_2 to X_0 (N). **(See Figure 13-7)**

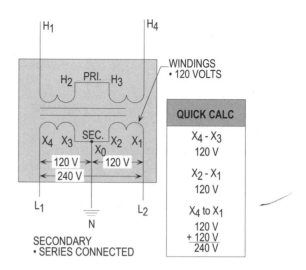

Figure 13-7. Connecting the secondary terminals of a transformer to derive a single-phase, 120/240 volt system.

THREE-PHASE, CLOSED DELTA CONNECTED SYSTEM

When connecting the secondary terminals of a three-phase closed delta system transformer, the first transformer winding X_6 is connected in series with a jumper to X_1 of the third transformer, the first transformer winding X_5 is connected in series with a jumper to X_4 of the second transformer, and the second transformer winding X_3 is connected in series with a jumper to X_2 of the third transformer.

The neutral connection can be tapped from the center of any 240 volt winding. 120 volts-to-ground is derived from the outside lines of the tap which is connected to ground. 208 volts-to-ground (high-leg) is derived from the current that must travel through one full-winding and one-half of the other winding to ground. The windings of a 120/240 volt, four-wire closed delta transformer are rated at 240 volts each with the high-leg rated at 208 volts. **(See Figure 13-8)**

THREE-PHASE, OPEN DELTA CONNECTED SYSTEM

When connecting the secondary terminals of a three-phase open delta (system) transformer, the first transformer winding X_3 is connected in series with a jumper to X_2 of the second transformer. 120 volts-to-ground is derived from either one of the 240 volt windings that are tapped. 240 volts is derived from phase-to-phase voltage that is connected from L_1 to L_2, L_1 to L_3, and L_2 to L_3. 208 volts-to-ground is derived from the high-leg. **(See Figure 13-9)**

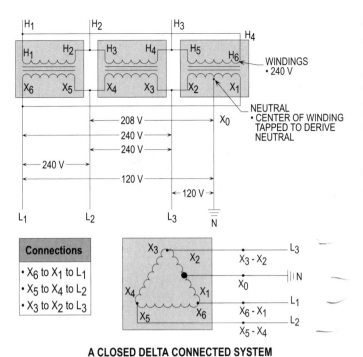

A CLOSED DELTA CONNECTED SYSTEM

Figure 13-8. Connecting the secondary terminals of a transformer to derive a three-phase, closed delta system.

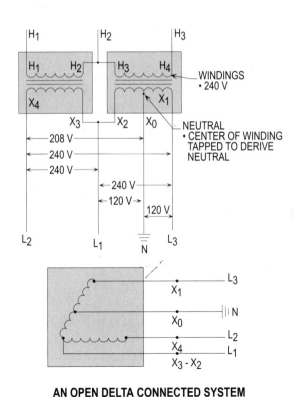

AN OPEN DELTA CONNECTED SYSTEM

Figure 13-9. Connecting the secondary terminals of a transformer to derive a three-phase, open delta system.

THREE-PHASE, WYE SYSTEM

When connecting the secondary terminals of a three-phase, wye (system) transformer, the first winding X_5 is connected in series with a jumper to X_3 of the second transformer and to X_1 of the third transformer, making the neutral connection. The connections in a wye (system) transformer, X_6, X_4, and X_2 will be Phases 1, 2, and 3. The neutral conductor connected to ground is X_0.

120 volts-to-ground is derived from L_1 to N, L_2 to N, and L_3 to N. 208 volts is derived from L_1 to L_2, L_1 to L_3, and L_2 to L_3. **(See Figure 13-10)**

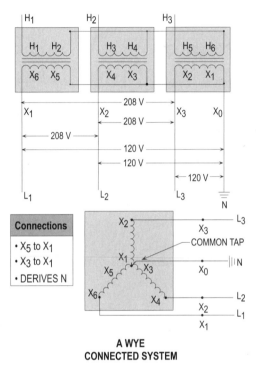

A WYE CONNECTED SYSTEM

Figure 13-10. Connecting the secondary terminals of a transformer to derive a three-phase wye system. Each winding is rated at 120 volts for a 120/208 volt, three-phase, wye-connected system.

THREE-PHASE, CORNER GROUNDED DELTA CONNECTED SYSTEM

When tapping any one of the ungrounded phase conductors of a corner grounded delta system, the grounded conductor is derived. The phase-to-phase voltage would be 480 volts, if the voltage of the windings is 480 volts. The voltage-to-ground from the grounded phase conductor is 0 volts.

A color of white or gray identification must be used for the grounded phase conductor per 200.7. The grounded phase conductor must never be fused per **230.90(B)**, except for motor circuits per **430.36**. Where the grounded phase conductor enters into a panelboard or switchgear, it must be connected to ground at the service equipment location only. **(See Figure 13-11)**

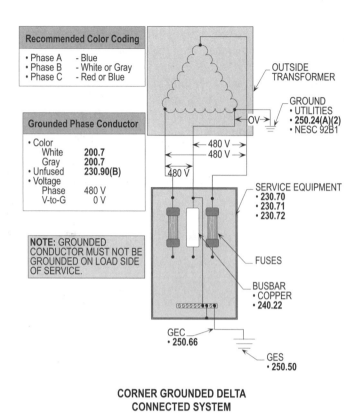

Recommended Color Coding

- Phase A - Blue
- Phase B - White or Gray
- Phase C - Red or Blue

Grounded Phase Conductor

- Color
  - White **200.7**
  - Gray **200.7**
- Unfused **230.90(B)**
- Voltage
  - Phase 480 V
  - V-to-G 0 V

NOTE: GROUNDED CONDUCTOR MUST NOT BE GROUNDED ON LOAD SIDE OF SERVICE.

OUTSIDE TRANSFORMER

GROUND
- UTILITIES
- **250.24(A)(2)**
- NESC 92B1

480 V
480 V
480 V

SERVICE EQUIPMENT
- **230.70**
- **230.71**
- **230.72**

FUSES

BUSBAR
- COPPER
- **240.22**

GEC
- **250.66**

GES
- **250.50**

CORNER GROUNDED DELTA CONNECTED SYSTEM

Figure 13-11. Voltage relationships and secondary terminal connections of a three-phase, corner grounded delta connected system.

SEPARATELY DERIVED AC SYSTEMS 250.30

Low-voltage and high-voltage feeder-circuits are sometimes installed in a building with transformers installed on each floor to reduce the voltage to 120/240, 120/208, or 277/480 volts for general use lighting and receptacle loads in large building applications. When designing and installing the bonding and grounding of a transformer system, the secondary of a separately derived system is divided into three parts. Such grounding since the 1978 NEC can be installed either at the transformer or at the load served which is connected and supplied from the secondary side per **240.21(B) and (C)**. The following three parts must be designed and installed for a separately derived system per **250.30(A)(1), (2), (3) and (7)**:
(See Figure 13-12)

(1) System bonding jumper,
(2) Grounding electrode conductor, and
(3) Grounding electrode.

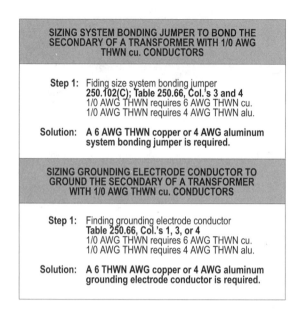

SIZING SYSTEM BONDING JUMPER TO BOND THE SECONDARY OF A TRANSFORMER WITH 1/0 AWG THWN cu. CONDUCTORS

Step 1: Fiding size system bonding jumper
250.102(C); Table 250.66, Col.'s 3 and 4
1/0 AWG THWN requires 6 AWG THWN cu.
1/0 AWG THWN requires 4 AWG THWN alu.

Solution: A 6 AWG THWN copper or 4 AWG aluminum system bonding jumper is required.

SIZING GROUNDING ELECTRODE CONDUCTOR TO GROUND THE SECONDARY OF A TRANSFORMER WITH 1/0 AWG THWN cu. CONDUCTORS

Step 1: Finding grounding electrode conductor
Table 250.66, Col.'s 1, 3, or 4
1/0 AWG THWN requires 6 AWG THWN cu.
1/0 AWG THWN requires 4 AWG THWN alu.

Solution: A 6 THWN AWG copper or 4 AWG aluminum grounding electrode conductor is required.

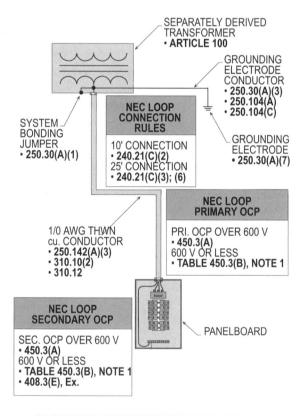

SEPARATELY DERIVED TRANSFORMER
- **ARTICLE 100**

GROUNDING ELECTRODE CONDUCTOR
- **250.30(A)(3)**
- **250.104(A)**
- **250.104(C)**

SYSTEM BONDING JUMPER
- **250.30(A)(1)**

NEC LOOP CONNECTION RULES
10' CONNECTION
- **240.21(C)(2)**
25' CONNECTION
- **240.21(C)(3); (6)**

GROUNDING ELECTRODE
- **250.30(A)(7)**

1/0 AWG THWN cu. CONDUCTOR
- **250.142(A)(3)**
- **310.10(2)**
- **310.12**

NEC LOOP PRIMARY OCP
PRI. OCP OVER 600 V
- **450.3(A)**
600 V OR LESS
- **TABLE 450.3(B), NOTE 1**

NEC LOOP SECONDARY OCP
SEC. OCP OVER 600 V
- **450.3(A)**
600 V OR LESS
- **TABLE 450.3(B), NOTE 1**
- **408.3(E), Ex.**

PANELBOARD

GROUNDING SEPARATELY DERIVED SYSTEMS NEC 250.30(A)(1); (2); (3)

Figure 13-12. The system bonding jumper and grounding electrode conductor is designed and installed based on the derived ungrounded phase conductors supplying the panelboard, switch, or other equipment connected from the secondary side of the transformer.

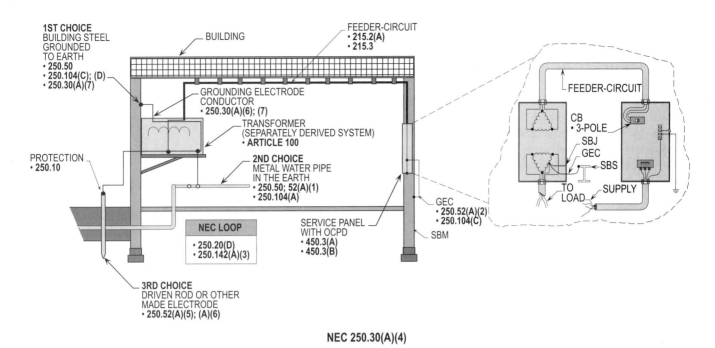

NEC 250.30(A)(4)

Figure 13-13. The grounding electrode conductor shall be as near as possible and preferably in the same area as the grounding electrode conductor connection to the system.

SYSTEM BONDING JUMPER
250.30(A)(1); (8)

The bonding jumper is designed and installed based on the derived ungrounded phase conductors supplying the panel, switch, or other equipment connected from the secondary side of the transformer and sized per **250.102(C)and (D)**. The bonding jumper must be sized per **250.66** from the ungrounded phase conductors up to 1100 KCMIL for copper and 1750 KCMIL for aluminum. The bonding jumper must be sized at least 12 1/2 percent (.125) of the area of the largest conductor where the service conductors are installed larger than 1100 KCMIL copper or 1750 KCMIL aluminum. The bonding jumper may be installed and connected at any point on the separately derived system from the source to the first system disconnecting means or overcurrent protection device. If the grounded phase conductors are larger than 1100 KCMIL for copper and 1750 KCMIL for aluminum, the bonding jumper will be larger than the grounding electrode conductor.

GROUNDING ELECTRODE CONDUCTOR
250.30(A)(3)

The grounding electrode conductor is designed and installed based on the derived ungrounded phase conductors supplying the panel, switch, or other equipment tapped from the secondary of the transformer and sized per **Table 250.66**. The grounding electrode conductor must be installed and connected at any point on the separately derived system

from the source to the first system disconnecting means or overcurrent protection device. When the KCMIL rating is greater than 1100 KCMIL for copper and 1750 KCMIL for aluminum, the grounding electrode conductor will usually be smaller than the bonding jumper.

GROUNDING ELECTRODE
250.30(A)(7)

The grounding electrode must be as near as possible and preferably in the same area as the grounding electrode conductor connection to the system. The following three choices must be selected in the order as they appear in the NEC. **(See Figure 13-13)**

(1) Nearest structural building steel,
(2) Nearest metal water pipe system, or
(3) Other electrodes as specified in **250.50** and **250.52**.

Note: Metal water pipe located in the area must be bonded to the grounded conductor per **250.104(A)(4)**.

The grounding electrode conductor does not have to be installed larger than 3/0 AWG copper or 250 KCMIL for aluminum when connecting to the nearest structural building steel or nearest metal water pipe system. The grounding electrode conductor does not have to be installed larger than 6 AWG copper or 4 AWG aluminum, when connecting to a driven rod or other made electrodes.

SIZING GROUNDING ELECTRODE CONDUCTOR
250.30(A)(3); TABLE 250.66

The procedure for selecting the grounding electrode conductor (GEC) to ground a separately derived system to the building steel is determined by the size of the ungrounded phase conductors in the feeder-circuit (connected conductors) between the panelboard and transformer.

For example: What size copper GEC is required to ground a separately derived system to the structural building steel when supplied by 500 KCMIL copper conductors?

Step 1: Finding the GEC
250.30(A)(3); Table 250.66
500 KCMIL cu. = 1/0 AWG cu.

Solution: The size GEC is 1/0 AWG copper.

The procedure for selecting the grounding electrode conductor to ground a separately derived system to a metal water pipe is determined by the size of the connected conductors from the secondary of the transformer. **(See Figure 13-14)**

For example: What size copper GEC is required to ground a separately derived systems to a metal water pipe supplied by 250 KCMIL copper conductors?

Step 1: Finding size GEC
250.30(A)(3); Table 250.66
250 KCMIL cu. = 2 AWG cu.

Solution: The size GEC is 2 AWG copper.

In cases where there are no other electrodes available, a separately derived system can be grounded with a driven rod or plate per **250.52 and 53**. A driven rod with a resistance of 25 ohms or less is considered low enough to allow the grounded system to operate safely and function properly. The grounding electrode conductor is not required to be larger than 6 AWG copper or 4 AWG aluminum where connected to electrodes such as driven rods.

For example: What is the current flow in a 6 AWG copper grounding electrode conductor (GEC) connecting the common grounded terminal bar in a separately derived system to a driven rod? (The supply voltage is 120/208 volt, three-phase system)

Step 1: Finding amperage
250.56
$I = 120 \text{ V} \div 25 \text{ R}$
$I = 4.8 \text{ A}$

Solution: The normal current flow is about 4.8 amps.

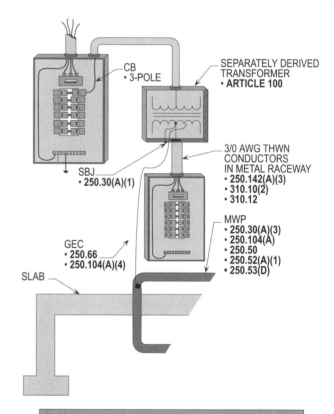

NEC 250.66
TABLE 250.66

Figure 13-14. The procedure for selecting the grounding electrode conductor to ground a separately derived system to the metal water pipe is determined by the size of the transformer's secondary conductors.

TROUBLESHOOTING TRANSFORMER WINDINGS

The winding of a transformer can be tested by taking resistance readings with an ohmmeter.

To check the X_1 winding, touch the case of the transformer with one lead of the ohmmeter and with the other lead touch the lead terminal of X_1. If a low resistance is measured, the winding is defective. X_2 and X_3 windings can be tested using the same procedure. **(See Figure 13-15)**

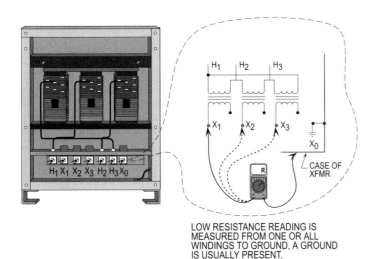

LOW RESISTANCE READING IS MEASURED FROM ONE OR ALL WINDINGS TO GROUND, A GROUND IS USUALLY PRESENT.

Figure 13-15. If a low resistance is read from one or all of the secondary windings to ground, measuring from the transformer's case to each individual winding, a ground is usually present. Note that primary windings can be tested using the same procedures.

Note: For further information on procedures for troubleshooting dry-type transformers, see Table 10 in Appendix A in the back of this book.

CONNECTING SECONDARY WINDINGS IN A WYE CONFIGURATION

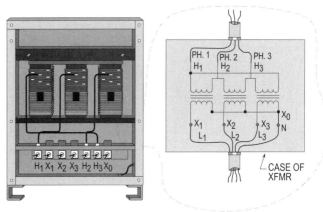

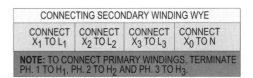

CONNECTING SECONDARY WINDING WYE			
CONNECT X_1 TO L_1	CONNECT X_2 TO L_2	CONNECT X_3 TO L_3	CONNECT X_0 TO N
NOTE: TO CONNECT PRIMARY WINDINGS, TERMINATE PH. 1 TO H_1, PH. 2 TO H_2 AND PH. 3 TO H_3.			

CONNECTING SECONDARY WINDINGS IN A PRIMARY CONFIGURATION

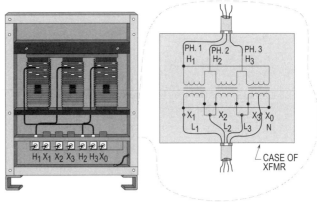

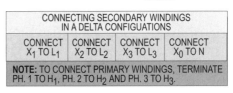

CONNECTING SECONDARY WINDINGS IN A DELTA CONFIGUATIONS			
CONNECT X_1 TO L_1	CONNECT X_2 TO L_2	CONNECT X_3 TO L_3	CONNECT X_0 TO N
NOTE: TO CONNECT PRIMARY WINDINGS, TERMINATE PH. 1 TO H_1, PH. 2 TO H_2 AND PH. 3 TO H_3.			

Chapter 13: Windings and Components

Section	Answer	

T   F   1. The induced voltage in the primary and secondary windings would be in opposite directions for transformer windings in additive polarity.

T   F   2. A measured voltage would be equal to the primary minus the secondary for transformer windings, if connected in subtractive polarity.

T   F   3. The letter H terminal and number is located at the left of the secondary side of the transformer, if the windings are connected with additive polarity.

T   F   4. When grounding any one of the ungrounded phase conductors of a corner grounded delta system the grounded conductor is derived.

T   F   5. Bonding jumpers for a separately derive system must be sized at least 125 percent of the area of the largest conductor where the service conductors are installed larger than 1100 KCMIL copper or 1750 aluminum.

T   F   6. The grounding electrode conductor for a separately derived system must always be connected to the nearest metal water pipe system.

T   F   7. The grounding electrode conductor to ground a separately derived system must be sized per **Table 250.122**.

T   F   8. A separately derived system can be grounded with a driven rod or plate when no other electrodes are available.

T   F   9. Grounding electrode conductors for a separately derived system must not be required to be larger than 6 AWG copper where connected to made electrodes such as driven rods.

10. The induced voltage and current in the primary and secondary windings would be in the same direction for transformers connected in _____ polarity.

11. The voltage would be equal to the primary and secondary added together for transformer windings that are _____ connected.

12. H_1 on the primary side will align with X_1 on the secondary side of the transformer if the windings are connected with _____ polarity.

13. Three-phase, closed delta secondary terminals are connected:

   (a) X_6 to X_1; X_5 to X_4; X_3 to X_2 **(c)** X_6 to X_5; X_4 to X_3; X_2 to X_1

   (b) X_6 to X_1; X_5 to X_2; X_4 to X_3 **(d)** X_6 to X_4; X_5 to X_3; X_3 to X_1

14. Three-phase, open delta secondary terminals are connected:

   (a) X_1 to X_2 **(b)** X_3 to X_2 **(c)** X_3 to X_1

_____ _____ **15.** What size bonding jumper is required to bond the secondary of a transformer with 2/0 AWG THWN copper conductors?

_____ _____ **16.** What size grounding electrode conductor is required to ground the secondary of a transformer with 2/0 AWG THWN copper conductors? (Use building steel)

_____ _____ **17.** What is the voltage when additive polarity is used?

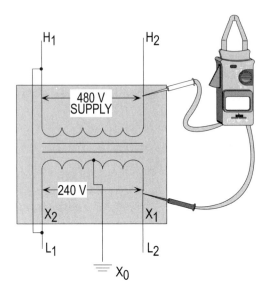

ADDITIVE POLARITY

_____ _____ **18.** What is the voltage when subtractive polarity is used?

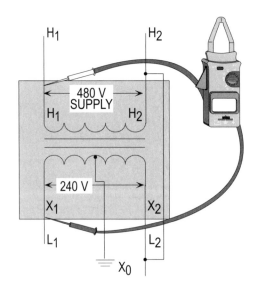

SUBTRACTIVE POLARITY

19. What is the voltage from phase-to-phase?   _____   _____ 13-13

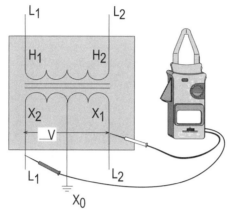

**PHASE-TO-PHASE
VOLTAGE**

20. What is the voltage from phase-to-ground?   _____   _____

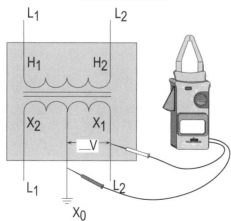

**PHASE-TO-GROUND
VOLTAGE**

Motors
Part Three

No day passes without the discovery of new ways to use the most efficient and most important device ever invented, the electric **motor**. Without it, the wheels of industry would grind to a halt and millions of time and labor saving devices would be rendered useless.

AC electrical motors are designed and selected by finding the three currents that make up the circuits that supply the power to the motors. The first current to be found is the full-load amps (FLA) from **Table 430.248** of the NEC for single-phase and **Table 430.250** for three-phase. The second current to be determined is the nameplate amps found on the motor. And the third current is the locked-rotor current (LRC), in amps, from **Table 430.7(B)** for motors with code letters and **Tables 430.251(A) and (B)** for motors with design letters. Both tables are found in the NEC.

AC electrical motors are designed and installed in a wide variety of sizes, types, and styles, ranging from tiny fractional horsepower units to very large machines of 20,000 HP and larger. These types of motors can be either of single-phase or three-phase construction based on the HP and voltage.

Part Three covers motor theory, types of motors, and the regulations of the NEC that pertain to the design and installation of these motors.

Motor Theory

For the operation of a motor, electricity and magnetism play a major role for producing power to the field windings, called *poles*. These windings induce magnetic lines of force from north to south poles. The rotor is connected in the motor to the load so that the rotor can drive the load. Circulating currents induced in the conducting material of the rotor as it cuts through the magnetic flux lines of the magnetic field are called eddy currents and must be circulated properly. Note that the foundation of motor operation is the attracting of unlike poles and the repelling of the like poles.

REGULAR MAGNETS

The earth is a permanent magnet with the north and south poles connected by an invisible field of magnetic force. If a piece of soft iron is placed within the field of a magnet, it becomes energized. The piece of soft iron is magnetized by the field of the permanent magnet. The piece of soft iron, when placed in the field of the permanent magnet, does not have to touch to become magnetized, and takes on the same characteristics as the permanent magnet. This type of action is called induction and is essential for motor operation. **(See Figure 14-1)**

The poles of two permanent magnets either attract or repel each other. Like poles repel each other, while unlike poles attract. By suspending a permanent magnet from a string, the suspended permanent magnet will rotate by attracting or repelling each end of a second magnet. This type of action illustrates one of the major principles used in the operation of electric motors. The attracting and repelling action by the field poles causes the rotor to rotate through the magnetic field and drive its connected load. **(See Figure 14-2)**

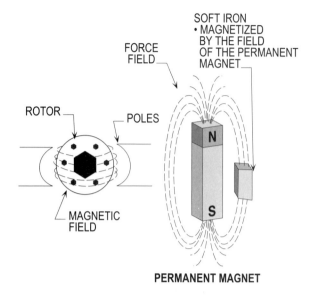

REGULAR MAGNETS

Figure 14-1. A permanent magnet becomes energized when a piece of soft iron is placed in the force field.

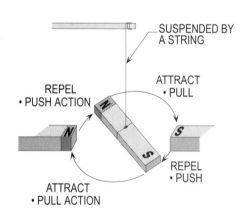

ATTRACTING AND REPELLING OF POLES

Figure 14-2. The poles of two permanent magnets either attract each other for unlike poles or repel each other for like poles.

ELECTROMAGNETS

A single insulated conductor wound around a soft iron core will produce an electromagnet that is much stronger than a permanent magnet. The strength of a magnetic field around a straight conductor carrying current is relatively weak. A strong magnetic field is produced by the number of turns in the winding (coil). A weak magnetic field is produced by fewer number of turns in the coil.

POLES

The polarity of its poles in an electromagnet is changed by reversing the current flow. Reversing the current flow in an electromagnet in one direction produces a south pole at one end and produces a north pole when current flow is in the opposite direction. Alternating current changes the poles of an electromagnet from north to south due to the flow of current changing direction. **(See Figure 14-3)**

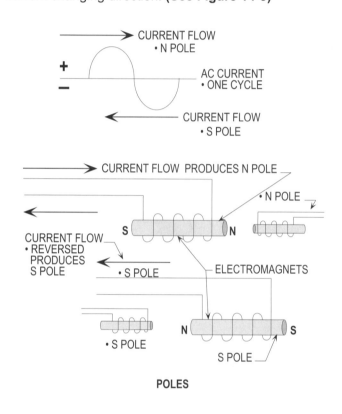

POLES

Figure 14-3. Reverse current flow causes the alternating current to change the poles of an electromagnet from a north pole to a south pole. (Note, see **Figure 14-6**.)

BASIC INDUCTION MOTORS

A basic induction motor consists of a fixed section called a stator and a rotating section called a rotor. A stator is cut into thin sections of soft iron or steel (laminations) and assembled in a sandwich-like manner to reduce eddy current losses. Circulating currents induced in the conducting material of the rotor when it cuts through the magnetic flux lines of the magnetic field are called eddy currents.

A heating effect is produced when eddy currents flow in a solid piece of metal. Eddy currents will only flow in sections of metal that is sandwiched or cut. Therefore, the heating effect is decreased by this process.

A stator that is equipped with two or more field poles with insulated wire wound around them and connected together will create two or more electromagnets. When applying 60 hertz (cycles) of alternating current to electromagnets, the magnetic poles of the electromagnet reverse their polarity 120 times per second. This occurs every time the current reverses direction and alternates. **(See Figure 14-4)**

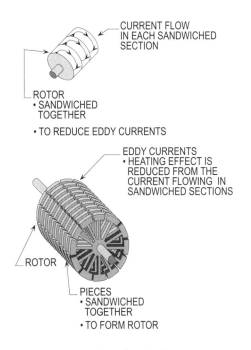

BASIC INDUCTION MOTORS

Figure 14-4. Eddy currents are reduced by a rotor being sandwiched together and laminated.

FIELD POLES

Induction motors operate on the following types of systems:

    (1) Single-phase,
    (2) Two-phase, or
    (3) Three-phase voltage.

Single-phase AC motors operate by the rotating magnetic field being produced by splitting the phases and shifting the AC power applied to the stator field poles. A means of starting must be provided for the rotor in a single-phase motor. The magnetic field alternates at such a fast rate (60 times a second) that the rotor cannot follow the alternating field. The rotor must start and turn fast enough to catch the rotating field. By using your hand or by using a starting winding, a rotor can be caused to rotate, and it will try to catch the magnetic field.

The current changes in the stator poles from north to south as the current alternates from positive to negative in a single-phase, 120 volt motor.

The phase displacement of different voltages is used when installing polyphase AC motors. The voltage in polyphase AC motors are one of the following:

(1) In-phase currents that rise and fall simultaneously.

(2) Out-of-phase currents 180 degrees out-of-phase have one current that rises past zero as the other falls past zero. Currents 90 degrees out-of-phase have one current reaching a peak while the other is at zero.

See Figure 14-5 for the rotating magnetic field of the stator using two voltages 90 degrees out-of-phase.

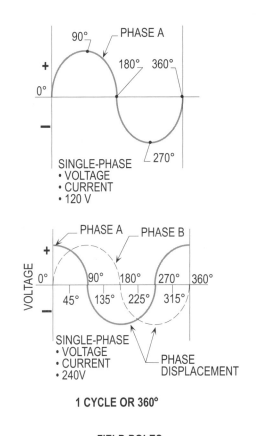

1 CYCLE OR 360°

FIELD POLES

Figure 14-5. Relationship of single-phase currents that are 90 degrees out-of-phase.

Figure 14-5 shows the relationship of single-phase voltage and current. Phase A current flow is at the 0 degree position when at maximum and Phase B current flow is at zero. Phase A windings in the stator will be at maximum value and so will its magnetic field. Phase B windings will produce a magnetic field that will be at zero. Note that Phase A and Phase B currents are of equal values at the 45 degree position.

Phase A current is at zero while phase B is at maximum at the 90 degree position. Phase B windings produce a magnetic field that is at maximum value while phase A windings are at zero. Phase A and phase B currents flowing at the 45 and 225 degree position are equal. The rotor continues to turn with the rotating magnetic field until it completes 360 degrees. By placing the two poles (windings) at right angles to each other in the stator, the rotating magnetic field can be accomplished with voltages that are 90 degrees out-of-phase.

ROTOR

The rotor is made of slotted sections that are cut and sandwiched together to reduce eddy current losses. The rotor is embedded with copper or aluminum bars and welded together by a ring. The flow of current travels a path provided by the ring through the bars of the rotor. Insulation between the bars and rotor is not required since the voltage induced into these bars is low.

The rotor may be placed between two or more stator poles. A power source of 60 hertz AC is applied to the stator poles. The stator poles build up a magnetic field and collapse for each alternation. The poles of the stator change from south-to-north or north-to-south from these alternations. The rotor has a push and pull action through the rotating magnetic field of the stator poles. **(See Figure 14-6)**

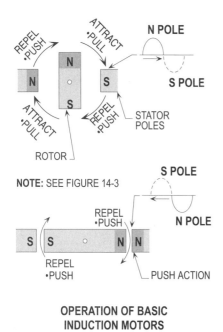

**OPERATION OF BASIC
INDUCTION MOTORS**

Figure 14-6. The operation of a basic induction motor.

Current is induced into the bars of the rotating rotor from the expanding and collapsing fields of the stator poles. This is accomplished by the rotor cutting the magnetic lines of force produced by the stator poles or field windings.

> **Motor Theory Tip:** The magnetic field of the stator field poles is opposite to the magnetic field in the rotor.

DESIGNS OF MOTORS

The following designs of motors are to be considered when designing speed regulation, starting torque, and full-voltage starting current:

(1) Class B,
(2) Class C,
(3) Class D, or
(4) Class E per 2002 NEC.

CLASS B MOTORS

A speed regulation based on 2 to 5 percent slip is used when designing and installing Class B motors. Class B motors are designed to drive loads such as blowers, fans, and centrifugal pumps.

A Class B motor has a starting torque of about 150 percent times the full-load torque rating of the motor. The full-voltage starting current is approximately 600 percent to 725 percent of the full load running current in amps of the motor.

Class B motors have a normal starting torque and normal starting current. Class B motor are used on loads that are started and reversed infrequently. Class B motors are the most used motors in the electrical industry. They are almost equal to motors marked with a code letter B.

CLASS C MOTORS

A speed regulation based on 2 percent to 5 percent slip is used when designing and Class C motors. Class C motors are designed to drive hard-to-start loads such as compressors, conveyors, reciprocating pumps, and crushers.

A Class C motor has a starting torque of about 225 percent of the full-load torque rating of the motor. The full-voltage starting current is approximately 600 percent to 650 percent of the running current of the motor. Class C motors have high starting torque and normal starting current. They are capable of starting loads that are hard to start and accelerate up to their running speed.

CLASS D MOTORS

A speed regulation based on 5 percent to 13 percent slip is used when designing with Class D motors. Class D motors are known as high-slip motors. Class D motors are designed to drive very hard-to-start loads that are started and reversed frequently such as cranes, hoists, elevators, cyclical loads, and punch presses.

A Class D motor has a starting torque of about 275 percent of the full-load torque rating of the motor. The full-voltage starting current is approximately 525 percent to 625 percent of the running current of the motor. Class D motors have high starting torque and low starting current.

CLASS E MOTORS

Due to a new law and energy conservation act, it is very clear that the need to install high energy efficiency motors will become more urgent. In fact, its essential that electrical personnel learn as much as possible about high energy efficient motors. For in the near future, such motors will be the only type available. For example, Tables 430.251(A) and (B) only recognize Design letter motors. See Table 430.7(B) for code letter motors. Note that high-efficiency motors per 2005 NEC are NEMA Design B.

THE LAW

In October 1992, the Energy Policy Act was accepted. The new law required standard efficiency motors will no longer be built after October 1997. After this date, only high-efficiency motors were to be manufactured. The motor industry seems to apply the term premium efficiency (PE) to identify high-efficiency motors.

The law basically requires all NEMA induction design motors of 200 HP or less that are single speed to comply with this rule. This law applies to motors as follows:

(1) Mainly Design A and B, continuous rated, and operating at 230 / 460 volt, 60 hertz.

(2) General purpose T-frame with the following characteristics:

(a) Single speed,
(b) Foot-mounted,
(c) Polyphase,
(d) Squirrel-cage,
(e) Induction motors, and
(f) 200 HP or less with exceptions

HIGH-EFFICIENCY MOTORS

To improve motor efficiency, the manufacturer must reduce motor losses. There are three categories of motor losses and they are as follows:

(1) I^2R losses,
(2) Mechanical losses, and
(3) Core losses.

I^2R LOSSES

I^2R losses account for 20 to 30 percent of the total loss of the motor. The I^2R losses of the windings depends on the current, and the winding resistance.

There are many conditions which will affect resistance and current. For example, such conditions as temperature, load, excitation, magnetic influences will affect the resistance in the windings. Since the current that a motor draws is primarily a function of the load, steps must be taken to improve the motor's power factor. This can be done by reducing the reactive component of the total motor current. Note that the way to reduce I^2R is to reduce the resistance of the windings. This is done by improving the motor's stator resistance by increasing both the size and number of conductors.

Rotor losses may be minimized by increasing the size of the conductor bars and using low-loss laminated steel.

For larger motors, copper bars can be utilized in rotors for best efficiency. For maximum efficiency, copper rotor bars are custom-fitted, brazed, formed, and designed and fitted into the rotor slots.

By reducing the air gap between the stator and rotor, winding losses can be reduced. This diminishes the reactive element of the motors total current. In general, the needed excitation current is reduced which improves the overall power factor. Due to the better designed steel core, the magnetic field does not have to be as large to provide the same amount of motor performance. Note that there is less current needed to produce the magnetic field.

MECHANICAL LOSSES

Mechanical losses are produced in motor operation from friction and windage that takes place within the motor. Windage losses are losses created by moving parts of the motor. For example, windage losses can be developed by the fan blade if such blade is not designed to help alleviate such unwanted friction. Another component of the motor that creates friction losses is the bearings. To help diminish bearing losses, high quality bearings are used and due to such close tolerance operation, the air gap variation between the stator and rotor greatly reduces these types of losses.

CORE LOSSES

Core losses are caused by two elements and they are as follows:

(1) Hysteresis, and
(2) Eddy currents.

The hysteresis loss is the element of the core loss in a magnetic circuit which can be reduced by using high quality steel to build the core.

Eddy current losses are reduced by laminating the core with thin sheets of steel that are insulated from each other. Thin sheets of steel are sandwiched together which will reduce eddy current problems even further. Eddy currents are circulating currents in the core of the stator poles. These currents are caused by varying magnetic fields in the core. Due to the I^2R losses in the resistance of the core material, unwanted heat is produced. Note that these currents do no useful work, they just create heating effects in the motor.

Motor Theory Tip: Stray load losses are dependent on the loading of the motor, and they increase as the load is applied. These losses are due to the location of the conductors, in the motor, skin effect, poor laminations, etc. All such factors are lumped into this catch-all category.

See Figure 14-7 for a detailed illustration of a high-efficiency motor.

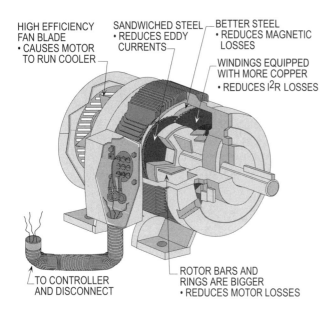

HIGH EFFICIENCY FAN BLADE
• CAUSES MOTOR TO RUN COOLER

SANDWICHED STEEL
• REDUCES EDDY CURRENTS

BETTER STEEL
• REDUCES MAGNETIC LOSSES

WINDINGS EQUIPPED WITH MORE COPPER
• REDUCES I^2R LOSSES

TO CONTROLLER AND DISCONNECT

ROTOR BARS AND RINGS ARE BIGGER
• REDUCES MOTOR LOSSES

HIGH EFFICIENCY MOTORS

Figure 14-7. The above is an illustration of a high-efficiency motor with elements that are designed to provide better efficiency.

MOTOR FACTS

The following items must be taken into consideration when designing and selecting a motor:

(1) Operating voltage,
(2) Operating current,
(3) Operating torque,
(4) Operating slip,
(5) Power factor, and
(6) Frequency

OPERATING VOLTAGE AND CURRENT

Single-phase voltage of 120 volts is like one person riding a bicycle. Only one stroke is produced that will peak and produce power from the one person riding the bicycle. Single-phase voltage of 208 or 240 volts is like two people riding a bicycle. Two power-producing strokes are provided by one rider leaving a power-producing stroke (peak) and the other rider entering the peak and producing power to drive the bicycle. This type of example shows that a 208 or 230 volt, single-phase motor is more efficient than a 120 volt, single-phase motor.

Three-phase voltage is compared to three riders on a bicycle. Three power-producing strokes are provided by the first rider leaving the peak stroke and the second rider entering at the peak. The third rider enters the peak as the second rider leaves the peak stroke. This type of example shows that a three-phase motor will produce more power because there are three different phases that are peaking and providing a smooth and continuous power to drive the rotor and load at its operating speed. **(See Figure 14-8)**

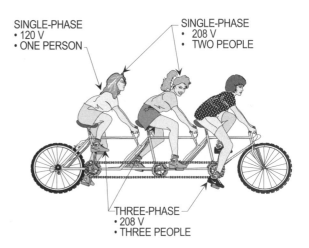

SINGLE-PHASE
• 120 V
• ONE PERSON

SINGLE-PHASE
• 208 V
• TWO PEOPLE

THREE-PHASE
• 208 V
• THREE PEOPLE

OPERATING VOLTAGE AND CURRENT

Figure 14-8. Single-phase and three-phase voltages and currents operate in comparison to riders on a bicycle.

The three phases of voltage and current supply one of the three separate pairs of poles. The first phase (peak stroke) delivers the greatest power. The second phase enters the peak as the first phase leaves the peak, delivering its greatest stroke of power. The third peak phase enters the peak as the second phase leaves the peak, and the process repeats itself. **(See Figure 14-9)**

Refer to **Figure 14-10** and notice that between poles A and D the greatest power stroke and magnetic field is produced for Phase 1. Between poles B and E the greatest power stroke is produced for Phase 2. Between poles C and F the greatest power stroke is produced for Phase 3. The voltage and current of the three phases is displaced 120 degrees on the stator of the motor. **(See Figure 14-10)**

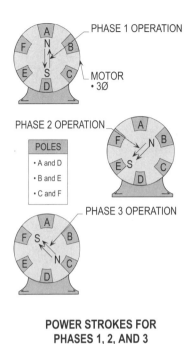

**POWER STROKES FOR
PHASES 1, 2, AND 3**

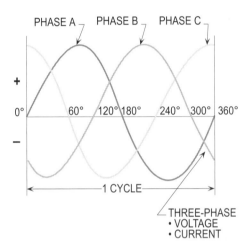

**THREE PHASES OF
VOLTAGE AND CURRENT**

Figure 14-9. The three phases (A, B, and C) of voltages supply one of the three separate pairs of poles.

Figure 14-10. Between poles A and D the greatest power stroke and magnetic field is produced for Phase 1. Between poles B and E the greatest power stroke is produced for Phase 2. Between poles C and F the greatest power stroke is produced for Phase 3.

OPERATING TORQUE AND SLIP

Good starting torque and slip is provided for the motor by dual bars installed in the rotor. The outer bars close to the surface have a high-resistance winding, and the inner bars have a low-resistance winding. Good starting is provided by the outer bars. More current is allowed to flow at the running speed of the motor from the inner bars. **(See Figure 14-11)**

The percentage of slip desired for each motor is designed by placement of these bars shown in **Figure 14-11**. The magnetic lines of force cutting across these bars embedded in the rotor produce the slip of the motor. Voltage is induced in the rotor only when these copper bars cut the magnetic lines of force created by the alternations of current in the stator field. Note that the rotor will never rotate at the same speed as the rotating magnetic field. This difference in rotating speed is the slip of the motor. **(See Figure 14-12)**

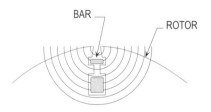

**CLASS B, C, OR D MOTOR IS
PRODUCED BY
ARRANGEMENT
OF BARS IN ROTOR**

Figure 14-11. Different classes of motors with different torque ratings can be produced by the bars in the rotor.

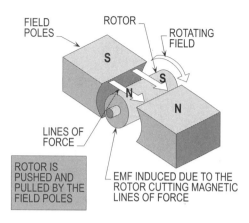

ROTOR SLIP

Figure 14-12. The rotor will never rotate at the same speed as the alternations of the current and rotating field.

The greater the slip a rotor has, resulting from a driven load, the more lines of force are cut during rotation and the slower the rotor will turn. The actual running speed of a motor is designed to have 5 percent slip which will allow the rotor to rotate at less than the synchronous speed created by the alternating current. **(See Figure 14-13)**

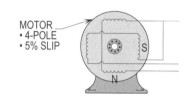

Finding Synchronous Speed And Actual Speed
Step 1: Finding synchronous speed in stator Syn. Sp = 120 x Frequency ÷ # of poles Syn. Sp = 120 x 60 cps ÷ 4 Syn. Sp = 1800 RPM
Step 2: Using alternate method Syn. Sp = cps x freq. ÷ pair of poles Syn. Sp = 60 cps x 60 cps ÷ 2 Syn. Sp = 1800 RPM
Step 3: Finding actual speed of rotor Slip (in rpm) = syn. speed x slip % Slip (in rpm) = 1800 x 5% Slip (in rpm) = 90 rpm Act. sp = syn. speed - slip (in rpm) Act. sp = 1800 - 90 rpm Act. sp = 1710 rpm
Solution: **The synchronous speed is 1800 RPM and the actual speed is 1710 RPM.**

FINDING ACTUAL RUNNING SPEED

Figure 14-13. Finding the synchronous and actual speed of a motor due to its slip characteristics.

Fewer lines of force are cut when the rotor turns at a faster speed through the field. The voltage and magnetic lines of force in the rotor becomes weaker and causes the rotor to slow down.

The rotor has 100 percent slip when the rotor is at rest and no lines of magnetic force are cut. However, the rotor begins to turn with the stator field when power is applied to the stator poles.

The rotor current tries to reach the peak of the alternation of current creating the magnetic lines of force between the stator field poles. As the rotor turns, the percentage of slip begins to decrease (usually 2 to 5 percent) until the designed amount of slip is reached as the motor drives the load. **(See Figure 14-14)**

POWER FACTOR

The ratio of the actual power used in a circuit to the apparent power drawn from the line is the power factor. The true power used to produce heat or work is the actual power. Actual power is also known as true, real, or useful power. A wattmeter is used to measure the actual power in watts and kW. A voltmeter and ammeter is used to measure the apparent power in VA and kVA. When measuring the voltage and current waveforms, they may be in-phase or out-of-phase. Note that the degree (0-90°) of shift indicates power factor. When the actual and apparent power are the same value, the power factor is 100 percent.

Less actual power is consumed for circuits with motors and transformers having windings producing magnetic fields. These circuits have a power factor that is less than 100 percent. The inductance of the windings causes the inequality between the actual power and apparent power. Note that the actual power never exceeds the apparent power.

REACTIVE POWER

When the kVA exceeds the kW a reactive power exists. The operating current consists of true current (in-phase) and reactive current. The current drawn by an inductive load is used to develop magnetic fields required for operation. This is reactive current. **(See Figure 14-15)**

APPARENT POWER

By multiplying the volts times amps the apparent power can be found. Apparent power can be equal to or greater than actual power. When apparent power and actual power are equal, their ratio is 1 to 1, 1.0, or 100 percent. For example, if the apparent power is 2000 W and the power consumed is 800 W, the ratio is .4 or 40 percent (PF = 800 W ÷ 2000

W = .4). A 40 percent power factor is low. A phase angle of 66 degrees is equal to cosine Ø of 40 and is listed in the trigonometric charts. **(See Figure 14-16)**

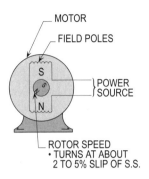

PERCENTAGE OF SLIP

Figure 14-14. In an induction motor the rotor speed usually turns at about 2 to 5 percent slip.

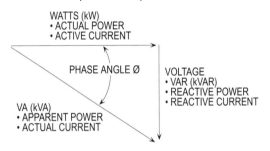

POWER FACTOR RELATIONSHIP

Figure 14-15. The above illustration shows the right angle relationship of terms used to demonstrate power factor.

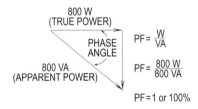

CIRCUIT WITH HIGH POWER FACTOR

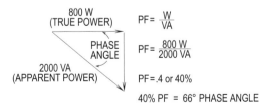

CIRCUIT WITH LOWER POWER FACTOR

Figure 14-16. Circuits with watts (or kW) closer to the VA (or kVA) rating of the apparent power have a higher power factor. Circuits with watts (or kW) less than the VA (or kVA) rating of the apparent power have a lower power factor. See trigonometric charts on page 15-6 to match power factor to the phase angle.

ACTUAL POWER

By multiplying the volts times amps in a pure resistance circuit the actual power can be found.

> **For example:** What are the watts for a 240 volt, single-phase load rated at 20 amps?
>
> **Step 1:** Finding W
> W = V x A
> W = 240 V x 20 A
> W = 4800
>
> **Solution: The actual power for the load is 4800 watts.**

By multiplying the volts times amps times cosine Ø (PF), the actual power can be found.

> **For example:** What are the watts for a 240 volt, single-phase load rated a 20 amps and a power factor of 70 percent?
>
> **Step 1:** Finding W
> W = V x A x PF
> W = 240 V x 20 A x 70%
> W = 3360
>
> **Solution: The actual power for the load is 3360 watts.**

By multiplying the volts times amps times cosine Ø (PF) for a three-phase circuit, the actual power can be found.

> **For example:** What are the watts for a 480 volt, three-phase motor with an FLC of 40 amps having a power factor of 80 percent?
>
> **Step 1:** Finding W
> W = V x $\sqrt{3}$ x A x PF
> W = 480 V x 1.732 x 40 x 80%
> W = 26,592
>
> **Solution: The actual power for the motor is 26,592 watts.**

OPERATING 230 V MOTORS ON 208 V SUPPLY CIRCUIT

Slight changes in voltage, torque, slip and current occur where 230 V, 3Ø, 60 Hz motors operate on 208 V, 3Ø, 60 Hz supply circuits. For example, the full-load current for a motor operating ten percent plus above rated voltage is seven percent below normal. **(See Figure 14-17)**

NEMA standards allow motor terminal voltage to vary 10 percent below or above rated voltage. For example, a 230 volt motor can operate between 207 volts (230 V x 10% = 230 V - 23 V = 207 V) and 253 volts (230 V x 110% = 253 V).

CHARACTERISTIC	+10%	-10%
TORQUE	UP 21%	DOWN 19%
FULL-LOAD SPEED	UP 1%	DOWN 2%
POWER FACTOR	DOWN 4%	UP 3%
FULL-LOAD CURRENT	DOWN 7%	UP 11%
TEMPERATURE	DOWN 10%	UP 17%
MAXIMUM OVERLOAD	UP 21%	DOWN 19%
EFFICIENCY	UP 1%	DOWN 2%

NOTE: FOR MORE FACTS, SEE FIGURE 16-13.

Figure 14-17. Characteristic of induction motors operating 10% above or 10% below supply voltage.

TORQUE

The torque of a motor varies with the square of the voltage. The starting torque and maximum running torque of motors running on 208 volts are determined by squaring voltage (208 V) and dividing by the voltage of the motor (230 V) squared. For example, a motor with a starting torque of 150 lb. ft. has a starting torque of 123 lb. ft. [(208 V/230 V)^2 = .82 x 150 lb. ft. = 123 lb. ft.)].

SLIP

The slip of induction motors varies inversely with the square of the voltage. The slip of a 230 volt motor operating on a 208 volt supply is 1.22 times the slip of the 230 volt rating on the motor's nameplate [(230 V/208 V)^2 = 1.22]. For example, if the synchronous speed is 1800 RPM and the actual speed is 1725 RPM, the slip is 75 RPM (1800 RPM – 1725 RPM = 75 RPM). The new slip of RPM for a 230 volt motor operating on 230 volts is 91.5 RPM [(230 V/208 V)^2 = 1.22 x 75 RPM = 91.5 RPM].

CURRENT

Motors are more efficient at higher voltage because they draw less current and run slightly cooler. Motors operating at lower voltage pull more current and are less efficient.

Motors rated at 230 volts and operating on a 208 volt supply draw approximately 11 percent more current than when operating at 230 volts. For example, a 230 volt motor operating at 30 amps on 208 volts draws approximately 33.3 amps (30 A x 111% = 33.3 amps).

MEASURING OPERATING AMPS

The operating amps of a motor must not exceed the nameplate rating of the motor for the motor to have normal operating life. The amp rating is read with an ampmeter for an accurate measurement. **(See Figure14-18)**

MEASURING OPERATING VOLTAGE

Two measurements are required to determine the operating voltage of a motor. The first measurement is for the system voltage to the motor. This voltage reading can be 10 percent above or 10 percent below the operating voltage. **(See Figure 14-19)** The second measurement is for the unbalanced voltage from each phase to ground, and you must average the measurements to obtain a percentage. The percentage must not exceed 1 percent. **(See Figure 14-20)**

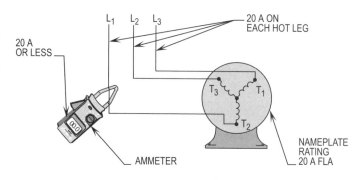

Figure 14-18. An amp reading of 20 amps or less indicates the motor should have a normal life of twenty years or less (SEE NEMA 1 MOTORS AND GENERATORS).

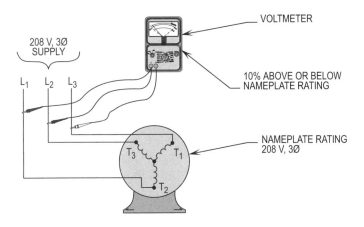

Figure 14-19. The operating voltage can read within 10 percent above or below the operating voltage listed on the nameplate of the motor.

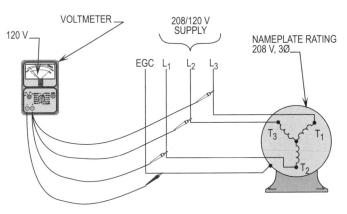

Figure 14-20. Unbalanced voltage (if measured) must not exceed 1 percent, or the running current of the nameplate must be derated to compensate for the percentage that exceeds 1 percent.

Chapter 14: Motor Theory

Section   **Answer**

_____   T   F   **1.** The earth is a permanent magnet with the north and south poles connected by an invisible field of magnetic force.

_____   T   F   **2.** The like poles of two permanent magnets attract each other.

_____   T   F   **3.** The rotor is made of a slotted section that is cut and sandwiched together to reduce eddy current losses.

_____   T   F   **4.** A speed regulation based on 2 to 10 percent slip is used when designing Class B motors.

_____   T   F   **5.** A Class D motor has a starting torque of about 250 percent of the full-load torque rating of the motor.

_____   T   F   **6.** Mechanical losses must be reduced to improve the efficiency of a motor.

_____   T   F   **7.** Good starting torque and slip can be provided by installing dual bars in the rotor of the motor.

_____   T   F   **8.** The ratio of actual power used in a circuit to the apparent power drawn from the line is the power factor.

_____   T   F   **9.** When kVA exceeds kW, a reactive power exists.

_____   T   F   **10.** Multiplying the volts times amps in a pure resistance circuit yields the reactive power.

_____ _____   **11.** If a piece of soft iron is placed within the field of a _____ it becomes magnetized.

_____ _____   **12.** The poles of two permanent magnets either _____ each for unlike poles or _____ each other for like poles.

_____ _____   **13.** A single energized insulated conductor wound around a soft iron core will produce an _____.

_____ _____   **14.** The polarity of poles in an electromagnet is changed by _____ the current flow.

_____ _____   **15.** A basic induction motor consists of a fixed section called a _____ and a rotating section called a _____.

_____ _____   **16.** The voltage in a polyphase motor's stator are 120° out of _____.

_____ _____   **17.** A Class B motor has a starting torque of about _____ percent times the full-load torque rating of the motor.

18. A Class C motor has a starting torque of about _____ percent times the full-load torque rating of the motor.

19. A speed regulation based on _____ percent to _____ percent slip is used when designing Class D motors.

20. I²R losses account for _____ to _____ percent of the total loss of the motor.

21. Which of the following elements are causes of core losses:

    (a) Hysteresis   (c) All of the above
    (b) Eddy currents (d) None of the above

22. The actual running speed of a motor is designed to have a slip which allows the motor to rotate at about _____ percent less than the synchronous speed.
    (a) 3   (c) 7
    (b) 5   (d) 10

23. A rotor has _____ percent slip when the rotor is at rest and no lines of magnetic force are cut.
    (a) 80   (c) 100
    (b) 90   (d) 110

24. Where actual and apparent power are the same, the power factor is _____ percent.
    (a) 100   (c) 150
    (b) 125   (d) 250

25. A Class D motor has a full-voltage starting current that is approximately _____ percent to _____ percent of the running current of the motor.
    (a) 600 to 725   (c) 525 to 625
    (b) 600 to 650   (d) 525 to 725

26. What are the synchronous speed and actual speed for a four-pole motor with 5 percent slip?

27. What are the watts (actual power) for a 240 volt, single-phase water heater with a 30 amp heating element?

28. What are the watts for a 240 volt, single-phase motor with a 30 amp FLC having a power factor of 70 percent?

29. What are the watts for a 480 volt, three-phase motor with an FLC of 50 amps having a power factor of 80 percent? (use 831 V for the three-phase calculation)

Types of Motors

Alternating current (AC) is used in the United States to power the majority of motors installed. These motors are usually connected by journeyman electricians and maintained by maintenance personnel. These motors are designed to operate on a single or dual voltage, based on the connections of the windings of the stator. These motors are equipped with either single-phase or three-phase windings.

Of all the different types of motors, the squirrel-cage induction motor is designed to have fewer parts, is less expensive, and requires less maintenance than wound rotor, synchronous, or direct-current (DC) motors. (See **Table 430.52** of the NEC for a listing of the different types of motors.)

SINGLE-PHASE MOTORS

Single-phase motors can operate from a single-phase lighting or power circuit. The following are the most commonly used single-phase motors:

(1) Split-phase,
(2) Capacitor-start,
(3) Capacitor start-and-run,
(4) Permanent split-capacitor,
(5) Shaded-pole, and
(6) Universal

SPLIT-PHASE MOTORS

The split-phase motor is an AC motor of fractional-horsepower size and is used to operate such devices as washing machines, oil burners, and small pumps. The motor consists of the following four main parts:

(1) A rotor (rotating part),
(2) Stator (a stationary part),
(3) Brackets (fastened to the frame of the stator by means of screws or bolts), and
(4) A centrifugal switch (located inside the motor).

Two windings are provided on the stator for a split-phase motor. These two windings consist of the main (running) winding and the auxiliary (starting) winding. The torque needed to start turning the load is produced from the starting windings which are placed about 30 degrees from the running windings.

A centrifugal switch (starting switch) is placed in series with the starting winding and the starting winding is connected in parallel with the running winding. The centrifugal switch contacts are closed when the rotor is at rest. The contacts open when the motor starts and begins to accelerate up to its running speed. The starting winding is taken out of the circuit by the centrifugal switch, at about 75 to 80 percent of the motor's running speed. The split-phase motor operates as a single-phase induction motor when the starting winding is disconnected from the circuit. **(See Figure 15-1)**

Split-phase motors have a high resistance in the starting windings which produces a high inrush current when starting. Starting windings consist of small wire and have many turns, which have a greater resistance than the running winding. Running (field) windings consist of larger size wire with fewer turns which have a lower resistance. **(See Figure 15-2)**

WITH CAPACITORS

The capacitor is used as the starting device in some motors. These are called capacitor-start motors. The capacitor motor operates on alternating current and is made in sizes ranging from 1/20 HP to 10 HP. It is mainly used to operate such machines as refrigerators, compressors, etc. A so-called capacitor motor is a split-phase motor with the addition of a capacitor connected in series with the starting or auxiliary winding. The capacitor is usually mounted on top of the motor but may be mounted in other external positions or inside the motor housing. The added capacitor provides higher starting torque with lower starting current than the regular split-phase motor. **(See Figure 15-3)**

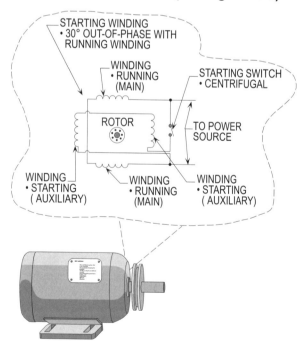

SPLIT-PHASE MOTOR

WINDINGS	RESISTANCE
RUNNING	LOW
STARTING	HIGH

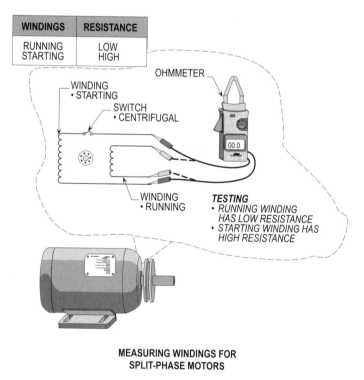

MEASURING WINDINGS FOR
SPLIT-PHASE MOTORS

Figure 15-1. The starting windings of a split-phase motor help start the motor and are disconnected from the circuit at about 75 to 80 percent of its running speed.

Figure 15-2. The running windings and starting windings are measured using an ohmmeter. Running windings have less resistance than starting windings.

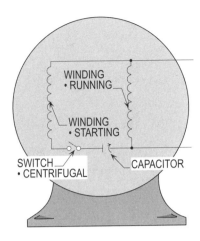

CAPACITOR-START MOTOR

Figure 15-3. Connections of a capacitor-start motor.

STARTING WINDINGS

There are three separate windings in the split-phase motor. The following are the three types of windings:

(1) Squirrel-cage winding (located in the rotor),

(2) Stator run winding (located at the bottom of the stator poles and known as the running or main winding), and

(3) Starting or auxiliary winding.

At the start, the current flowing through both the running and starting windings, which are connected in parallel, causes a magnetic field to be formed inside the motor. This magnetic field rotates and induces a voltage in the rotor winding, which in turn causes another magnetic field. These magnetic fields combine in such a manner as to cause rotation of the rotor. The starting winding is necessary at the start in order to produce the rotating field. After the motor is running the starting winding is no longer needed and is cut out of the circuit by means of the centrifugal switch.

Use these methods to check the starting winding if the motor fails to start:

(1) Turn the shaft of the motor by hand. If it starts, the trouble is in the starter winding circuit.

(2) Disassemble the motor and check the starting winding for an open circuit.

(3) Use a test light or an ohmmeter to determine that the winding is complete. **(See Figure 15-4 and Table 1 in the Appendix)**

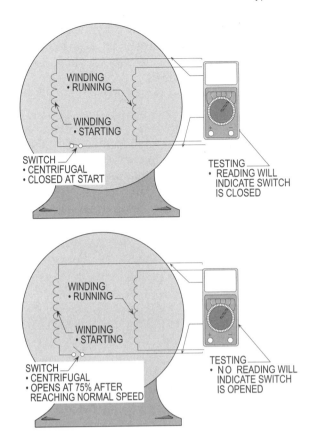

SPLIT-PHASE MOTOR

Figure 15-4. This split-phase motor is started by a centrifugal switch closing and opening the starting winding .

REVERSING DIRECTION

The flow of current in a split-phase motor is changed by reversing the flow of current through the running or starting windings. The rotor will rotate in a counterclockwise direction when the flow of current in the starting winding and the running winding are in the same direction. The rotor will rotate in a clockwise direction when the flow of current in the starting winding and running winding are in the opposite direction. **(See Figure 15-5)**

IDENTIFICATION OF LEADS

While older motors are usually tagged M_1 and M_2 for the running winding, S_3 and S_4 for starting winding, or R_1 and R_2 for the running winding, new motor types dictate a different color coding of the starting and running windings. These newer motors follow a typical color coding of red for T_1, black for T_2 yellow for T_3, and blue for T_4.

THERMAL PROTECTION

An additional switch is sometimes provided for split-phase motors to protect them from overheating. Overheating can be caused by lack of ventilation or by high temperatures.

Ventilation problems can be caused by the motor's inlets and outlets being covered with lint or dirt. High temperatures can also develop in the windings from a stuck bearing in the motor or on the driven load.

Overcurrent protection devices (bi-metal disks or strip composed of dissimilar metals) are connected in series with the running winding. The amount of current flow and temperature rise of the windings is moderated by the overload protector. The overload protector will open the circuit if the current flow and temperature rise exceeds the predetermined (set) value. The overload can be designed to connect the power supply and start the motor when the running winding temperature decreases. **(See Figure 15-6)**

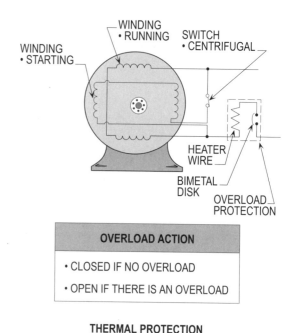

OVERLOAD ACTION
• CLOSED IF NO OVERLOAD
• OPEN IF THERE IS AN OVERLOAD

THERMAL PROTECTION

Figure 15-6. Overload protection is provided for the running winding by the bi-metal disk or strip.

Motor Theory Tip: A motor will shut down and not re-start until the problem is corrected. Note that the overload protector provides this protection.

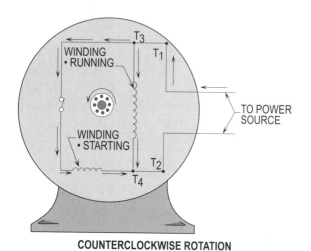

COUNTERCLOCKWISE ROTATION

CLOCKWISE ROTATION

REVERSING DIRECTION

Figure 15-5. The rotation of the split-phase motor is reversed by changing the flow of current through the running winding as shown above.

CAPACITOR START MOTORS

A capacitor start motor creates a greater starting torque when a capacitor is connected in series with the starting winding and the centrifugal switch. The capacitor causes the current in the starting winding to lead, by almost 90°, the current in the running winding. This condition produces a revolving magnetic field in the stator, which in turn induces a current in the rotor winding. As a result, the magnetic field acts in such a manner as to produce rotation of the motor.

ELECTROLYTIC CAPACITORS

Many capacitor motors employ the electrolytic capacitor. This type consists of two sheets of aluminum foil that are separated by one or more layers of gauze. The gauze has previously been saturated with a chemical solution called an electrolyte. The electrolyte forms a film, which acts as the insulating medium of the electrolytic capacitor. These layers are rolled together and fitted into an aluminum container. Electrolytic capacitors should not be kept in a circuit more than a few seconds at a time because they are designed for only intermittent operation. **(See Figure 15-7)**

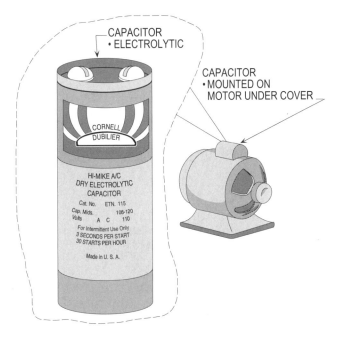

ELECTROLYTIC CAPACITOR

Figure 15-7. The above is an illustration of an electrolytic capacitor.

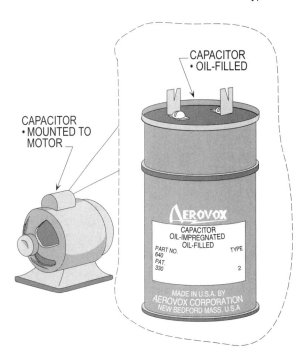

OIL-FILLED CAPACITOR

Figure 15-8. The above is an illustration of an oil-filled capacitor.

OIL-FILLED CAPACITORS

Some capacitors are made with paper that has been impregnated with oil and then inserted in a container that is filled with oil. This is done to increase the insulating quality of the paper and to help keep the capacitor from overheating. **(See Figure 15-8)**

SELECTING CAPACITORS

The capacitor acts essentially as a storage unit; that is, it has the capacity to store electricity and release electricity when needed.

All capacitors have this quality, and all are electrically the same. They differ only in mechanical construction as detailed above.

USING CHARTS

To correct an existing power factor when the existing power factor for a motor is known and the power factor is low, you may find the size kVAR capacitor needed by using the manufacturer's capacitor chart in **Figure 15-9.**

REVERSING DIRECTION

Reversing the direction of a capacitor start motor can be achieved by reversing the flow of current through the running or starting winding. If the flow of current in the starting winding and running winding are the same, the rotor will rotate in a counterclockwise direction. If the current flow of the starting and running winding is in opposite directions, the rotor will rotate in a clockwise direction. **(See Figure 15-10)**

CAPACITOR START-AND-RUN MOTORS

In capacitor start-and-run motors, the capacitor is utilized during starting and remains in operation during running. The capacitor start-and-run motor is quiet and smooth running. It is similar to the capacitor-start motor, except that the starting winding and capacitor are connected in the circuit at all times. The capacitor is connected in series with the starting winding and connected in parallel with the running winding. **(See Figure 15-11)**

A high starting torque is provided with a capacitor start motor. When the motor reaches its running speed, the centrifugal switch opens the circuit and drops out the starting capacitor. The running capacitor is left in the running circuit to provide a higher running torque and improve the running power factor while the motor is in operation.

CAPACITOR CALCULATING CHART

%	80	81	82	83	84	85	86	87	88	89	90	91	92	93	94	95	96
50	0.982	1.003	1.034	1.060	1.086	1.112	1.139	1.165	1.192	1.220	1.248	1.276	1.306	1.337	1.369	1.403	1.442
51	.937	.962	.989	1.015	1.041	1.067	1.094	1.120	1.147	1.175	1.203	1.231	1.261	1.292	1.324	1.358	1.395
52	.893	.919	.945	.971	.997	1.023	1.050	1.076	1.103	1.131	1.159	1.187	1.217	1.248	1.280	1.314	1.351
53	.850	.876	.902	.928	.954	.980	1.007	1.033	1.060	1.088	1.116	1.144	1.174	1.205	1.237	1.271	1.308
54	.809	.835	.861	.837	.913	.939	.966	.992	1.019	1.047	1.075	1.103	1.133	1.164	1.196	1.230	1.267
55	.769	.795	.821	.847	.873	.899	.926	.952	.979	1.007	1.035	1.063	1.090	1.124	1.156	1.190	1.228
56	.730	.756	.782	.808	.834	.860	.887	.913	.940	.968	.996	1.024	1.051	1.085	1.117	1.151	1.189
57	.692	.718	.744	.770	.796	.822	.849	.875	.902	.930	.958	.986	1.013	1.047	1.079	1.113	1.151
58	.655	.681	.707	.733	.759	.785	.812	.838	.865	.893	.921	.949	.976	1.010	1.042	1.076	1.114
59	.618	.644	.670	.696	.722	.748	.755	.801	.828	.856	.884	.912	.939	.973	1.005	1.039	1.077
60	.584	.610	.636	.662	.688	.714	.741	.767	.794	.822	.850	.878	.905	.939	.971	1.005	1.143
61	.549	.575	.061	.627	.653	.679	.706	.732	.759	.787	.815	.843	.870	.904	.936	.970	1.008
62	.515	.541	.567	.593	.619	.645	.672	.698	.725	.753	.781	.809	.836	.870	.904	.936	.974
63	.483	.509	.535	.561	.587	.613	.640	.666	.693	.721	.749	.777	.804	.838	.870	.902	.942
64	.450	.476	.502	.528	.544	.580	.607	.633	.660	.688	.716	.744	.771	.805	.837	.871	.909
65	.419	.445	.471	.497	.523	.549	.576	.602	.629	.657	.685	.713	.740	.744	.806	.840	.878
66	.398	.414	.440	.466	.492	.518	.545	.571	.598	.626	.654	.682	.709	.743	.775	.809	.847
67	.358	.384	.410	.436	.462	.488	.515	.541	.568	.596	.624	.652	.679	.713	.745	.779	.817
68	.329	.355	.381	.417	.433	.459	.486	.512	.539	.567	.595	.623	.650	.684	.716	.750	.788
69	.299	.325	.351	.377	.403	.429	.456	.482	.509	.537	.565	.593	.620	.654	.686	.720	.758
70	.270	.296	.322	.348	.374	.400	.427	.453	.480	.508	.536	.564	.591	.625	.657	.691	.729
71	.242	.268	.294	.320	.346	.372	.399	.425	.452	.480	.508	.536	.563	.597	.629	.633	.701
72	.213	.239	.265	.291	.317	.343	.370	.396	.423	.451	.479	.507	.534	.568	.600	.634	.672
73	.186	.212	.238	.264	.290	.316	.343	.369	.396	.424	.452	.480	.507	.541	.573	.607	.645
74	.159	.185	.211	.237	.263	.289	.316	.342	.369	.397	.425	.453	.480	.514	.546	.580	.618
75	.132	.158	.184	.210	.236	.262	.289	.315	.342	.370	.398	.426	.453	.487	.519	.553	.591
76	.105	.131	.157	.183	.209	.235	.262	.288	.315	.343	.371	.399	.426	.460	.492	.526	.564
77	.079	.105	.131	.157	.183	.209	.236	.262	.289	.317	.345	.373	.400	.434	.466	.500	.538
78	.053	.079	.105	.131	.157	.183	.210	.236	.263	.291	.319	.347	.374	.408	.440	.474	.512
79	.026	.052	.078	.104	.130	.156	.183	.209	.236	.264	.292	.320	.347	.381	.413	.447	.485
80	.000	.026	.052	.078	.104	.130	.157	.183	.210	.238	.266	.294	.321	.355	.387	.421	.459

For Example: A power factor of 65% is present in an existing circuit to a motor. How many kvar will it take to correct the power factor to 96%? (The motor is 100 HP, 208 V, three-phase)

Step 1: Finding existing VA
VA = V x √3 x A
VA = 208 V x 1.732 x 273 A (360 V x 273 A)
VA = 98,280

Step 2: Applying existing PF; Finding Watts
W = VA x PF
W = 98,280 x 65%
W = 63,882

Step 3: Applying multiplier in chart to find kVAR
kvar = 63,882 W ÷ 1000 x .878
kvar = 56.09

Solution: it will take 56.09 kvar to correct the power factor.

USING CHART TO DETERMINE kvar

Smaller capacitor rated in microfarads (μf) is determined by applying the following formula.

$$\mu f = \frac{159,300}{hertz} \times \frac{amps}{volts}$$

For example, the microfarads of a 15-horse-power, 208-volt, three-phase motor pulling 46.2 amps can be found by the following procedures.

Step 1: $\mu f = \frac{159,300}{hertz} \times \frac{amps}{volts}$

$\mu f = \frac{159,300}{60} \times \frac{46.2 \ A}{208 \ V \times 1.732}$

$\mu f = 340$

Solution: 340 μf

Note: See page 20-13 for a similar calculation.

USING FORMULA TO DETERMINE MICROFARADS

Figure 15-9. The kVAR's are selected from the capacitor calculating chart based on the existing power factor of the motor and the existing VA is multiplied by this value to correct power factor problems. The procedure for determining microfarads for a smaller capacitor is also shown.

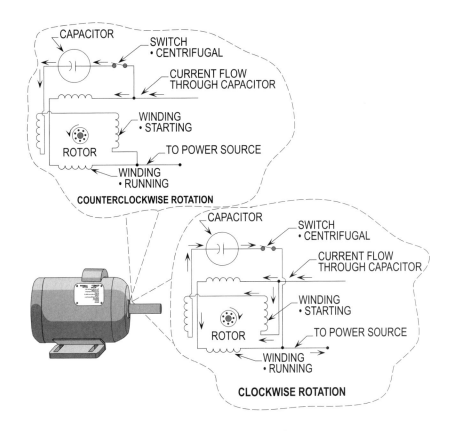

REVERSING DIRECTIONS

Figure 15-10. The rotation of a rotor is determined by the direction of current flow through the capacitor and windings in the motor.

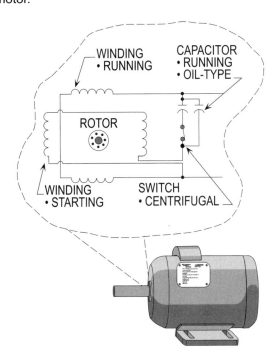

CAPACITOR START-AND-RUN MOTOR

Figure 15-11. Starting torque can be increased by providing a properly sized capacitor.

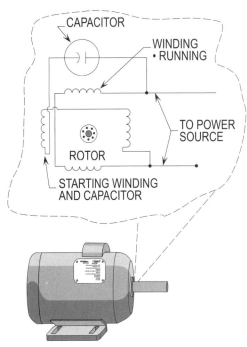

PERMANENT SPLIT CAPACITOR MOTOR

Figure 15-12. A centrifugal switch is not required for permanent split capacitor motors. Note that the capacitor is never moved from the circuit.

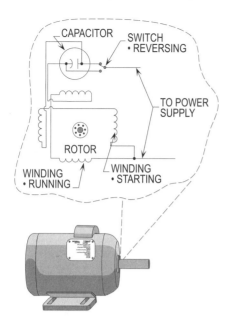

REVERSING DIRECTION

Figure 15-13. A reversing switch is used to reverse a permanent split capacitor motor.

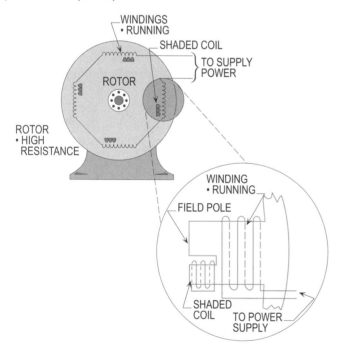

SHADED POLE MOTOR

Figure 15-14. The shaded coil of a shaded pole motor forms a closed (loop) circuit with the running windings.

REVERSING DIRECTION

To reverse the rotation of a capacitor start-and-run motor, the terminal cover must be removed and the leads of the starting or running winding must be reversed. Reversing switches may also be used. **(See Figure 15-10)**

PERMANENT SPLIT CAPACITOR MOTORS

Permanent split capacitor motors are similar in all respects to the capacitor-start motor except that it does not contain a centrifugal switch.

By checking the resistance of the starting and running winding, the windings of a permanent split capacitor motor can be identified. **(See Figure 15-2)** These motors are commonly called *single-value motors*. The low value of the capacitor results in a motor of medium starting torque. Consequently, this motor can only be used for oil burners, voltage regulators, fans, etc. **(See Figure 15-12)**

REVERSING DIRECTION

Permanent split capacitor motors may be reversed by reversing the terminal leads or by using a reversing switch. However, most permanent split capacitor motors are used in equipment, such as fans and blower motors, that requires a reversing switch. **(See Figure 15-13)**

SHADED POLE MOTORS

The trailing edge of each pole for a shaded-pole motor is wound with a shaded coil. Torque is provided to start and run the load when the shaded coil produces a slip. Equipment that requires a high starting torque cannot be operated with shaded pole motors. Shaded pole motors provide a very low starting torque. Each field pole is cut with a slot containing the shaded coil. The coil forms a closed circuit (loop) with the running windings wound around each field pole. **(See Figure 15-14)**

A magnetic field is set up between the poles and the rotor when power is applied to the running windings. An out-of-phase condition is created with the flux lines when the shaded coil cuts through a portion of the magnetic field. A two-phase magnetic field is created and the phase shifting provides the torque needed to start and rotate the driven equipment. **(See Figure 15-15)**

Additional starting torque is provided for shaded pole motors by using a high-resistance rotor. A higher slip and poor speed regulation is created by using a high-resistance rotor. Shaded pole motors are available for 115 volt, 230 volt, and dual-voltage operation.

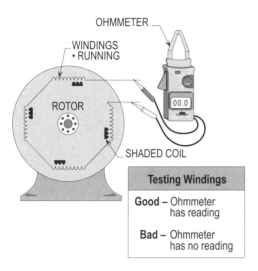

**TESTING WINDINGS OF A
SHADED POLE MOTOR**

Figure 15-15. The resistance of the winding in a shaded pole motor is measured by an ohmmeter.

REVERSING DIRECTION

By using one of the following methods, shaded pole motors can be reversed:

(1) By placing the rotor in the motor housing in the opposite direction.

(2) By two sets of field windings used with each shaded coil.

(3) By a switch that can open or close the circuit to the correct windings for the direction of rotation. The rotor will always rotate toward the shaded coil. **(See Figure 15-16)**

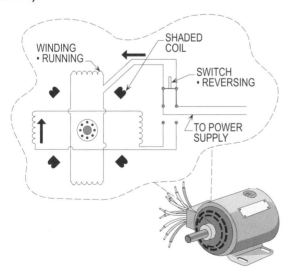

REVERSING DIRECTION

Figure 15-16. Two sets of field (running) windings are used to reverse shaded pole motors.

REGULATING SPEEDS

A shaded pole motor can have different speeds which are provided by tapping a coil. For example, a coil can be used to provide three speeds: low (L), medium (M), and high (H). Low speed is produced by tapping all the windings. Medium speed is produced by tapping half of the windings. High speed is produced by tapping none of the windings. **(See Figure 15-17)**

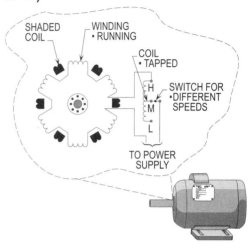

REGULATING SPEEDS

Figure 15-17. Different speeds for shaded pole motors are obtained from a tapped coil.

A multiple (six) winding provides two motor speeds with the windings located in separate slots. The same size wire is used for five of the windings, while smaller wire with more turns is used for the sixth winding. When all six windings are used in the motor, the motor has a low-speed operation. When only five of the windings are used, the motor has a high-speed operation. **(See Figure 15-18)**

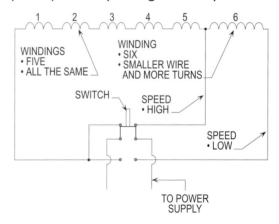

REGULATING SPEEDS

Figure 15-18. When all six windings are used in the motor, the motor has a low-speed operation. When only five of the windings are used in the motor, the motor has a high-speed operation.

UNIVERSAL MOTORS

A universal motor is one that can be operated on either direct current or single-phase alternating current at approximately the same speed. All windings are series connected and the motor has high starting torque and a variable speed characteristic. They are usually designed and built in sizes varying from 1/150 to 3/4 HP but are obtainable in much larger sizes for special applications.

Universal motors are equipped with field windings and an armature with brushes and a commutator. The commutator keeps the armature turning through the magnetic field of the field windings. It also changes the flow of current in relation to the field windings and armature so there is a push-and-pull action. This push-and-pull action is created by the north and south poles of the field windings and armature. **(See Figures 15-19 and 15-20)**

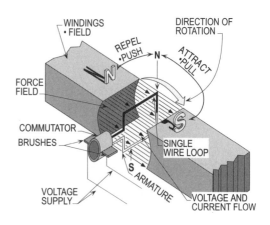

UNIVERSAL MOTOR

Figure 15-20. The above illustrates a simple universal motor with all of its parts.

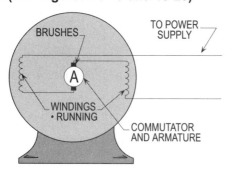

UNIVERSAL MOTOR

Figure 15-19. Universal motors are equipped with field windings and an armature with brushes and a commutator.

The north pole of the field windings pulls the south pole of the armature (loop) into the main strength of the magnetic field (field force). The commutator and brushes reverse the current flow through the armature, creating a north pole in the loop. The north pole of the field winding then repels the north pole of the armature. This push-and-pull action rotates the armature through the magnetic field of the field windings, establishing motor operation.

When the universal motor operates on AC voltage, the current is constantly changing direction in the field windings. Both the armature and field windings have their current reversed simultaneously. Therefore, the motor operates similar to an inductive motor. The field windings of a universal motor are connected in series with the brushes and armature. **(See Figure 15-20)**

REVERSING DIRECTION

Changing the flow of current through the armature by interchanging the leads on the terminals or the use of a reversing switch will reverse the rotation of the motor.

REGULATING SPEED

Resistance determines the speed of a motor. The higher the resistance, the lower the speed. By using a variable resistor, the speed of a universal motor can be controlled.

To obtain three speeds, one of the field windings must be tapped. For slow speed, tap all of the winding, for medium speed, half of the winding, and for fast speed the entire winding is bypassed. **(See Figure 15-21)**

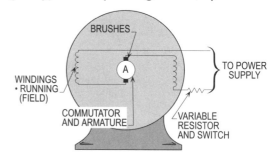

REGULATING SPEED

Figure 15-21. The above illustrates that high resistance in a circuit that is used for slower motor speeds and low resistance in a circuit is used for higher motor speeds. Resistance is produced by using a variable resistor.

REPULSION MOTORS

Repulsion motors are divided into three distinct classifications. The following are the types of repulsion motors:

(1) Standard repulsion,
(2) Repulsion-start induction, and
(3) Repulsion-induction.

These classifications are often confused because of the similarity of names. But each is different, having its own characteristics and applications. However, one feature common to all is that each has a rotor containing a winding that is connected to a commutator. These motors generally operate from a single-phase lighting or power circuit, depending on the size of the motor.

STANDARD REPULSION MOTORS

The standard repulsion motor is a single-phase motor, often called an inductive series motor. It starts and runs on the induction principle and is a varying-speed type of motor. This motor is a brush-riding type and does not have any centrifugal mechanism. It starts and runs on the repulsion principle. **(See Figure 15-22)**

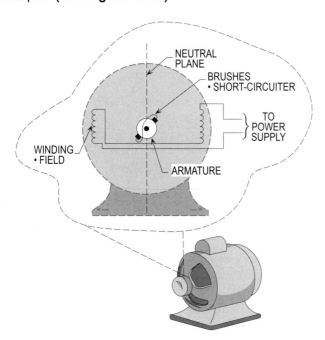

STANDARD REPULSION MOTOR

Figure 15-22. A standard repulsion motor has field windings and a wound rotor with brushes and a commutator.

REPULSION-START INDUCTION MOTORS

The repulsion-start induction motor and the standard repulsion motor's starting procedures are the same, but at a predetermined speed, a special device is actuated which short-circuits all the commutator windings. From this point on, the motor operates as a single-phase induction motor. This device is typically called a *short-circuiter*.

Since an induction motor functions on the magnetic-induction principle, where fields of both the stator and armature rotate in the same direction, the higher the speed at which the *short-circuiter* operates, the less line current is drawn by the motor. (**See Figure 15-23**)

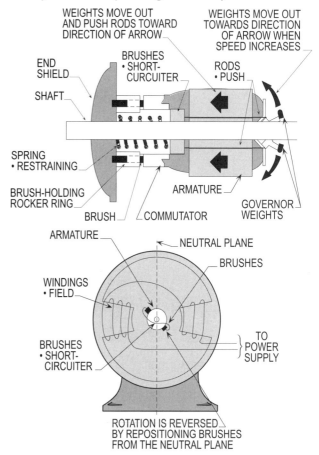

REPULSION-START INDUCTION MOTOR

Figure 15-23. A repulsion-start induction motor develops a centrifugal force which operates the short-circuiter and brush-lifting mechanisms to cause a normal induction motor operation.

REPULSION INDUCTION MOTORS

As in the standard and repulsion-start induction motor, the repulsion-inductor motor has the same starting principle, but no mechanism is included in its construction. It combines instead, a repulsion and squirrel-cage winding in its armature. Both windings are always in operation while the armature rotates. **(See Figure 15-24)**

All starting torque is provided by the repulsion winding. However, once the armature begins its rotation, the voltage induced in the squirrel-cage winding produces some torque within its winding.

The characteristics of the repulsion-induction motor are very similar to that of the standard type. Direction of rotation and setting of the neutral are the same, In fact, it is sometimes difficult to differentiate between the two. An easy way to determine which is, remove the load, start the motor and then remove the brushes. If the motor continues to run, it is a repulsion-induction type.

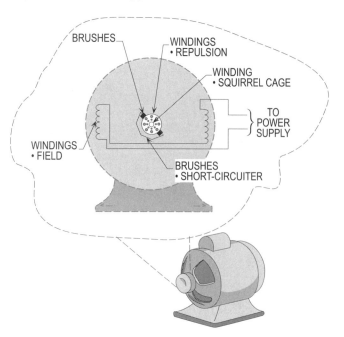

REPULSION INDUCTION MOTOR

Figure 15-24. A repulsion induction motor has a squirrel-cage rotor with a wound armature and a commutator with short-circuiter brushes.

REVERSING ROTATION

A repulsion motor is reversed by shifting the brush holder to either side of the neutral position. Its speed can be decreased by moving the brush holder further away from the neutral position. **(See Figure 15-25)**

THREE-PHASE MOTORS

Three-phase motors vary from fractional-horsepower sizes to several thousand horsepower ratings. These motors have a fairly constant speed characteristic and are made in designs giving a variety of torque characteristics. They are made for practically every standard voltage and frequency and are almost always dual-voltage motors.

The operating principles of a two-phase motor apply to the three-phase motor. For the three-phase motor, however, the generated magnetic fields are 120 degrees out-of-phase with each other. An additional starting winding is not required for three-phase motors to start and run. An induction motor will always have a peak phase of current. This is due to alternating current reversing its direction of flow. In other words, when the alternating current of one phase reverses

its direction of flow, a peak current will be developed on one phase and as current reverses direction again, a second phase will peak, etc. Three-phase motors provide a smooth and continuous source of power once they are started and driving their load. Three-phase motors are used to drive machine tools, pumps, elevators, fans, hoists, and many other machines. **(See Figure 15-26)**

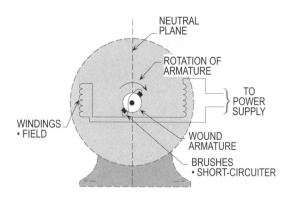

ARMATURE WILL ALWAYS ROTATE TOWARDS BRUSHES

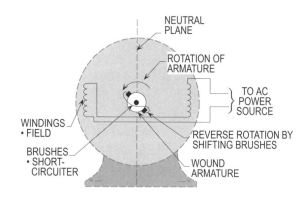

REVERSING ROTATION

Figure 15-25. The armature of a repulsion motor always rotates toward the position of the short-circuiter brushes from the neutral plane.

SQUIRREL-CAGE INDUCTION MOTORS

A squirrel-cage motor is an induction motor and so called because of its construction. The rotating (stator) magnetic field induces voltages in the rotor which in turn causes the rotor to turn. The rotor consists of an iron core mounted on a concentric shaft. Copper or brass bars run the entire length of this core and are set into slots on the core. At each end of the core, end rings are welded to the copper or brass bars so that a complete short-circuit exists within the rotor. The entire assembly resembles the type of cage within which squirrels, etc. are placed to run through various tests. In effect, the rotor acts as the secondary winding of transformer while the stator acts as the primary winding. **(See Figure 15-27)**

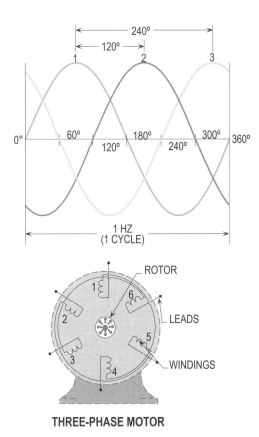

THREE-PHASE MOTOR

Figure 15-26. Voltage from 2 to 1 is 120° behind 1 and voltage from 3 to 1 is 240° behind 1.

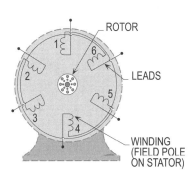

SQUIRREL-CAGE INDUCTION MOTOR

Figure 15-27. The above illustrates a stator (field poles) and a rotor equipped in a squirrel-cage motor.

CONNECTING LEADS

All three-phase motors are wound with a number of coils, which are placed in slots in the stator. These coils are so connected as to produce three separate windings called phases, and each of which must have the same number of coils. The number of coils in each must be one-third the total number of coils in the stator. Therefore, if a three-phase motor has 36 coils, each phase will have 12 coils. These phases are listed as phase 1, phase 2, and phase 3.

All three-phase motors have their phases arranged in either a wye, sometimes called a star connection (⋏) or a delta, sometimes called a triangle connection (Δ). Either of these connections are so connected that only three leads come from the stator, making the line connections very simple.

SIX LEAD MOTORS

The windings of a motor can be designed with six leads to connect the windings to the three-phase supply lines. A six lead motor used for a delta connection has the winding leads connected so that 1 and 2 close one end of the delta (triangle), 5 and 6 close one end of the delta, and 3 and 4 close one end of the delta to form a closed-delta connection of the motor windings. **(See Figure 15-28)**

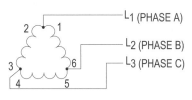

DELTA CONNECTION

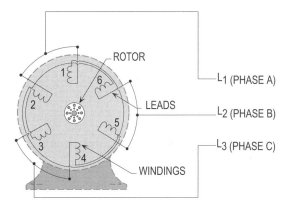

DELTA-CONNECTED SIX LEAD MOTOR

Figure 15-28. The above illustrates a six lead squirrel-cage induction motor with internal windings connected for delta operation.

The six leads can be wye-connected with one lead for each winding being connected to form the wye or star connection. The three remaining leads are connected to the three-phase supply lines L_1, L_2, and L_3. **(See Figure 15-29)**

Note that delta connected windings in a motor will usually always run cooler than windings connected in a wye configuration.

NINE LEAD MOTORS

The windings of a motor can be designed with nine leads to connect the windings to the three-phase supply. The nine leads are connected to the internal windings for delta operation. A closed delta is formed by connecting six internal windings together. The three windings are marked 1-4-9, 2-5-7, and 3-6-8. A nine lead motor is used to operate as a closed delta system with the windings connected for single or dual voltage operation. **(See Figure 15-30)**

The nine leads can be wye-connected with three leads of its windings connected to form a wye with three remaining leads (7-8-9). The three remaining windings are numbered 1-4, 2-5, and 3-6. The windings are connected to operate on low or high-voltage. Windings are connected in parallel for low-voltage and windings are connected in series for high-voltage. This type of connection applies for either wye or delta connected windings. **(See Figure 15-31)**

REVERSING DIRECTION

By interchanging any two of the three-phase leads the rotation can be reversed for any three-phase squirrel-cage induction motor. Using windings 1, 2, and 3 as a reference, the 3 winding will follow the 2 winding rather than the 1 winding. The rotating field will rotate in the opposite direction when reversing the polarity through the windings, carrying the rotor with it. **(See Figure 15-32)**

REGULATING SPEED

The speed of a squirrel-cage motor depends upon the following four conditions:

(1) Load,
(2) Applied voltage,
(3) Frequency, and
(4) Number of poles within the stator.

In the squirrel-cage motor, we are more interested in the amount the rotor speed lags the speed of the rotating field. This difference in speed, called slip, is entirely dependent on the load. The greater the load, the greater the amount of slip, and the slower the speed of the rotor. However, this slip is such a small fraction of the synchronous speed, that the squirrel-cage motor is used widely as a constant-speed type.

Because of their constant-speed characteristics, they are more often used in such items as larger types of fans, conveyor-belt applications, presses, etc.

SYNCHRONOUS MOTORS

Synchronous motors are available in a wide range of sizes and types which are designed to run at synchronous speeds. The following are two types of synchronous motors that are available:

(1) Nonexcited
(2) Direct-current excited

A DC source of excitation is required. The torque required to turn the rotor for a synchronous motor is produced when the DC current of the rotor field locks in with the magnetic field of the stator AC current. **(See Figure 15-33)**

Synchronous motors are made in sizes varying from approximately 20 HP to hundreds of horsepower and are used wherever it is necessary or desirable to obtain constant speed. In many cases, synchronous motors are used to improve the power factor of the electrical system of a plant or factory. Many small synchronous motors are also made for clocks, but are constructed differently from the large ones.

STARTING METHOD

The synchronous motor does not start by itself, some kind of starting action must be supplied to bring the rotor up to synchronous speed.

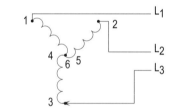

WYE OR STAR CONNECTION

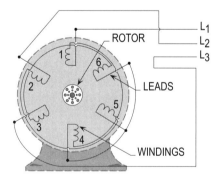

WYE-CONNECTED SIX LEAD MOTOR

Figure 15-29. The above illustrates a six lead squirrel-cage induction motor with internal windings connected for wye operation.

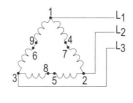

DELTA CONNECTION

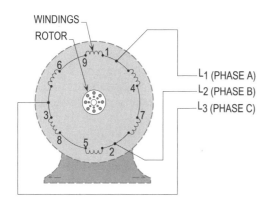

DELTA-CONNECTED NINE LEAD MOTOR

Figure 15-30. The above illustrates a nine lead squirrel-cage induction motor with internal windings connected for delta operation.

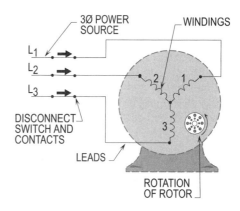

COUNTERCLOCKWISE ROTATION

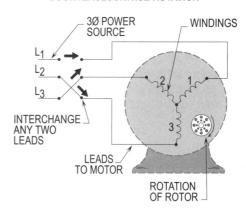

CLOCKWISE ROTATION

REVERSING DIRECTION

Figure 15-32. By interchanging any two leads of a three-phase squirrel-cage induction motor the rotation is reversed.

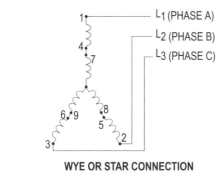

WYE OR STAR CONNECTION

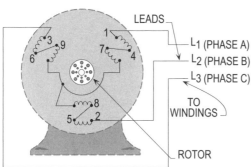

WYE-CONNECTED NINE LEAD MOTORS

Figure 15-31. The above illustrates a nine lead squirrel-cage induction motor with internal windings connected for wye operation.

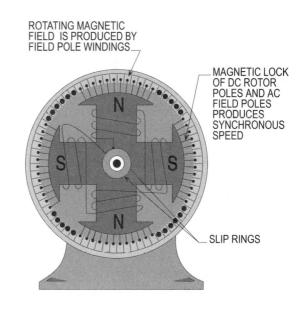

SYNCHRONOUS MOTOR

Figure 15-33. The torque required to turn the rotor of a synchronous motor is produced when the DC current of the rotor field locks in with the magnetic field of the stator AC current.

WITH MOTORS

One type of starting uses another motor, either DC or induction, with a high starting torque. This auxiliary motor brings the synchronous motor up to almost full speed and is automatically disconnected and the synchronous motor comes up to full speed under its own power when DC excitation is applied.

WITH DAMPER WINDINGS

Another method of starting may be provided by a adding common winding to the rotor DC winding. The added winding is an induction type or squirrel-cage construction. The voltage induced in this winding by the rotating stator field produces poles of opposite polarity in the rotor. Opposite poles attracting the rotating magnetic field in the stator, provides the necessary starting torque. At some point slightly below synchronous speed, the rotor DC voltage is fed in to the rotor and the motor reaches full operating speed.

WOUND-ROTOR MOTORS

Wound-rotor motors are classified as three-phase induction motors and have two sets of leads. One set is the main leads to the motor windings (stator or field poles) and the other set is the secondary leads to the rotor. The secondary leads are connected to the rotor through the slip rings, while the other end of the leads are connected through a controller and a bank of resistors.

Wound-rotor motors operate on the same principle as the squirrel-cage induction motor. The major difference is in construction. Where the squirrel-cage motor has copper or brass bars that are permanently short-circuited, the wound rotor has insulated windings in their place. These windings are not permanently short-circuited.

PRINCIPLES OF STARTING

The starting principle of the wound-rotor motor is identical to that of the squirrel-cage motor, but the wound-rotor is not permanently short-circuited; instead, the currents produced in the rotor are fed into slip rings mounted on the end of the rotor. These currents, in turn, are coupled through slip ring brushes (carbon or graphite) into an external control device which is either a variable resistor or a variac, usually the latter.

SPEED REGULATION

Because of the external control, the starting torque, the

current, the operating speed, and the acceleration of the motor up to full-load speed can be varied. However, even though it is possible to control the wound-rotor motor's operation, a big disadvantage results. Any loss in motor speed results in a loss of efficiency. Therefore, the wound-rotor motor is used mostly on heavy equipment which requires a high starting torque and smooth acceleration up to full-rated load, or where variable speed is essential.

REVERSING DIRECTION

Reversing direction of three-phase motors may be done very simply. The entire procedure consists of disconnecting any two of the three stator leads from the line, interchanging them, and then reconnecting them to the line.

Typical connection illustrations for three-phase motors are illustrated. Note how the stator leads are changed to achieve a reversal in rotation.

See Figure 15-34 for a detailed illustration of the parts making up a wound-rotor motor.

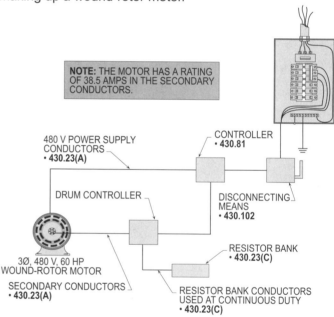

WOUND-ROTOR MOTOR

Figure 15-34. The above illustrates the parts of a wound-rotor motor.

NEMA TYPE ENCLOSURES

The Underwriter's Laboratories (UL) has defined the requirements for protective enclosures according to the hazardous conditions and the National Electrical Manufacturers Association (NEMA) has standardized enclosures from these requirements.

The correct selection and installation of an enclosure for a particular application can contribute considerably to the length of life of a motor.

When selecting and installing a motor enclosure, it is always necessary to consider carefully the conditions under which the motor must operate. Note that there are many applications where a general-purpose enclosure does not afford protection.(See **Table 430.91** of the NEC.)

NEMA TYPE 1 ENCLOSURES

This enclosure is suitable for general-purpose applications indoors where it is not exposed to unusual service conditions. A NEMA 1 enclosure serves as protection against limited falling dirt, dust and light indirect splashing water.

NEMA TYPE 2 ENCLOSURES

This enclosure is intended to provide suitable protection against specified weather hazards. A NEMA 2 enclosure is intended for indoor use, primarily to provide a degree of protection against limited amounts of falling water and dirt.

NEMA TYPE 3 ENCLOSURES

This enclosure protects against interference in operation of the contained equipment due to rain and resists damage from exposure to weather. A NEMA 2 enclosure is intended for outdoor use, primarily to provide a degree of protection against rain, sleet, windblown dust; and damage from external ice formation.

NEMA TYPE 3R ENCLOSURES

This enclosure is intended for outdoor use, primarily to provide a degree of protection against rain, sleet; and damage from external ice formation, and must have a drain hole.

NEMA TYPE 3S ENCLOSURES

This enclosure is intended for outdoor use primarily to provide a degree of protection against rain, sleet, windblown dust; and to provide for operation of external mechanisms when ice laden.

NEMA TYPE 4 ENCLOSURES

This enclosure is intended for indoor or outdoor use, primarily to provide a degree of protection against windblown dust and rain; splashing water, hose directed water; and damage from external ice formation.

NEMA TYPE 4X ENCLOSURES

This enclosure is intended for indoor or outdoor use, primarily to provide a degree of protection against corrosion, windblown dust and rain, splashing water, hose directed water; and damage from external ice formation.

NEMA TYPE 5 ENCLOSURES

This enclosure is intended for indoor use, primarily to provide a degree of protection against settling airborne dust, falling dirt, and dripping noncorrosive liquids.

NEMA TYPE 6 ENCLOSURES

This enclosure is intended for indoor or outdoor use, primarily to provide a degree of protection against hose directed water, the entry of water during occasional temporary submersion at a limited depth; and damage from external ice formation.

NEMA TYPE 6P ENCLOSURES

This enclosure is intended for indoor or outdoor use, primarily to provide a degree of protection against hose-directed water, the entry of water during prolonged submersion at a limited depth; and damage from external ice formation.

NEMA TYPE 7 ENCLOSURES

This enclosure is intended for indoor use in locations that are classified as Class I, Groups A, B, c, or D, as defined in the National Electrical Code.

NEMA TYPE 8 ENCLOSURES

This enclosure is for indoor or outdoor use in locations that are classified as Class I, Groups A, B, C, or D, as defined in the National Electrical Code.

NEMA TYPE 9 ENCLOSURES

This enclosure is intended for indoor use in locations that are classified as Class II, Groups E, F, and G, as defined in the National Electrical Code.

NEMA TYPE 10 ENCLOSURES

This enclosure is constructed to meet the applicable requirements of the Mine Safety and Health Administration.

NEMA TYPE 12 ENCLOSURES

This enclosure is intended for indoor use, primarily to provide a degree of protection against circulating dust, falling dirt, and dripping noncorrosive liquids.

NEMA TYPE 12K ENCLOSURES

This enclosure, with knockouts, are intended for indoor use, primarily to provide a degree of protection against circulating dust, falling dirt, and dripping noncorrosive liquids.

NEMA TYPE 13 ENCLOSURES

This enclosure is intended for indoor use, primarily to provide a degree of protection against dust, spraying of water, oil, and noncorrosive coolant.

For more information about enclosures, see the NEMA Standards Publication No. 250, Enclosures for Electrical Equipment (1000 Volts Maximum) or other third party certification standards for specific requirements for product construction, testing and performance such as Underwriters Laboratories Inc.©, Standard UL 50 "Standard for Enclosures for Electrical Equipment", and UL 886 "Outlet Boxes and Fittings for use in Hazardous (Classified) Locations".

Chapter 15: Types of Motors

Section	Answer		

_____ T F **1.** Split-phase motors have a low-resistance in the starting windings which produce a high inrush current when starting.

_____ T F **2.** Split-phase motors are manufactured with a squirrel cage winding.

_____ T F **3.** The current running through the starting or running winding is changed when the direction of rotation is reversed in split-phase motors.

_____ T F **4.** Overcurrent protection devices (bi-metal discs or composed of dissimilar metals) are connected in series with the running winding.

_____ T F **5.** A capacitor start motor creates a greater starting torque and is connected in series with the starting winding and centrifugal switch.

_____ T F **6.** The capacitor of a capacitor start-and-run motor is connected in series with the starting winding and in series with the running winding.

_____ T F **7.** Reversing switches may not be used on capacitor start-and-run motors.

_____ T F **8.** Equipment that requires a high starting torque cannot be used with shaded pole motors.

_____ T F **9.** Two sets of field windings can be used to reverse shaded pole motors.

_____ T F **10.** A standard repulsion motor has field windings and a wound rotor with brushes and a commutator.

_____ _____ **11.** The windings of a squirrel-cage induction motor can be designed with six leads to connect the windings to the three-phase supply lines.

_____ _____ **12.** By interchanging any two leads of a three-phase, squirrel-cage induction motor the rotation of the motor is reversed.

_____ _____ **13.** Synchronous motors are able to start without the aid of a starting action being supplied to bring the rotor up to synchronous speed.

_____ _____ **14.** Wound-rotor motors are classified as three-phase induction motors and have three sets of leads.

_____ _____ **15.** Reversing direction of wound-rotor motors is accomplished by disconnecting any two of the three stator leads from the line, interchanging them, and then reconnecting them to the line.

_____ _____ **16.** The running winding and starting winding of a split-phase motor is measured for resistance using an _____.

_____ _____   **17.** After a split-phase motor is running, the starting winding is no longer needed and is disconnected from the circuit by means of the _____ switch.

_____ _____   **18.** The rotation of a capacitor start motor is usually determined by the direction of current flow through the _____ and running winding.

_____ _____   **19.** A _____ switch is not required for a permanent split capacitor motors.

_____ _____   **20.** A _____ switch is used to reverse a permanent split capacitor motor.

_____ _____   **21.** The shaded coil of a shaded pole motor forms a _____ circuit with the running winding.

_____ _____   **22.** Different speeds for a shaded pole motor is obtained from a _____ transformer or coil.

_____ _____   **23.** A universal motor is equipped with a field winding and an _____ with brushes and a commutator.

_____ _____   **24.** When a _____ resistor is used to add higher resistance in a running winding, slower universal motor speeds are produced, while lower resistance added in a running winding produces faster speeds.

_____ _____   **25.** The running torque required to turn the rotor for a synchronous motor is produced when the _____ current of the rotor field locks in with the rotating magnetic field of the stator AC current.

Design Letters and Code Letters

Motor circuits must be designed to provide protection for motor windings and components when motors are starting, running, and driving loads. Motor windings are protected by overcurrent protection devices which are selected according to the type that are used, based upon the amount of starting current required. Overcurrent protection devices are sized by percentages based on the type motor, starting method, design, or code letter. Starting methods are to be selected based on the amount of current required to start and run the motor or the amount that is be reduced by utilizing a starting method.

This chapter adresses these motors and their many different characteristics and why it's sometimes desirable to choose one over the other, based on the requirements of the driven load or equipment.

TYPES OF MOTORS
TABLE 430.52

The following are five types of motors that must be considered when sizing OCPD's to allow motors to start and run:

(1) Single-phase AC squirrel-cage,
(2) Three-phase AC squirrel-cage,
(3) Wound-rotor,
(4) Synchronous, and
(5) DC

SINGLE-PHASE SQUIRREL-CAGE MOTORS

Squirrel-cage motors are known in the electrical industry as induction motors. An induction motor operates on the same principles as the primary and secondary windings of a transformer. When power energizes the field windings they serve as the primary by inducing voltage into the rotor which serves as the secondary windings. Squirrel-cage motors have two windings on the stator, one winding is the run winding and the other is the starting winding. This additional starting winding on the stator is required for split-phase, single-phase, induction motors to provide the capacity to start and run. The starting winding has a higher resistance than the running winding which creates a phase displacement between the two windings. It is this phase displacement between the two windings that gives split-phase motors the power to start and run.

The phase displacement is about 18 to 30 degrees in an angular phase displacement, which provides enough starting torque, called twist or force, to start the motor. The motor operates on the running winding after the rotor starts running and has established a running speed at about 75 to 80 percent of the motor's synchronous speed. The starting winding is disconnected by a centrifugal switch that is installed in the circuit of the starting winding. **(See Figure 16-1)**

motors to start and run. An induction motor will always have a peak phase of current. This is due to alternating current reversing its direction of flow. In other words, when alternating current of one phase reverses its direction of flow, a peak current will be developed on one phase and as current reverses direction again, a second phase will peak, etc. Three-phase motors provide a smooth and continuous source of power once they are started and driving the load. **(See Figure 16-2)**

WOUND-ROTOR MOTORS

Wound-rotor motors are classified as three-phase induction motors and are similar in design to squirrel-cage induction motors. They are three-phase motors having two sets of leads. One set is the main leads to the motor windings called field poles and the other set is the secondary leads to the rotor. The secondary leads are connected to the rotor through the slip rings, while the other end of the leads are connected through a controller and a bank or resistors. The speed of the motor varies with the amount of resistance added in the motor circuit. The rotor will turn slower when the resistance is greater in the rotor, and vice versa. The resistance may be incorporated in the controller or the resistor banks may be separate from the motor. **(See Figure 16-3)**

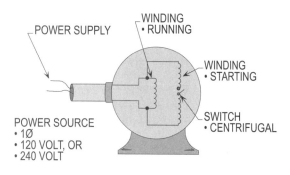

NEC TABLE 430.52

Figure 16-1. The above illustrates an example of a single-phase squirrel-cage motor which is listed in **Table 430.52** of the NEC.

THREE-PHASE SQUIRREL-CAGE MOTORS

Three-phase squirrel-cage motors have three separate windings per pole on the stator that generate magnetic fields that are 120 degrees out-of-phase with each other. An additional starting winding is not required for three-phase

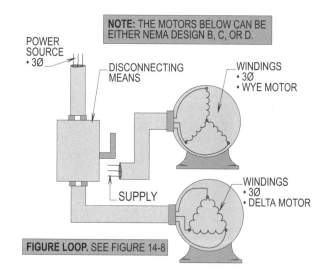

NEC TABLE 430.52

Figure 16-2. The above is an example of a three-phase squirrel-cage motor which is listed in **Table 430.52** of the NEC.

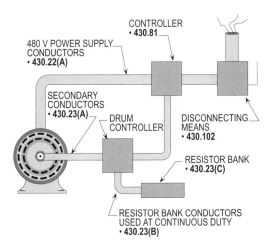

NEC TABLE 430.52

Figure 16-3. The above is an example of a three-phase wound-rotor motor which is listed in **Table 430.52** of the NEC.

SYNCHRONOUS MOTORS

The following are two types of synchronous motors that are available:

(1) Nonexcited, and
(2) Direct-current excited.

Synchronous motors are available in a wide range of sizes and types which are designed to run at designed speeds. A DC source is required to excite a direct-current excited synchronous motor. The torque required to turn the rotor for a synchronous motor is produced when the DC current of the rotor field locks in with the magnetic field of the stator's AC current. **(See Figure 16-4)**

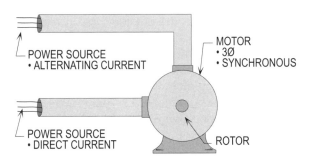

NEC TABLE 430.52

Figure 16-4. The above is an example of a three-phase synchronous motor which is listed in **Table 430.52** of the NEC.

DC MOTORS

Direct-current only is used to operate DC related motors. A DC motor is designed with the following two main parts:

(1) The stator, and
(2) The rotor or armature

The stationary frame of the motor is called the stator. The armature mounted on the drive shaft is known as the rotor. By applying direct-current to the rotor, the speed may be adjusted for a DC motor which drives the driven load at a specific speed. **(See Figure 16-5)**

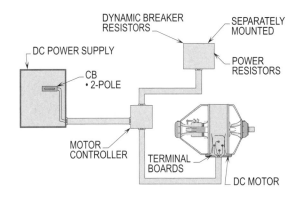

NEC 430.29
TABLE 430.29

Figure 16-5. The above is an example of a DC motor which is listed in **Table 430.52** of the NEC.

SERIES DC MOTORS

A very high starting torque of 300 percent to 375 percent of the full-load torque is provided when using series DC motors. Loads that are required to be driven with high torque and low speed regulate use this type of motor. Depending on the load requirements, the speed varies. Series DC motors are used in such installations such as traction work, where the speed varies depending on the load on the hoist. The armature and fields are connected in series. **(See Figure 16-6)**

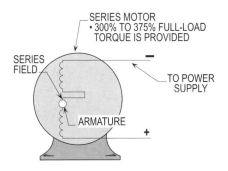

SERIES DC MOTOR

Figure 16-6. A series DC motor provides a very high starting torque of 300 percent to 375 percent of the full-load torque.

SHUNT DC MOTORS

A high torque of 125 to 200 percent of the full-load torque is provided when using shunt DC motors. Loads that are required to be driven with constant or adjustable speeds and loads that do not require high starting torque use this type of motor. Loads such as woodworking machines, printing presses, and papermaking machines use shunt DC motors. **(See Figure 16-7)**

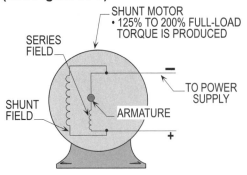

SHUNT DC MOTOR

Figure 16-7. A shunt DC motor provides a medium starting torque of 125 percent to 200 percent of the full-load torque.

COMPOUND DC MOTORS

A high torque of 180 to 260 percent of the full-load torque is provided when using compound DC motors. A fairly constant speed is obtained when using this type of motor. The compound DC motor is equipped with a series and shunt winding. A series winding is connected in series with the armature and the shunt winding is connected in parallel with the armature. This type of motor has the characteristics of both a series and shunt motor during operation. Loads such as crushers, reciprocating compressors, and punch presses use compound DC motors. **(See Figure 16-8)**

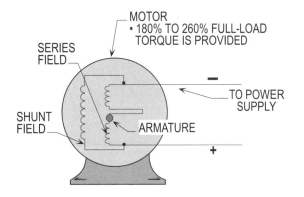

COMPOUND DC MOTOR

Figure 16-8. A compound DC motor provides a high torque of 180 percent to 260 percent of the full-load torque.

CALCULATING TORQUE

To accelerate and drive a piece of equipment the motor must be capable of producing a torque. Torque is called the turning or twisting force of the motor and is measured in foot-pounds or pound-feet.

FULL-LOAD TORQUE

The full-load torque of a motor is determined by dividing the horsepower times 5252, the rpm of the motor.

> **Motor Tip:** The value of 5252 is found by dividing 33,000 foot-pounds per minute by 6.2831853 (33,000 ÷ 6.2831853 = 5,252), which is found by multiplying π (3.14159265) by 2. **(See Figure 16-9)**

STARTING TORQUE

The starting torque of a motor varies with the classification of the motor. Motors are classified by NEMA as Design B, C or D motors. These types of standardized motors are the most used motors in the electrical industry. Other types of motors classified by NEMA are Design F or G motors.

A different rotor design is offered for each class of motors which will create a different value of starting torque. A different value of torque, speed, current, and slip to start and drive the various types of loads is produced when using Design B, C or D motors classified by NEMA. The design motor to be selected and used depends on the starting torque of the driven load and the running torque required to drive the load. **(See Figure 16-13)**

CLASS B MOTORS

The most used motor in the electrical industry is Class B design motors. For example, the starting torque of an induction motor will increase by 150 percent of the full-load torque when using Class B design motors. The starting torque of an induction motor is usually increased less than 150 percent by most designers when using Class B motors to start and run loads. **(See Figure 16-10)**

CLASS C MOTORS

The starting torque of a squirrel-cage induction motor will increase about 225 percent of the full-load torque when using Class C design motors. However, to keep from overloading the starting torque of a motor, designers will often load a motor to a value less than 225 percent.

For example: What is the full-load torque and starting torque of a 40 HP, Class C design induction motor operating at 1725 rpm?

Step 1: Finding full-load torque
Torque = HP x 5252 ÷ rpm
Torque = 40 x 5252 ÷ 1725
Torque = 210,080 ÷ 1725
Torque = 121.8 ft. lbs.

Step 2: Finding starting torque
Full-load torque increased by 225%
Torque = 121.8 ft. lbs. x 225%
Torque = 274.05 ft. lbs.

Solution: The full-load torque is 122 ft. lbs. and the starting torque is 274 ft. lbs.

For example: What is the full-load torque and starting torque of a 50 HP, Class D design induction motor operating at 1725 rpm?

Step 1: Finding full-load torque
Torque = HP x 5252 ÷ rpm
Torque = 50 x 5252 ÷ 1725
Torque = 262,600 ÷ 1725
Torque = 152.2 ft. lbs.

Step 2: Finding starting torque
Full-load current increased by 275%
Torque = 152.2 ft. lbs. x 275%
Torque = 418.6 ft. lbs.

Solution: The full-load torque is 152.2 ft. lbs. and the starting torque is 418.6 ft. lbs.

CLASS D MOTORS

The starting torque of a squirrel-cage induction motor is increased about 275 percent of the full-load torque when using Class D design motors. However, to keep from overloading the starting torque of a motor, designers will often load a motor to a value less than 275 percent.

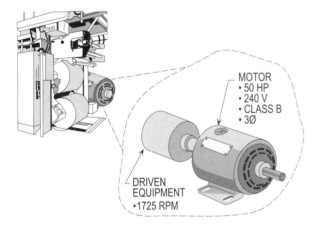

MOTOR
• 50 HP
• 240 V
• CLASS B
• 3Ø

DRIVEN EQUIPMENT
•1725 RPM

Finding Full-Load Torque
Step 1: Finding full-load torque Torque = HP x 5252 ÷ RPM Torque = 50 x 5252 ÷ 1725 Torque = 262,600 ÷ 1725 Torque = 152.2
Solution: The full-load torque is 152.2 ft. lbs.

CALCULATING FULL-LOAD TORQUE

Figure 16-9. To find the full-load torque of a motor, multiply the horsepower rating times 5252 and divide by rpm's.

CLASS E MOTORS PER 2002 NEC

When designing and installing a high-efficient motor, it is most important to know the starting and running torque of the load. The difference between the nominal and the minimum efficiency must also be determined. The motor must be sized to start and drive the load.

STARTING CURRENTS

Most high-efficiency motors do have higher starting currents and this presents a real problem where a standard motor is replaced with such. Nuisance tripping of the OCPD can occur during full-voltage start up.

There are some high-efficiency motors that have starting currents as high as 1500 percent of the full-load current. If 1700 percent per **Ex. 1 to 430.52(C)(3)** does not allow the motor to start and run, reduced voltage starting or use of modern electronic types of motor start/run technologies must be utilized. Note that starting currents of high-efficiency motors varies based on manufacturer and size. High efficiency motors must be selected with enough starting torque and break down torque to start and run the driven loads.

The nameplate on most motors will list the starting and running kVA of the motor. It is from these values and the manufacturer data that the OCPD and conductors are sized. The motor should be loaded based upon the minimum efficiency and not the motors nominal efficiency. Note that a high-efficiency motor is equipped on its nameplate with a nominal and minimum efficiency full-load rating.

TWO-SPEED MOTORS

The full-load torque of a motor is determined by the rpm of the motor. A motor turning at 1800 rpm produces less torque than motor turning at 1200 rpm.

For example: What is the full-load torque for a two-speed motor, 30 HP motor operating at either 1200 rpm or 1800 rpm?

Step 1: Finding full-load torque (1200)
Torque = HP x 5252 ÷ rpm
Torque = 30 x 5252 ÷ 1200
Torque = 157,560 ÷ 1200
Torque = 131.3 ft. lbs.

Step 2: Finding full-load torque (1800)
Torque = HP x 5252 ÷ rpm
Torque = 30 x 5252 ÷ 1800
Torque = 157,560 ÷ 1800
Torque = 87.5 ft. lbs.

Solution: The full-load torque for 1200 rpm is 131.3 ft. lbs. and the full-load torque for 1800 rpm is 87.5 ft. lbs.

RESISTOR OR REACTOR-REDUCED STARTING

To reduce the inrush starting current (LRC) of a motor, a resistor or reactor-reduce starting method can be used. The starting current is reduced to 65 percent by using either method. The starting torque will be reduced to 42 percent (65% x 65% = 42%) if the starting current is reduced. When selecting a reduced starting method, care must be taken to ensure that enough foot-pounds are provided to accelerate the load. **(See Figure 16-11)**

CODE LETTERS
TABLES 430.7(B) AND 430.251

Code letters are installed on motors by manufacturers for calculating the locked-rotor current (LRC) in amps based upon the kVA per horsepower which is selected from the motor's code letter. Overcurrent protection devices shall be set above the locked-rotor current of the motor to prevent the overcurrent protection device from opening when the rotor of the motor is starting. The following two methods can be used to calculate and select the locked-rotor current of motors:

(1) Utilizing code letters to determine LRC, and
(2) Utilizing horsepower to determine LRC.

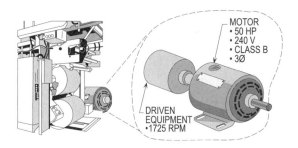

MOTOR	PERCENTAGE
B	150%
C	225%
D	275%

Finding Starting Torque

Step 1: Finding starting torque
Torque = HP x 5252 ÷ RPM
Torque = 50 x 5252 ÷ 1725
Torque = 262,600 ÷ 1725
Torque = 152.2

Step 2: Increasing torque by 150% for Class B
Torque = 152.2 x 150%
Torque = 228.3

Solution: The starting torque required is 228.3 ft. lbs.

CALCULATING STARTING TORQUE

Figure 16-10. To find the starting torque of a motor (Class B, C or D design), the full-load torque is multiplied by the percentages of the proper motor design letter.

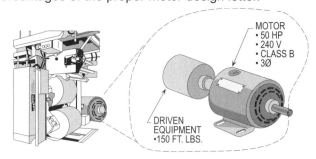

Finding Resistor Starting Torque

Step 1: Finding resistor starting torque
Torque = HP x 5252 ÷ RPM
Torque = 50 x 5252 ÷ 1725
Torque = 262,600 ÷ 1725
Torque = 152.2

Step 2: Increasing 150% for Class B
Torque = 152.2 x 150%
Torque = 228.3

Step 3: Reducing 42% for starting torque
Torque = 228.3 x 42%
Torque = 95.9

Solution: The reduced starting torque of 95.9 ft. lbs. for resistor starting will not start the driven load at 150 ft. lbs.

RESISTOR STARTING TORQUE

Figure 16-11. Resistor or reactor starting used to reduce the starting torque of a motor.

LOCKED-ROTOR CURRENT BASED UPON CODE LETTERS 430.7(B)

Code letters must be marked on the nameplate of existing motors and these letters are used for designing locked-rotor current. Locked-rotor current for code letters are listed in **Table 430.7(B)** in kVA (kilovolt-amps) per horsepower which is based upon a particular code letter.

For example: What is the locked-rotor current rating for a three-phase, 208 volt, 20 horsepower motor with a code letter B marked on the nameplate of the motor?

Step 1: Finding LRC amps
Table 430.7(B)
A = kVA per HP x 1000 ÷ V x 1.732
A = 3.54 x 20 x 1000 ÷ 208 V x 1.732
A = 70,800 ÷ 360 V
A = 197

Solution: The locked-rotor current is 197 amps. Note that Table 430.7(B) must be used to find LRC's of motors based on their code letters per 1996 NEC and earlier editions.

LOCKED-ROTOR CURRENT UTILIZING HP TABLES 430.251(A) AND (B)

The locked-rotor current of a motor may be found in **Tables 430.251(A) and (B)**. The locked-rotor current for single-phase and three-phase motors are selected from this Table based upon the phases, voltage, and horsepower rating of the motor. For motors with code letters A through G, round the nameplate current in amps up to an even number and multiply by 6 to obtain the LRC of the motor. Note that code letters can't be found in **Tables 430.251(A) and (B)** because they won't be listed on the motors nameplate anymore. Motors will be marked either as Design B, C or D to indicate which locked-rotor currents are to be selected from **Tables 430.251(A) and (B)** based on horsepower, phases, and voltages.

For example: What is the locked-rotor current rating for a three-phase, 460 volt, 50 horsepower, Design B motor?

Table method using Design letter

Step 1: Finding LRC amps
Table 430.251(B)
50 HP requires 363 A

Solution: The locked-rotor current is 363 amps.

For example: Consider a motor with a nameplate current of 63 amps and determine the LRC of the motor based upon code letters A through G?

Rule of thumb method using code letter

Step 1: Finding even unit of ten number
Table 430.7(B)
Round up 63 A to 70 A
and multiply by 6

Step 2: Calculating LRC
Table 430.7(B)
70 A x 6 = 420 A

Solution: The locked-rotor current is 420 amps. Note: This method can only be used for code letters A through G.

See **Figures 16-12(a) and (b)** for calculating and selecting the locked-rotor current of a motor.

Motor Tip: Engineers and electricians must select the locked-rotor current rating from **Tables 430.251(A) and (B)** when using Design B, C or D motors. The overcurrent protection device must be set above the locked-rotor current of the motor so the motor can start and run. See problem in **Figure 16-12(b)**.

When code letters are used, the locked rotor current must be calculated per **Table 430.7(B)** or the rule of thumb method applied, based on code letters A through G. See the problem and Quick Calc in **Figure 16-12(a)**.

See Figure 16-13 for a chart showing the different electrical characteristics for design type motors.

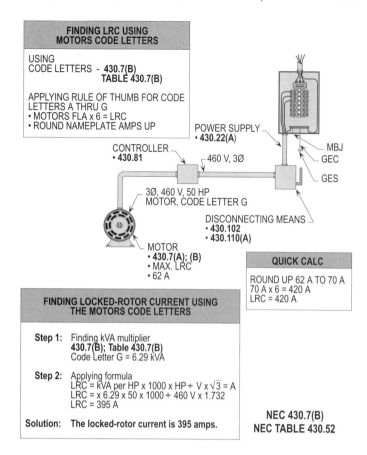

FINDING LRC USING MOTORS CODE LETTERS

USING
CODE LETTERS - **430.7(B)**
TABLE 430.7(B)

APPLYING RULE OF THUMB FOR CODE
LETTERS A THRU G
• MOTORS FLA x 6 = LRC
• ROUND NAMEPLATE AMPS UP

POWER SUPPLY
• **430.22(A)**

CONTROLLER
• **430.81**

460 V, 3Ø

MBJ
GEC
GES

3Ø, 460 V, 50 HP
MOTOR, CODE LETTER G

DISCONNECTING MEANS
• **430.102**
• **430.110(A)**

MOTOR
• **430.7(A); (B)**
• MAX. LRC
• 62 A

QUICK CALC

ROUND UP 62 A TO 70 A
70 A x 6 = 420 A
LRC = 420 A

FINDING LOCKED-ROTOR CURRENT USING THE MOTORS CODE LETTERS

Step 1: Finding kVA multiplier
430.7(B); Table 430.7(B)
Code Letter G = 6.29 kVA

Step 2: Applying formula
LRC = kVA per HP x 1000 x HP ÷ V x √3 = A
LRC = x 6.29 x 50 x 1000 ÷ 460 V x 1.732
LRC = 395 A

Solution: The locked-rotor current is 395 amps.

NEC **430.7(B)**
NEC **TABLE 430.52**

Figure 16-12(a). For motors having code letters instead of Design letters, the LRC must be calculated per **Table 430.7(B)** using the code letter of the motor.

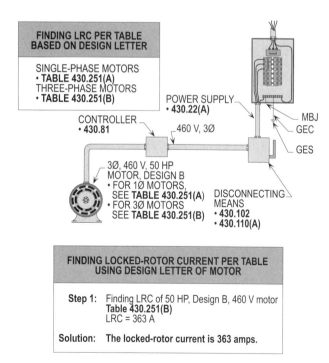

FINDING LRC PER TABLE BASED ON DESIGN LETTER

SINGLE-PHASE MOTORS
• **TABLE 430.251(A)**
THREE-PHASE MOTORS
• **TABLE 430.251(B)**

POWER SUPPLY
• **430.22(A)**

CONTROLLER
• **430.81**

460 V, 3Ø

MBJ
GEC
GES

3Ø, 460 V, 50 HP
MOTOR, DESIGN B
• FOR 1Ø MOTORS,
SEE **TABLE 430.251(A)**
• FOR 3Ø MOTORS
SEE **TABLE 430.251(B)**

DISCONNECTING
MEANS
• **430.102**
• **430.110(A)**

FINDING LOCKED-ROTOR CURRENT PER TABLE USING DESIGN LETTER OF MOTOR

Step 1: Finding LRC of 50 HP, Design B, 460 V motor
Table **430.251(B)**
LRC = 363 A

Solution: The locked-rotor current is 363 amps.

NEC TABLES **430.251(A) AND (B)**

Figure 16-12(b). Tables **430.251(A)** and **(B)** can be used to determine the LRC in amps for motors with Design letters.

NEMA Design	Starting Torque	Starting Current	Breakdown Torque	Full-Load Slip
A	Normal	Normal	High	Low
B	Normal	Low	Medium	Low
C	High	Low	Normal	Low
D	Very high	Low	-	High

FOR MORE MOTOR FACTS, SEE **FIGURE 14-17**.

Figure 16-13. The type of motor will determine the electrical characteristics of the design. Note that NEMA has designated the above designs for polyphase motors. Note that Design E motors are not listed.

For example: What size OCPD, using a CB, is required to allow the 3Ø, 460 V, 50 HP motor in Figure 16-12(a) and (b) to start and run?

Step 1: Finding FLC of motor
Table 430.250
50 HP = 65 A

Step 2: Finding percentage to size OCPD (CB)
Table 430.52
Percentage = 250%

Step 3: Performing math
65A x 250% = 162.5 A

Step 4: Selecting OCPD (CB)
430.52(C)(1), Ex. 1; 240.6(A)
162.5 A = 175 A CB

Solution: A 175 amp CB will hold about 525 A (175 A x 3 = 525 A) for 4 to 9 sec.

Motor Tip: Inverse time CB's (600 V or less) will hold about three times their rating for different periods of time, based on their frame size.

Motor Tip: It does not matter if the code letter or design letter is used to determine LRC (starting current). The size OCPD (CB), when used per Table 430.52, is large enought to hold such current and allow the motor to start and run.

Chapter 16: Design Letters and Code Letters

_____ T   F   **1.** Single-phase squirrel-cage induction motors are equipped with two windings on the stator.

_____ T   F   **2.** The stationary frame of a DC motor is called the rotor.

_____ T   F   **3.** A series DC motor produces a very high starting torque of 125 percent to 200 percent of the full-load torque.

_____ T   F   **4.** The full-load torque of a motor is determined by multiplying the horsepower times 5252 by the rpm of the motor.

_____ T   F   **5.** The starting torque of a squirrel-cage induction motor will be about 125 percent of the full-load torque for Design C motors.

_____ T   F   **6.** To reduce the inrush starting current (LRC) of a motor, a resistor starting method can be used.

_____ T   F   **7.** Code letters that are marked on the nameplate of motors are used for calculating locked-rotor current.

_____ T   F   **8.** The locked-rotor current of a three-phase motor can be found in **Tables 430.250(B)**.

_____ T   F   **9.** For motors with code letters A through G, round the nameplate current in amps up to the next number (unit of 10) and multiply by 6 to obtain the LRC of the motor.

_____ T   F   **10.** Code letters for locked rotor current of a motor can be found in **Table 430.7(B)**.

_____ _____ **11.** Squirrel-cage motors are known in the electrical industry as _____ motors.

_____ _____ **12.** Three-phase, squirrel-cage motors have three separate windings per stator pole that generate magnetic fields that are _____ degrees out-of-phase with each other.

_____ _____ **13.** A shunt DC motor produces a medium starting torque of _____ percent to _____ percent of the full-load torque.

_____ _____ **14.** A compound DC motor produces a high torque of _____ percent to _____ percent of the full-load torque.

_____ _____ **15.** The starting torque of an induction motor will be _____ percent of the full-load torque for Design B motors.

_____   _____

16. The starting torque of an induction motor will be _____ percent of the full-load torque of Design D motors.

_____   _____

17. When a high-efficiency Design E motor will not start and run, an instantaneous trip circuit breaker can be increased _____ percent of the full-load current.

_____   _____

18. When applying 65% resistor-reduced starting methods, the starting current is reduced to _____ percent.

_____   _____

19. When applying 65% reactor-reduced starting methods, the starting torque is reduced to _____ percent.

_____   _____

20. When code letters are marked on the nameplate, such letters are used for calculating locked-rotor current (LRC) in amps based upon the _____ per horsepower which is selected from the motor's LRC code letter.

_____   _____

21. What is the full-load torque and starting torque of a 50 HP, Design C motor operating at 1725 rpm? (Round up calculation).

_____   _____

22. What is the full-load torque and starting torque of a 40 HP, Design D motor operating at 1725 rpm?

_____   _____

23. What is the full-load torque for a two-speed motor, 40 HP motor operating at either 1200 rpm or 1800 rpm? (Calculate each speed.)

_____   _____

24. What is the lowest (42%) reduced-resistor starting torque for a 240 volt, 40 HP, three-phase, Design B motor operating at 1725 rpm?

_____   _____

25. What is the locked-rotor current rating for a three-phase, 208 volt, 40 HP motor with a LRC code letter B marked on the nameplate of the motor?

_____   _____

26. What is the locked-rotor current rating for a three-phase, 460 volt, 40 horsepower, Design B motor using the LRA listed in **Table 430.251(B)**?

_____   _____

27. Consider a motor with a nameplate current of 58 amps and calculate the LRC of the motor based upon code letters A through G. (Use the rule-of-thumb method.)

Starting Methods

The starting method of a motor must be considered when sizing the overcurrent protection device for the motor circuit. The starting method is determined and selected based on the amount of current required to be reduced. Overload protection for a circuit is used to allow a motor to start but will open if the motor develops overloads during operation. The starting methods are designed by using the external components in motor starters or the windings of the motor.

TYPES OF STARTING METHODS

The following seven starting methods must be considered when sizing the overcurrent protection device.

(1) Full-voltage starting,
(2) Reactor starting,
(3) Resistor starting,
(4) Autotransformer starting,
(5) Solid state starting,
(6) Wye-delta starting, and
(7) Part-winding starting.

Motor Starting Tip: The starting methods are no longer listed in **Table 430.52** of the NEC.

FULL-VOLTAGE STARTING

Full-voltage starting applies 480 volts directly to the motor's windings when the supply voltage from the utility company is three-phase, 480 volts. A disconnecting switch or circuit breaker is used as a single main switch for connecting a motor across the line. The coil in a magnetic starter may be controlled by a start and stop pushbutton station or other control devices to bridge the line to the load terminals of the motor. **(See Figure 17-1)**

High torque is produced when using full-voltage starting. Full-voltage starting has a starting torque per ampere of line current which is the highest of all starting methods. Full-voltage starting for motors have an inrush starting current that varies from 3 1/2 to 10 times the normal full-load running amps. The power system for full-voltage starting must be capable of delivering its starting current without a voltage dip. Full-voltage staring for the driven equipment must be designed to withstand heavy currents. When selecting a starting method, full-voltage starting is usually selected because of its lower cost and maintenance. However, integral horsepower motors or high-voltage systems use full-voltage starting because the starting current of the motor is low due to the high-voltage or lower horsepower.

For example: What is the full-voltage starting current, in amps, applied across the line to the windings of a 240 volt, three-phase, Design B, 50 horsepower motor?

Step 1: Finding FLA
Table 430.251(B)
50 HP = 725 A

Solution: The locked-rotor starting current is 725 amps.

By using the design letter of the motor, the locked-rotor starting current of the motor is 725 amps per **Table 430.251(B)**. By using the code letter of the motor, if available, and calculating the locked-rotor current of the motor, the LRC may be less or greater per **Table 430.7(B)**.

The LRC of a motor, when calculated, is used to select the components that make up a motor circuit. For example, a larger disconnect would be required for code letters selected above H due to the larger size fuses required to allow the motor start and run.

REACTOR STARTING

Reduced voltage starting is accomplished by placing a reactor in series with each phase of the motor. The insulation problems that are created, due to heat, when using resistor starting is not a problem when using reactor starting to reduce the voltage and current for starting a motor. Reactor starting is designed and installed mainly for motors with high-voltage systems. When using reactor starting, the torque efficiency is less than that of full-voltage starting. **(See Figure 17-2)**

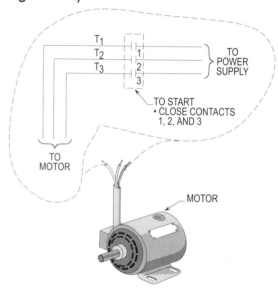

**FULL-VOLTAGE STARTING
NEMA 1**

Figure 17-1. Supply voltage and starting current are applied to the motor windings when using full-voltage starting.

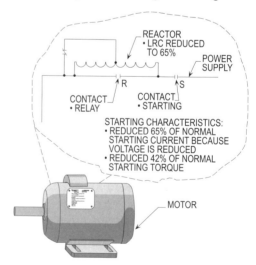

**REACTOR STARTING
NEMA 1**

Figure 17-2. A percentage of reduced voltage and starting current is applied to the motor when using reactor starting. Note that starting torque is also reduced.

The inrush starting current of a motor is reduced to about 65 percent when using reactor starting. The starting torque is reduced to about 42 percent of the normal starting torque. When applying either reactor or resistor starting, the starting current and starting torque of a motor will be reduced about the same. Reactor starting will affect the system's power factor. Note that high-voltage systems over 600 volts usually use reactor starting to start and run motor loads.

RESISTOR STARTING

Reduced voltage and current is accomplished by placing a resistor in series with each phase of a motor. When using resistor starting, the torque efficiency is less than that of full-voltage starting. **(See Figure 17-3)**

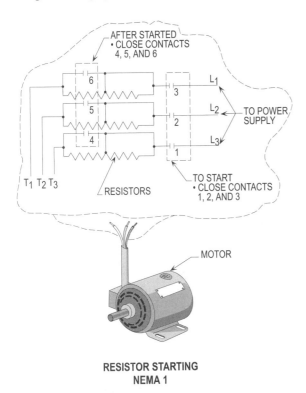

RESISTOR STARTING
NEMA 1

Figure 17-3. A percentage of reduced voltage and starting current is applied to the motor when using resistor starting.

Resistor starting is the least expensive reduced starting method. This type of starting method reduces the starting torque and provides smooth starting and acceleration up to the motor's running speed.

Resistor starting does not provide as high a starting torque and does not reduce the starting current to the same value as an autotransformer or a solid state starter.

Motor Starting Tip: When using resistor starting the starting current cannot be limited to the value of autotransformer starting.

The normal inrush starting current is reduced to about 65 percent LRC for resistor starting. The normal starting torque is reduced to about 42 percent of the starting torque. **(See Figure 17-4)**

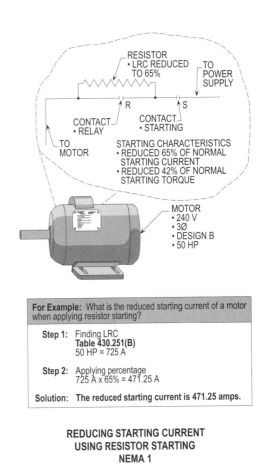

REDUCING STARTING CURRENT
USING RESISTOR STARTING
NEMA 1

Figure 17-4. By applying resistor starting, the inrush starting current is reduced to an acceptable level.

Care must be exercised to ensure that the amount of reduced torque will start the motor and driven load. When designing and selecting a reduced voltage starting method, the amount of starting torque required to start the motor and driven load and accelerate it to running speed must be considered.

Motors operating on low-voltage systems that are rated 600 volts or less are usually installed with resistor starting methods. Note that the starting torque of a motor using resistor starting is reduced to about 42 percent of the full load torque. The reduced starting torque, using a one step acceleration, may not be high enough to accelerate the motor up to its running speed. However, a resistor starting method with two steps of acceleration can be selected when this type of problem occurs. **(See Figure 17-5)**

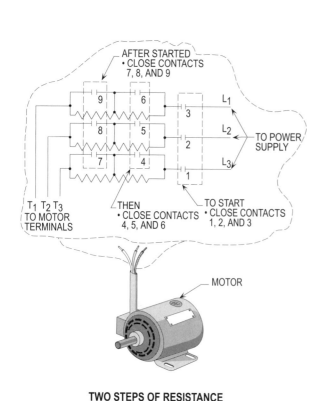

**TWO STEPS OF RESISTANCE
USING RESISTANCE STARTING**

Figure 17-5. Two steps of acceleration is selected to reduce starting voltage and current during acceleration.

AUTOTRANSFORMER STARTING

Autotransformer starting is designed and selected by providing taps to start the motor at 50, 65, or 80 percent of the applied line voltage. A tap of 50 percent can be provided to the line voltage to start a motor rated above 50 horsepower. When designing and installing autotransformer starting, an autotransformer with step-down taps and a switching device to start the motor are provided. Once started, the switching device switches the autotransformer out of the circuit and the motor is connected directly to the line. This type of reduced starting has the same effect as full-voltage starting because it provides good torque efficiency.

When determining the torque efficiency, the starting torque of the motor is divided by the locked-rotor current. The inrush starting current is reduced by switching the autotransformer into the motor circuit by contacts that connect to the desired tap. The autotransformer is switched or transferred out of the motor circuit when the motor accelerates up to its running speed. **(See Figure 17-6)**

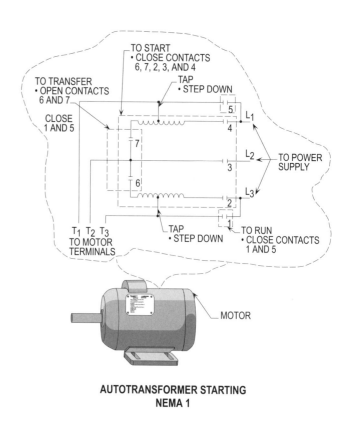

**AUTOTRANSFORMER STARTING
NEMA 1**

Figure 17-6. The above illustrates the percentage applied for reducing voltage and starting current to a motor using autotransformer starting.

Inrush starting current of a motor can be reduced by 50, 65, or 80 percent of the applied line voltage from the taps when using autotransformer starting. The percentages for autotransformer starting are based on voltage taps produced from 25 to 64 percent of the full-load starting torque. For example, 725 amps of locked-rotor starting current is listed for a 230 volt, three-phase, Design B, 50 horsepower motor, per **Table 430.251(B)**.

The percentage of the tap is squared to determine the starting torque (50% x 50% = 25%). When installing an autotransformer with a 50 percent tap, the starting torque will be reduced by 25 percent.

See Figure 17-7 for a detailed illustration for estimating the reduced starting torque and inrush current in a motor circuit using an autotransformer reduced starting method.

When designing and installing an autotransformer starting method, care must be exercised when selecting the percentage of tap required to reduce the starting current. When selecting a tap, the tap must be sized large enough to provide starting torque for the driven load and accelerating up to its running speed. For example, 725 amps of locked-rotor starting current per **Table 430.251(B)** are needed for a 230 volt, three-phase, Design B, 50 horsepower motor. The locked-rotor current (starting current) is altered per **Table 430.7(B)** when using the code letter on the nameplate of the motor. An autotransformer designed and installed with a 50 percent tap reduces the starting torque to 25 percent of the original value.

When applying a 50 percent voltage tap on an autotransformer with a (delta) motor starting current of 725 amps, the starting current is reduced to 210.25 amps for the transformation line-to-winding current (725 A x 50% x .58 = 210.25 A). Therefore, the current on the line conductors will be reduced to 210.25 amps. The current to the motor windings will be reduced to about 362.5 amps (725 A x 50% = 362.5 A). **(See Figure 17-8)**
Note, for the 58% rule, see **Figures 17-8** and **17-9**.

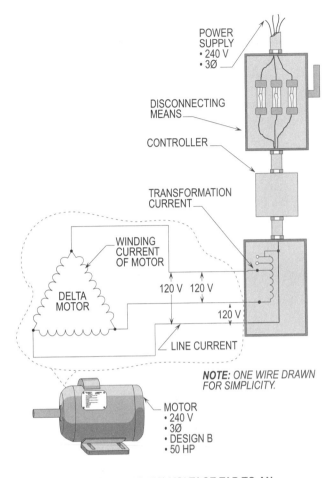

NOTE: ONE WIRE DRAWN FOR SIMPLICITY.

MOTOR
• 240 V
• 3Ø
• DESIGN B
• 50 HP

APPLYING 50% VOLTAGE TAP TO AN AUTOTRANSFORMER NEMA 1

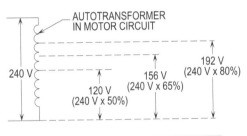

TAP	VOLTAGE	STARTING CURRENT	STARTING TORQUE
50%	120 V	50%	25%
65%	156 V	65%	42%
80%	192 V	80%	64%

NEMA 1-MOTORS AND GENERATORS

Figure 17-7. The reduced voltage for autotransformer starting is determined by using voltage taps.

An autotransformer with a 65 percent tap is squared to determine the starting torque (65% x 65% = 42%). The starting current of a (delta) motor with 725 amps is reduced to 273.325 amps (725 A x 65% x 58% = 273.325 A). The locked-rotor current of 725 amps is multiplied by the 65 percent tap on the autotransformer to determine the winding current. Therefore, the motor's winding current is about 471.25 amps. (725 A x 65% = 471.25 A)

An autotransformer with a 80 percent tap is squared to determine the starting torque (80% x 80% = 64%) of the motor. The conductor's line current for the delta motor is found by multiplying 725 amps by 80 percent (725 A x 80% x 58% = 336.4 A). The locked-rotor current of 725 amps is multiplied by the 80 percent tap on the autotransformer to determine the motor's winding current. Therefore, the motor's winding current is about 580 amps (725 A x 80% = 580 A).

For Example: What is the (delta) motor's winding current and the conductor's line current?

Step 1: Finding motor's winding current (WC)
WC = 725 A x 50%
WC = 362.5 A

Step 2: Finding conductor's line current (LC)
LC = 725 A x 50% x 58%
LC = 210.25

Solution: The motor's winding current is 362.5 amps and the conductor's line current is 210.25 amps.

Figure 17-8. Autotransformer starting having a 50 percent voltage tap will have 50 percent winding and line current.

SOLID STATE STARTING

Solid state reduced starters use silicon controlled rectifiers (SCR's) to control voltage and current flow that is directed through the SCR's by a gate (terminal) to start and accelerate motors up to their running speeds. A low-voltage signal is applied to the gate which switches the voltage and current ON and OFF through the SCR's.

When applying AC power to the circuit, a signal is sent to the gate, allowing current to flow. This current flows in one direction only though the SCR. The SCR's will turn OFF for each half cycle. The flow of current can be traced by using an AC waveform as the gate switches the flow of current through the SCR. **(See Figure 17-9)**

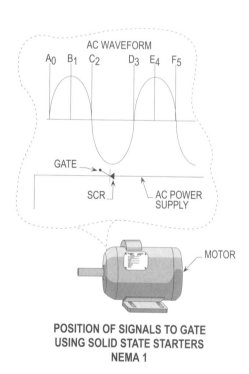

POSITION OF SIGNALS TO GATE
USING SOLID STATE STARTERS
NEMA 1

Figure 17-9. The above illustrates the use of an AC waveform. B_1 sends a signal to the gate and current flows. At C_2, the flow of current changes direction. E_4 sends a signal to the gate and current flows for the last half of the cycle.

B_1 sends a signal to the gate that turns the gate ON and current flows. At C_2, the flow of current changes direction. E_4 sends a signal to the gate and current flows for the last half of the cycle until F_5 turns the signal to the gate OFF.

When installing two SCR's, each SCR is connected in parallel to the motor in opposite directions. The voltage supply to the motor is controlled by signals that are sent to the gate. The current flow through the windings are changed by SCR's each half cycle and the circuit detects alternating current. By controlling the signals to the gate, the starting current and voltage can be adjusted to the desired level.

The circuit is energized to the motor windings when the power source is connected and the starting contacts close. By switching the signals to the gates and controlling the SCR's, acceleration of the motor is controlled. By turning ON the SCR's with gate signals, different voltage and current levels are obtained. The motor windings are connected directly to the supply line when the running contacts close, which brings the motor up to its running speed. The motor runs at full-voltage when the SCR's are disconnected from the line by the opening of the starting contacts. **(See Figure 17-10)**

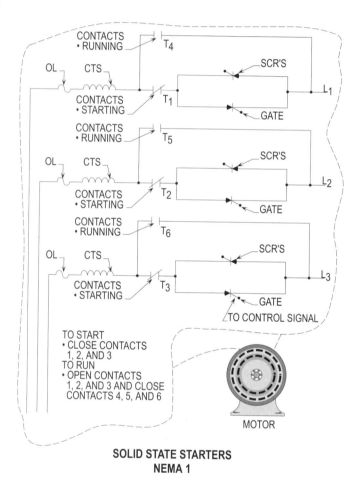

SOLID STATE STARTERS
NEMA 1

Figure 17-10. The motor is started by closing contacts 1, 2, and 3. When the motor reaches running speed, contacts 1, 2, and 3 open and contacts 4, 5, and 6 close to run the motor.

The starting current of a motor is approximately 100 percent to 400 percent of the motor's full-load current rating when using solid state starters. **(See Figure 17-11)**

Note that solid state starters provide increments of starting torque that allow for a motor to start and run its load in a very smooth operation.

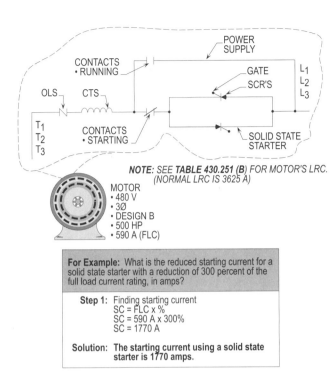

Step 1: Finding starting current
SC = FLC x %
SC = 590 A x 300%
SC = 1770 A

Solution: The starting current using a solid state starter is 1770 amps.

REDUCING STARTING CURRENT USING A SOLID STATE STARTER

Figure 17-11. The above is an illustration using solid state starting to reduce starting current for a motor.

For example: What is the reduced starting current for a solid state starter with a reduction of 300 percent of the full-load current rating for a 480 volt, three-phase, Design B, 400 horsepower motor? (See **Table 430.250** for FLC).

Step 1: Finding starting current
SC = FLC x %
SC = 477 A x 300%
SC = 1431 A

(Normal LRC = 2900 A per **Table 450.251(B)**)

Solution: **The starting current using a solid state starter is 1431 amps.**

Motor Starting Tip: When using the rule-of-thumb method, the starting current for a motor without the use of a solid state starter is approximately 2862 amps (477 A x 6 = 2862 A).

Overloads can be set in a solid state starter to sense any amount of overload current higher than the running current of the motor. Therefore, all types of overload conditions are provided with closer protection for the windings. Overloads in a solid state starter can protect special motors by disconnecting the motor from the power supply where overloads exist. The power supply is connected again to the solid state starter and motor after the overload condition is corrected.

During the starting period, the starting torque will vary with the percentage of starting current. The motor's starting torque will be reduced to approximately 60 percent if the starting current is reduced to 400 percent, 40 percent if reduced to 300%, and 30 percent if reduced to 200 percent.

See Figure 17-12 for a detailed illustration of a solid state starter used to reduce the starting current and torque of a motor.

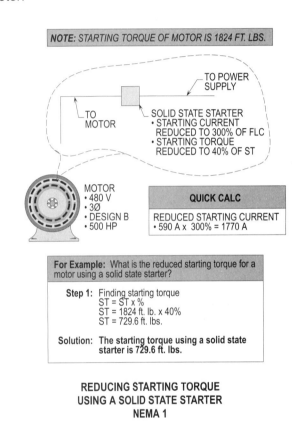

Step 1: Finding starting torque
ST = ST x %
ST = 1824 ft. lb. x 40%
ST = 729.6 ft. lbs.

Solution: The starting torque using a solid state starter is 729.6 ft. lbs.

REDUCING STARTING TORQUE USING A SOLID STATE STARTER NEMA 1

Figure 17-12. The above illustrates a solid state starter where the starting torque is reduced to 40 percent if the starting current is reduced to 300 percent.

ADJUSTABLE FREQUENCY DRIVES

A basic adjustable frequency drive system consists of the following:

(1) AC squirrel-cage induction motor,
(2) Inverter, and
(3) Operator's control station.

Motor Starting Tip: A control station may be installed on an inverter cabinet, if necessary. **(See Figure 17-13)**

Adjustable frequency drives for AC squirrel-cage induction motors are used to control the speed by varying the frequency of the power supply to the motors.

A variety of sizes are available to give designers a broad selection of adjustable speed applications for an economical installation. Adjustable frequency drives have become a popular method by which designers and installers control the speed of a motor.

AC INDUCTION MOTOR

Class B, 460 volt, three-phase, AC squirrel-cage induction motors are usually used with adjustable frequency drives that are connected to a three-phase, power supply. Note that adjustable frequency drive systems can be fitted for any size AC squirrel-cage induction motor.

By reducing the applied frequency to a value of 2 hertz or less, an adjustable frequency drive system will start an AC squirrel-cage motor. The inrush current is reduced to approximately 150 percent of the rated current of the motor when using such a low frequency.

For example: What is the inrush current for a 460 volt, three-phase, Design B, 100 horsepower AC squirrel-cage induction motor?

> **Step 1:** Finding A
> **Table 430.250; Table 430.251(B)**
> 100 HP = 124 A of FLC
> 100 HP = 725 A of LRC
>
> **Step 2:** Calculating A
> 124 A x 150% = 186 A
>
> **Solution: The inrush current would be 186 amps using an adjustable frequency drive system.**

Motor Starting Tip: By adjusting the frequency between 2 hertz or less, the AC squirrel-cage induction motor can be started at 186 amps and slowly brought up to the desired running speed.

See Figure 17-14 for a detailed illustration showing the varying percentages of torque and current due to the amount of frequency applied to the controller during starting.

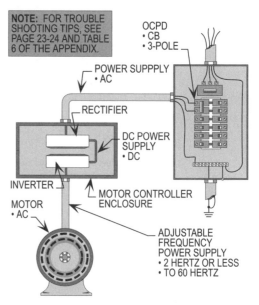

**ADJUSTABLE FREQUENCY DRIVES
USED TO CONVERT AC TO DC
NEMA 1**

Figure 17-13. To start and run AC motors, adjustable frequency drives are installed to convert AC power to DC power. Rectifiers are used to convert AC to DC. The level of frequency is regulated by the use of an inverter to start and drive the AC squirrel-cage induction motor.

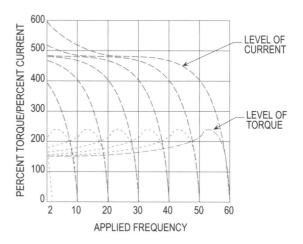

**ADJUSTABLE FREQUENCY DRIVES WITH
PERCENTAGES OF TORQUE AND CURRENT
NEMA B**

Figure 17-14. The percentage of torque and current vary with the amount of frequency applied by the controller.

INVERTERS

Inverters are solid state power conversion units which convert AC power to DC power or DC power to AC power. The following two stages of power conversion are used:

(1) Controlled or uncontrolled rectifier section (AC to DC), and
(2) Inverter (DC to AC).

A three-phase, 480 volt power supply at 60 hertz is used when inverters are installed. The AC squirrel-cage induction motor rotates at its maximum speed when the full 60 hertz is applied to the inverter by the controller. The AC squirrel-cage induction motor rotates slower when the value of frequency applied is less than 60 hertz to the inverter by the controller. Torque would be less when applying more than 60 hertz from an inverter listed for the purpose, which would produce a higher output speed (rpm's) from the motor.

When designing and installing an inverter, care must be exercised for sizing and matching it to the AC squirrel-cage induction motor. Generally, an inverter and AC squirrel-cage induction motor can be used with the same horsepower rating. The motor rating in horsepower must equal the amount of current required to drive the load. This method is used when designing and selecting an AC squirrel-cage induction motor. The inverter is designed and selected based on this rating.

An inverter must be sized for the amount of current required for an AC squirrel-cage induction motor at the maximum operating torque when the motor is oversized to provide a wider range of speed control. The inverter output current rating must be equal to or greater than the current listed on the nameplate of the AC squirrel-cage induction motor if operating at full-load current.

A maximum ambient temperature of 40°C is used on most inverters. However, 50°C is used on some inverters. Inverters are oversized by one size for areas with a higher ambient temperature. Check with manufacturer before designing and installing to verify this.

CONTROL STATION

The operator's control station is equipped with start and stop pushbuttons which are normally open or normally closed contacts which are designed and installed to start and stop the motor circuit. The rotating speed of an AC squirrel-cage induction motor is adjusted by using a speed-setting potentiometer. A potentiometer (rheostat or resistor) has three terminals with one or more sliding contacts that are adjustable and act as an adjustable voltage divider.

STARTING TORQUE

An AC squirrel-cage induction motor has an inrush current of approximately 600 percent of the motors full-load current rating when started across the line at full-voltage and full-frequency. AC squirrel-cage induction motors marked with code letters A through H have 600 percent starting currents. Inrush current (IC) for AC squirrel-cage induction motors marked with code letters J through K must be calculated on individual code letters, per Table 430.7(B).

> **For example:** What is the inrush current for a 460 volt, three-phase, 100 horsepower AC squirrel-cage induction motor with the code letter T?
>
> **Step 1:** Finding IC
> IC = kVA x 1000 x HP x ÷ V x 1.732
> IC = 19.99 x 1000 x 100 HP ÷ 460 x 1.732
>
> IC = 2508 amps
>
> **Solution: The inrush current is 2508 amps.**

EDDY-CURRENT DRIVES

Eddy-current drives consist of soft iron bars (electromagnets) shaped like a U which are magnetized by applying DC voltage to the coil of insulated wire around the bases. An eddy-current clutch is developed by a solid ring or soft iron (drum assembly) which is added to encircle the poles of the electromagnets. **(See Figure 17-15)**

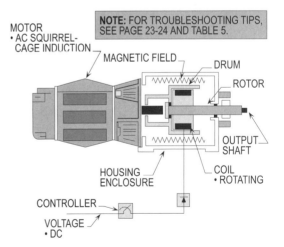

BASIC EDDY-CURRENT DRIVE

Figure 17-15. The above illustrates a basic eddy-current drive system which consists of soft iron bars (electromagnets), a coil of insulating wire, and an iron ring (drum assembly).

A wide range of stepless, adjustable speeds are obtained from eddy-current drives when used from AC power supply lines operating at standard frequencies. Eddy-current drives are designed and installed to consist of an AC squirrel-cage induction motor and a magnetic eddy-current clutch. Equipment requiring a variety of speed control or regulated torque is where eddy-current drives are used.

AC SQUIRREL-CAGE INDUCTION MOTORS

When designing and installing eddy-current drives, the AC supply power is converted to rotational power using AC squirrel-cage induction motors. AC squirrel-cage induction motors are designed and installed with basic output speeds of 3600, 1800, and 1200 rpm. The full-load speeds of AC squirrel-cage induction motors are 2 to 5 percent less than synchronous speeds.

For example: What is the actual speed if the synchronous speed is 1800 rpm and the motor is operating at 3 percent slip?

Step 1: Finding rpm
rpm = 1800 rpm x 3% (.03)
rpm = 54
rpm = 1800 rpm - 54 rpm
rpm = 1746

Solution: The actual speed of the motor is 1746 rpm.

Note: By increasing DC voltage to the coil, the output shaft will speed up and slow down when the voltage is decreased. (See Page 17-11.)

The AC squirrel-cage induction motor has an eddy-current clutch added in which the motor drives to obtain variable output speeds.

PURPOSE OF CLUTCH

The following are the three main components of the eddy-current clutch:

(1) Drum,
(2) Rotor, and
(3) Rotating coil.

The drum (steel drum) of an eddy-current clutch is the input member that is driven by the AC squirrel-cage induction motor. The rotor of an eddy-current clutch is the output

member and is free to rotate in the clutch drum. The rotating coil in an eddy-current clutch is wound around the rotor clutch and is supplied with DC voltage to produce a flux pattern through the drum and rotor.

Poles are cast in each section of the rotor which develop a north and south pole in each section of the rotor when the field coil is excited. The polarity is opposite that of the other section in each section of the rotor. Magnetic lines of force will flow through the north poles of the rotor into the drum and through the south poles of the rotor when the field coil is excited and then return to the field assembly.

The motion between the rotating drum and the rotor generate eddy-currents in the drum. Circulating currents are induced in a conducting material when they cut the magnetic flux lines which are eddy currents. These small currents are produced by the voltage through the conducting material. Eddy currents also produce a second magnetic field. The rotor rotates in the same direction as the drum when the magnetic field, generated by the eddy-currents, interacts with the magnetic field, generated by the field coil.

The rotor and drum will rotate freely, with no rotation of the output shaft, when no voltage is applied to the coil. The output shaft will pick up speed when voltage is applied and continue to increase until it is rotating slightly less than the motor. The output shaft will not rotate at the same speed as the motor due to the percentage of slip generated by the difference in speed between the drum and rotor.

By adding or subtracting the amount of DC voltage applied to the coil, the output shaft speed can be varied. The speed of the output shaft will slow down when the voltage to the coil is decreased and will speed up when voltage to the coil is increased. **See Figure 17-15** for a detail illustration of a basic eddy-current drive.

CONTROLLER

The eddy-current controller changes the strength of the magnetic field on the rotating drum by the excitation voltage (which varies) to the clutch field coil. The output shaft of the clutch will speed up when the controller is set up to add more DC voltage to the coil. The output shaft of the clutch will slow down when the controller has less excitation voltage applied to the coil which causes the magnetic field to be weaker on the drum.

The motor can be protected from overload by the controller being set at a moderate amount of current flow. The motor can be programmed to be shut off by the controller and reverse the motor's rotation and then start again in the original rotation. This type of programming would be necessary where blockage could occur in the supply line.

Solid state transistorized boards are used with the controller, which helps to facilitate troubleshooting procedures and replacement. These boards are easy to replace and can be repaired and used again. **(See Figure 17-16)**

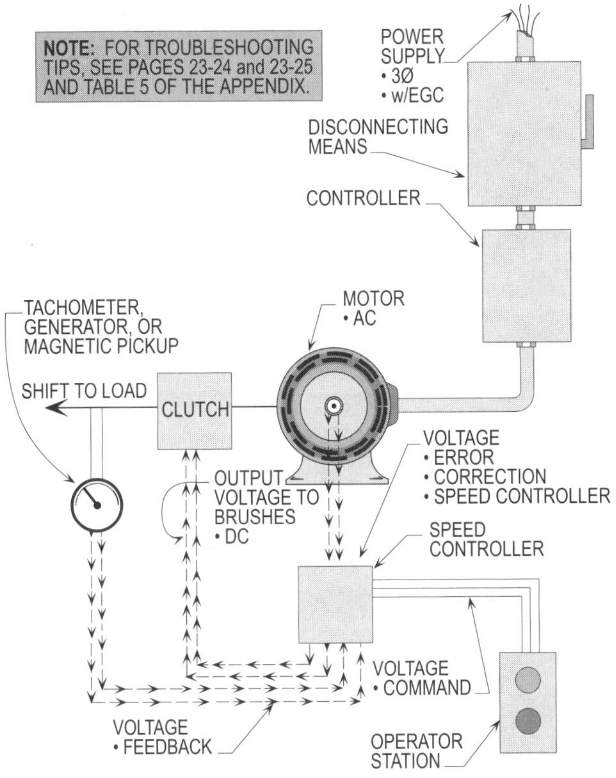

NOTE: FOR TROUBLESHOOTING TIPS, SEE PAGES 23-24 AND 23-25 AND TABLE 5 OF THE APPENDIX.

CONTROLLER ACTION

Figure 17-16. To increase or decrease the rotation speed of the eddy-current drive, the amount of excitation voltage is controlled by an operation station and controller .

TORQUE OUTPUT

The load will vary when the output speed of the clutch is increased or decreased by a fixed amount of excitation voltage applied to the coil.

By adjusting the level of excitation to the coil, the amount of torque transmitted from the AC squirrel-cage induction motor to the output shaft can be varied. The magnetic field will be greater on the drum and the faster the drum will rotate the output shaft as more excitation DC voltage is applied to the coil.

CONTROLLING SPEED OF THE MOTOR

A tachometer generator is designed to provide a signal that is proportional to the output speed of the shaft. The speed between the present speed of the controller and the actual speed of the load are realigned by this signal which is designed to adjust the excitation to the coil. The tachometer generator is mounted integrally with the output shaft.

A drive regulator is installed for realigning the output speed to the present speed. **(See Figure 17-17)**

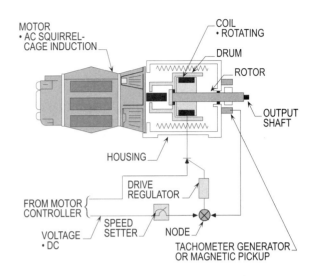

CONTROLLING SPEED

Figure 17-17. A tachometer generator and drive regulator are designed to adjust and correct the realignment of speed between the actual output speed of the output shaft to the present speed where the load is varied.

WYE-DELTA STARTING MOTORS

A special wound six lead motor is required for a wye-delta starting method. Note that the wye winding is used to start the motor and then switched to the delta winding for the run operation of the motor. Lower current is produced in wye windings due to lower voltages.

The current in wye windings is equal to the line current and not 58 percent times the line current. Each wye winding has 139 volts (240 V x 58% = 139 V) impressed across it, instead of 240 volts as in delta windings. A three-phase, 240 volt supply is used to derive these values of voltages to the motor.

Wye-connected windings have different values of phase voltage and line current. The line voltage is equal to 1.732 times the voltage-to-ground. For example, the line voltage is 208 volts (120 V x 1.732 = 208 V) when the winding voltage is 120 volts. However, the line current and phase current are the same value for wye systems. The phase voltage in wye-connected windings can be found by multiplying the phase voltage (208 V x 58% = 120 V). **(See Figure 17-18)**

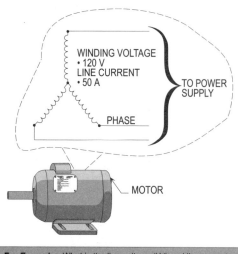

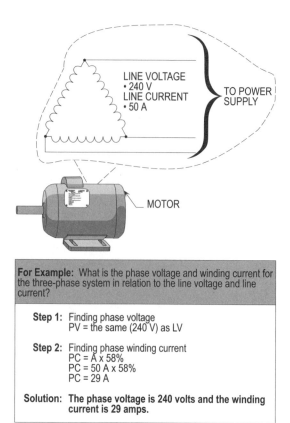

For Example: What is the line voltage (LV) and line current (LC) for the three-phase system in relation to the phase voltage (120 V) and line current (50 A)?

Step 1: Finding line voltage (Ph-to-Ph)
LV = V x √3
LV = 120 V x 1.732
LV = 208 V

Step 2: Finding line current
LC = the same (50 A) as PC

Solution: The line voltage is 208 volts and the line current is 50 amps.

WYE- SYSTEM

For Example: What is the phase voltage and winding current for the three-phase system in relation to the line voltage and line current?

Step 1: Finding phase voltage
PV = the same (240 V) as LV

Step 2: Finding phase winding current
PC = A x 58%
PC = 50 A x 58%
PC = 29 A

Solution: The phase voltage is 240 volts and the winding current is 29 amps.

DELTA SYSTEM

Figure 17-18. Wye-connected windings produce the same values of phase (winding) current (PC) and line current (LC).

Delta-connected windings produce the same value of phase voltage and line voltage. If the supplied phase voltage is 240 volts, the line voltage is 240 volts. If the line current is 50 amps, the phase current is multiplied by 58 percent (50 x 58% = 29 A). **(See Figure 17-19)**

> **Motor Starting Tip:** The reciprocal of the square root of 3 is found by dividing 1 by 1.732 to derive 58 percent (1 ÷ 1.732 = 58%).

The current flow into two phase windings connected to the line produces a different value between the phase current and line current. The phase current for delta-connected windings is found by multiplying the line current by 58 percent.

Note that the phase to phase voltage in a wye system is multiplied by 58% to derive the winding voltage and in the delta system, the winding current is multiplied by the phase current to determine the winding current.

Figure 17-19. The phase voltage (PV) and line voltage (LV) are the same value and the winding line current and phase current have different values for a delta-connected winding.

STARTING A MOTOR ON A WYE AND RUNNING ON A DELTA

Line currents and phase currents are the same currents in a wye connected winding. However, the phase voltage is 58 percent of the line voltage. The voltage value does not change in a delta-connected winding but the winding current value is 58 percent of the line current. The voltage and starting current are reduced when starting a motor on the wye winding. The circuit is automatically connected to the delta windings when the motor accelerates up to its running speed.

> **Motor Starting Tip:** After a motor has been started by a wye and reaches its running speed, the wye windings are cut out and it then runs on a delta winding.

See Figure 17-20 for a detailed illustration pertaining to the values of voltage and current when starting a motor on a wye winding and running it on a delta winding.

The inrush starting current for a motor started on a wye-connection is reduced to one-third the value of the locked-rotor line current. **(See Figure 17-21)**

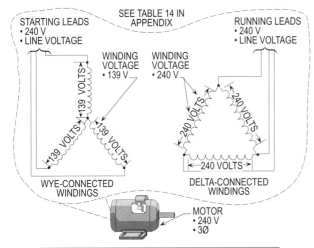

STARTING ON A WYE AND
RUNNING ON A DELTA

For Example: What is the voltage for the motor windings when using a wye-delta starting method?

Step 1: Finding delta voltage
DV = 240 V

Step 2: Finding multiplyier
M = 1 ÷ 1.732
M = .58

Step 3: Finding wye voltage
V = V x M
V = 240 V x .58
V = 139 V

Solution: The delta winding voltage is 240 volts and the wye-winding voltage is 139 volts.

Figure 17-20. The above illustration shows the values of voltage and current when starting a motor on a wye winding and running it on a delta winding. The phase voltage in wye windings is 58 percent of the line voltage.

For example: What is the inrush current for a 240 volt, three-phase, Design B, 50 horsepower motor with a wye-delta starting method?

Step 1: Finding Amps
Table 430.251(B)
50 HP = 725 A

Step 2: Calculating Amps
A = 725 A x .33
A = 239 A

Solution: The starting inrush current is 239 amps.

When designing and selecting the starting torque for a motor starting on a wye-connection and running on a delta connection, the starting torque is reduced to one-third.

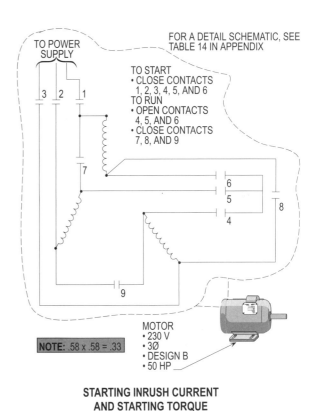

NOTE: .58 x .58 = .33

STARTING INRUSH CURRENT
AND STARTING TORQUE

For Example: What is the inrush current and starting torque for a wye-delta connected motor with a torque of 216 ft. lbs.?

Step 1: Finding A
Table 430.251(B)
50 HP = 725 A

Step 2: Calculating A
A = 725 A x .33
A = 239 A

Step 3: Finding ft. lbs.
216 ft. lbs. x .33 = 71 ft. lbs.

Solution: The starting inrush current is 239 amps and the starting torque is 71 ft. lbs.

Figure 17-21. The inrush starting current and torque for a wye-delta connected motor is calculated at 33 percent of the normal current and torque.

For example: What is the starting torque for a wye-delta connected motor with a torque of 216 foot-pounds?

Step 1: Finding ft. lbs.
216 ft. lbs. x .33 = 71 ft. lbs.

Solution: The starting torque is reduced to 71 foot-pounds.

PART-WINDING STARTING MOTORS

Part-winding starting is mostly used to reduce the voltage on weak power systems and prevent voltage disturbances. No voltage dip will occur when using a part-winding starting method during the starting and acceleration of the motor. Two separate parallel windings with two basic starting units is used for part-winding motors. Each individual starting unit is designed and selected for half the horsepower rating of the motor. When the motor is started, one winding of the motor is connected to the supply voltage. A the preset time, delay is used at a predetermined time to connect the second winding of the motor.

When using part-winding starting, only half of the motor's copper is utilized during the starting operation. The inrush starting current and starting torque is reduced by 65 percent, when designing and installing part-winding starting methods. **(See Figure 17-22)**

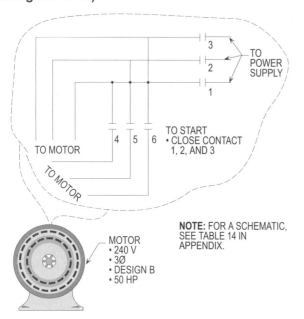

PART-WINDING STARTING

Figure 17-22. The inrush starting current and starting torque is reduced by 65 percent, when designing and installing part-winding starting methods.

OVERCURRENT PROTECTION
430.4

The requirements for sizing the overcurrent protection device for a part-winding motor are listed in **430.4**. When selecting the protective device per **430.52(C)(1)**, the percentages to be applied are found in **Table 430.52**. Since only half of the motor's horsepower is used for starting, only one half of the percentages listed in **Table 430.52** are used for selecting the overcurrent protection device. **(See Figure 17-23)**

> **Motor Starting Tip:** The size OCPD can also be found by multiplying the FLC of the motor by the percentages in **Table 430.52** and dividing the size OCPD by 2.

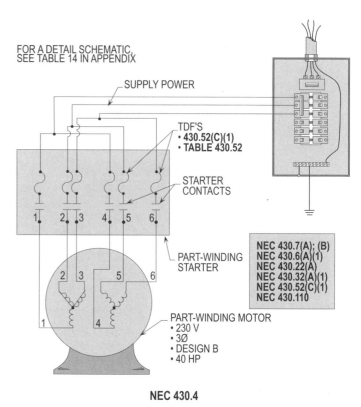

NEC 430.4

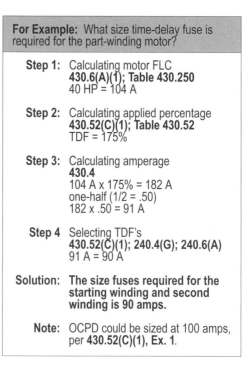

For Example: What size time-delay fuse is required for the part-winding motor?

Step 1: Calculating motor FLC
430.6(A)(1); Table 430.250
40 HP = 104 A

Step 2: Calculating applied percentage
430.52(C)(1); Table 430.52
TDF = 175%

Step 3: Calculating amperage
430.4
104 A x 175% = 182 A
one-half (1/2 = .50)
182 x .50 = 91 A

Step 4 Selecting TDF's
430.52(C)(1); 240.4(G); 240.6(A)
91 A = 90 A

Solution: **The size fuses required for the starting winding and second winding is 90 amps.**

Note: OCPD could be sized at 100 amps, per **430.52(C)(1), Ex. 1**.

Figure 17-23. Sizing time-delay fuses for a part-winding starting method to start and run a motor.

Chapter 17: Starting Methods

Section   Answer

_____   T   F   **1.**   Full-voltage starting has a starting torque per ampere for line current which is the highest of all starting methods.

_____   T   F   **2.**   Reduced voltage starting is accomplished by placing a reactor in series with each phase of the motor.

_____   T   F   **3.**   The inrush starting current of a motor can be reduced by using an autotransformer starting method.

_____   T   F   **4.**   Solid state reduced starters use silicon controlled rectifiers (SCR's) to control voltage and current.

_____   T   F   **5.**   When installing two gated SCR's, each SCR is connected in series with the motor and in opposite directions.

_____   T   F   **6.**   Adjustable frequency drives for AC squirrel-cage induction motors are used to control the speed by varying the frequency of the power supply to the motors.

_____   T   F   **7.**   Inverters are solid state power conversion units which convert AC power to DC power.

_____   T   F   **8.**   A special wound nine lead motor is required for a wye-delta starting method.

_____   T   F   **9.**   Wye-connected windings have the same values of phase current and line current.

_____   T   F   **10.**   The phase voltage and line voltage have different values and the line current and phase current have the same values for a delta-connected winding.

_____ _____   **11.**   Part-winding starting is used to reduce the voltage mostly on weak power systems and prevent voltage disturbances, due to high inrush current.

_____ _____   **12.**   One half of the motor's copper is utilized during the starting operation when using part-winding starting.

_____ _____   **13.**   The inrush starting current and torque for a wye-delta connected motor is calculated at 42 percent of the normal current and torque.

_____ _____   **14.**   Full-voltage starting applies _____ volts directly to the motor's windings when the supply voltage from the utility company is a 480 V, three-phase system.

_____ _____   **15.**   Full-voltage starting for motors have an inrush starting current that varies from _____ to _____ times the normal full-load running value.

——————— ———————   **16.**   The inrush starting current of a motor is reduced to about _____ percent when using the lowest reactor starting method to start a motor.

——————— ———————   **17.**   The normal starting torque is reduced to about _____ percent of the starting torque when using full resistor starting to start a motor.

——————— ———————   **18.**   A tap of _____ percent is applied to the line voltage to start a motor, 50 HP or higher, when autotransformer starting is used.

——————— ———————   **19.**   Autotransformer starting utilizing a 50 percent voltage tap has a transformation current of _____ percent.

——————— ———————   **20.**   The starting current of motors are reduced approximately _____ percent to _____ percent of the motor's full-load current rating when using solid state starters.

——————— ———————   **21.**   The motor speed varies with the amount of _____ applied by an adjustable frequency drive system.

——————— ———————   **22.**   Torque would be less when applying more than 60 hertz from an _____ listed for the purpose, which would produce a higher output speed (rpm's) from an A/C squirrel-cage induction motor.

——————— ———————   **23.**   Eddy-current drives consists of soft _____ bars shaped like a U, which are magnetized by applying DC voltage to the coil of insulated wire around the bases.

——————— ———————   **24.**   The rotating coil in an eddy-current clutch is wound around the rotor clutch and is supplied with _____ voltage to create a flux pattern through the drum and rotor.

——————— ———————   **25.**   A _____ generator or magnetic pick up is designed to provide a signal that is proportional to the output speed of the shaft.

——————— ———————   **26.**   In a delta-connected winding the current value is _____ percent in each winding of the line current.

——————— ———————   **27.**   The inrush current for a motor started on a wye-connection is reduced to _____ the value of the locked-rotor line current.

——————— ———————   **28.**   The inrush starting current and starting torque is reduced to _____ percent of normal for part-winding starting.

29. What is the full-voltage starting current supplied to its windings for a 240 volt, three-phase, 40 horsepower, Design B motor?

_____ _____

30. What is the reduced starting current of a 240 volt, three-phase, 40 horsepower, Design B motor when applying resistor starting and using 65% resistance?

_____ _____

31. What is the winding current, line current, and transformation current for 208 volt, three-phase, 50 horsepower, Design B motor when applying autotransformer starting with a 50 percent tap?

_____ _____

32. What is the reduced starting current for a solid state starter with a reduction of 200 percent of the full-load current rating for a 480 volt, three-phase, 300 horsepower, Design B motor?

_____ _____

33. What is the inrush current for a 460 volt, three-phase, 125 horsepower, Design B squirrel-cage induction motor? (Using an adjustable frequency drive system).

_____ _____

34. What is the inrush current for a 460 volt, three-phase, 125 horsepower, Code letter T, squirrel-cage induction motor? (Use code letter method.)

_____ _____

35. What is the actual speed if the synchronous speed for an induction motor is 1800 rpm and is operating at 5 percent slip where using eddy-current drive?

_____ _____

36. What is the inrush current for a 240 volt, three-phase, 40 horsepower, Design B motor with a wye-delta starting method?

_____ _____

37. What is the starting torque for wye-delta connected windings with a starting torque of 208 foot-pounds using wye-delta starting?

_____ _____

38. What size time-delay fuse is required for each winding of a 230 volt, three-phase 50 horsepower, Design B part-winding motor?

_____ _____

Overcurrent Protection for Individual Motors

The full-load current (FLC) in amps from **Table 430.248** for single-phase and **Table 430.250** for three-phase are used when designing and selecting the elements to make up circuits supplying power to motors. This current rating is used to size all the elements of the circuit except for the overload (OL) protection. **Table 430.7(B)** and **Tables 430.251(A) or (B)** are used to find the locked-rotor current (LRC) in amps. The OCPD must be sized large enough to hold the LRC in amps and allow the motor to start and run.

SHORT-CIRCUIT, GROUND-FAULT AND OVERLOAD PROTECTION
TABLE 430.52; 430.52(C)(1); (C)(3)

The motor branch-circuit overcurrent protection device shall be capable of carrying the starting current of the motor. Short-circuit and ground-fault currents are considered as being properly sized when the overcurrent protection device does not exceed the values in **Table 430.52** as permitted by the provisions of **430.52(C)(1)** with Exceptions.

Different percentages are selected for particular OCPD's from one of the four Columns listed in **Table 430.52**. The percentages are used to size and select the proper OCPD to allow a certain type of motor to start and run. The motor has a momentary starting current that is necessary for the motor to have power to start and drive the connected load. Note that the OCPD sized per **430.52(C)(1)** provides protection from short-circuits and ground-faults. Overload protection must be provided for conductors and windings per **430.32(A)(1)** or **430.32(C)**. **(See Figure 18-1)**

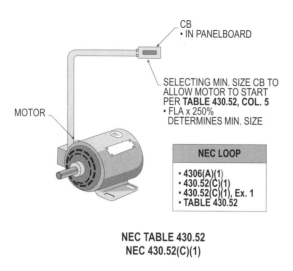

NEC TABLE 430.52
NEC 430.52(C)(1)

Figure 18-1. Selecting the percentages to size the minimum, next size, and maximum size circuit breaker to allow a motor to start and run.

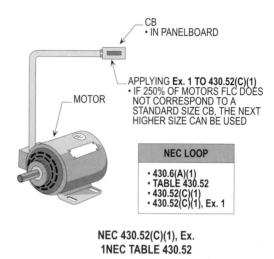

NEC 430.52(C)(1), Ex.
1NEC TABLE 430.52

Figure 18-2. Where the percentages of **Table 430.52** times the full-load current of motor in amps does not correspond to a standard size OCPD, the next higher size above this percentage may be used.

Motor Tip: In cases where the values for branch-circuit protective devices, determined by **Table 430.52**, do not correspond to the standard sizes or ratings of fuses, nonadjustable circuit breakers, thermal devices, or possible settings of adjustable circuit breakers adequate to carry the starting currents of the motor, the next higher size rating or setting may be used.

EXCEPTION 2

If the ratings listed in **Table 430.52** and **Ex. 1 to 430.52(C)(1)** are not sufficient for the starting current of the motor, the OCPD's with percentages shown can be used to start and run motors having high inrush starting currents: **(See Figure 18-3)**

APPLYING THE EXCEPTIONS 430.52(C)(1), Ex.'s 1 AND 2

There are exceptions permitting larger OCPD's to be used when the overcurrent protection devices specified in **Table 430.52** will not handle the starting current and allow the motor to start and run. Where the motor fails to start and run due to excessive inrush starting currents, one of the following exceptions can be applied:

EXCEPTION 1

If the values of the branch-circuit, short-circuit and ground-fault protection devices determined from **Table 430.52** do not conform to standard sizes or ratings of fuses, nonadjustable circuit breakers, or possible settings on adjustable circuit breakers, it does not matter if they are capable or not capable of adequately carrying the load involved, the next higher setting or rating shall be permitted. In other words, you can round up or round down the size of the overcurrent protection device automatically, by choice. **(See Figure 18-2)**

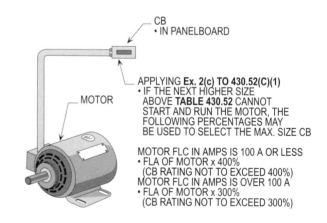

NEC TABLE 430.52
NEC 430.52(C)(1)
NEC 430.52(C)(1), Ex. 1
NEC 430.52(C)(1), Ex. 2(c)

Figure 18-3. When the percentages of **Table 430.52** and **430.52(C)(1), Ex. 1** won't allow the motor to start and run the driven load, the maximum size CB of **430.52(C)(1), Ex. 2(c)** may be used.

Anytime the percentages of **Table 430.52** and **Ex. 1 to 430.52(C)(1)** won't allow the motor to start and run its driven load, the following percentages can be applied for different types of OCPD's:

(1) When nontime-delay fuses are used and they do not exceed 600 amperes in rating, the fuse size can be increased up to 400 percent of the full-load current, but never over 400 percent.

(2) Time-delay fuses (dual-element) are not to exceed 225 percent of the full-load current, but they may be increased up to this percentage.

(3) Inverse time-element circuit breakers are permitted to be increased in rating. However, they shall not exceed:

    (a) 400 percent of the full-load current of the motor for 100 amperes or less, or

    (b) They may be increased to 300 percent where a full-load current is greater than 100 amperes.

See Figure 18-4 for a detailed illustration on selecting percentages for sizing OCPD's.

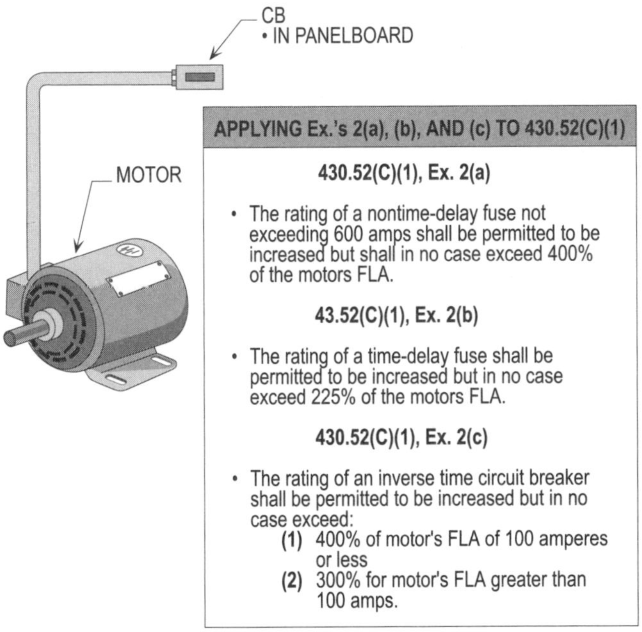

USING INSTANTANEOUS TRIP CB's 430.52(C)(3), Ex. 1

An instantaneous trip circuit breaker shall be used only if it is adjustable and is a part of a combination controller that has overcurrent protection in each conductor. Such combination, when used, must be approved, per **110.3**. An instantaneous trip circuit breaker is allowed to have a damping device to limit the inrush current when the motor is started.

If the specified setting in **Table 430.52** is found not to be sufficient for the starting current of the motor, the setting on an instantaneous trip circuit breaker may be increased, provided that in no instance it exceeds 1300 percent of the motor's full-load current ratings, in amps, for motors marked Class B, C, or D.

> **Motor Tip:** For Design E per 2002 NEC and Design B, high efficiency motors, the setting on the instantaneous circuit breaker can be adjusted up to 1700 percent to allow the motor to start and run.

See Figure 18-5 for adjusting the maximum trip settings on instantaneous trip circuit breakers to allow motors to start and accelerate their driven loads.

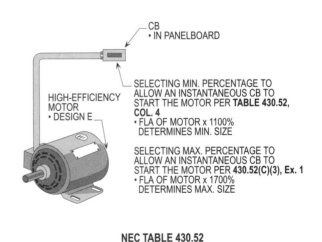

Figure 18-4. When the percentages of **Table 430.52** and **430.52(C)(1), Ex. 1** won't allow the motor to start and run, the maximum percentage of **430.52(C)(1), Ex.'s 2(a), (b), and (c)** may be applied.

Figure 18-5. Determining the minimum and maximum size instantaneous trip circuit breaker setting to allow the motor to start and run the driven load.

SIZING AND SELECTING OCPD'S TABLE 430.52, COLUMNS 2, 3, 4, AND 5

The overcurrent protection device must be sized for the starting current of the motor and selected to allow the motor to start and run. The OCPD per **Table 430.52** must protect the branch-circuit conductors from short-circuit and ground-faults. The following four overcurrent protection devices selected from **Table 430.52** will start most motors under normal starting conditions:

(1) Nontime-delay fuses - Column 2,
(2) Time-delay fuses - Column 3,
(3) Instantaneous trip circuit breakers - Column 4, and
(4) Inverse-time circuit breakers - Column 5.

NONTIME-DELAY FUSES
TABLE 430.52, COLUMN 2

Nontime-delay fuses are installed with instantaneous trip features to detect short-circuits and thermal characteristics for sensing slow heat buildup in the circuit. A nontime-delay fuse will hold 5 times (500 percent) its rating for approximately 1/4 to 2 seconds based upon the type used.

For example: What is the holding time in amps for a nontime-delay fuse of 150 amps?

Step 1: Finding holding amps
A = fuse rating x 500%
A = 150 A x 500%
A = 750

Solution: **The holding time in amps of a nontime-delay fuse is about 750 amps.**

Note: **This fuse will blow in 1/4 to 2 seconds so the motor will have to start and accelerate the load quickly.**

See Figure 18-6 for a detailed illustration of sizing nontime-delay fuses to allow a motor to start and run.

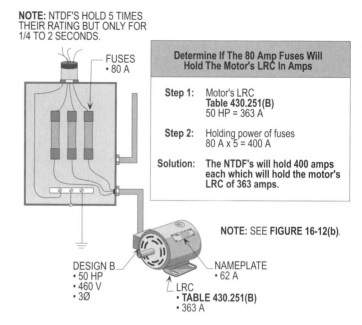

NOTE: SEE **FIGURE 16-12(b)**.

NEC TABLE 430.251(B)

Figure 18-6. Nontime-delay fuses will hold five times their rating and when this rating is above the locked-rotor current it should allow the motor to start and run, based upon LRC.

TIME-DELAY FUSES
TABLE 430.52, COLUMN 3

Time-delay fuses are also equipped with instantaneous features to detect short-circuits and thermal characteristics to sense slow heat buildup in the circuit. Time-delay fuses are used because of their time-delay action in allowing a motor to start. Time-delay fuses will hold 5 times (500 percent) their rating, which will permit most motors to start and accelerate the driven load. Note that time-delay fuses, which are size at 125 percent or less of the motor's nameplate FLC rating, in amps, provide running overload protection for the motor.

A time-delay fuse will hold 5 times its rating for ten seconds and this delayed action provides more acceleration time to allow the motor to start without tripping the OCPD.

Note that nontime delay fuses are to start a motor and the load that they are designed to drive, must be sized larger than time delay fuses because they do not have the characteristics of time delay fuses.

For example: What is the holding power in amps for a time-delay fuse of 150 amps?

Step 1: Finding holding amps
A = fuse rating x 500%
A = 150 A x 500%
A = 750

Solution: **The rating of the time-delay fuse is about 750 amps.**

Note: **This fuse holds five times its rating for ten seconds without blowing and opening the circuit.**

See Figure 18-7 for sizing time-delay fuses to hold the motor's locked-rotor current.

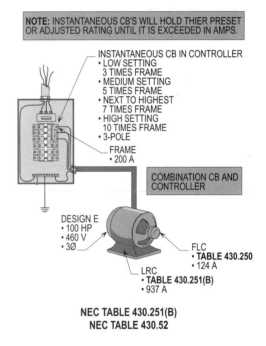

NOTE: TDF'S HOLD 5 TIMES THEIR RATING FOR 10 SECONDS.

FUSES
• 110 A

Determine If The 110 Amp Fuses Will Hold The Motor's LRC in Amps

Step 1: Motor's LRC
Table 430.251(B)
50 HP = 515 A

Step 2: Holding power of fuses
110 A x 5 = 550 A

Solution: **The TDF's will hold 550 amps each which will hold the motor's LRC of 515 amps.**

DESIGN E
• 50 HP
• 460 V
• 3Ø

NAMEPLATE
• 62 A

LRC
• **TABLE 430.251(B)**
• 515 A

NEC TABLE 430.151(B)

Figure 18-7. Time-delay fuses will hold five times their rating and when this rating is above the locked-rotor current of the motor, it should allow the motor to start and run.

INSTANTANEOUS TRIP CIRCUIT BREAKER
TABLE 430.52, COLUMN 4

Instantaneous trip circuit breakers are installed with instantaneous values of current to respond to short-circuits only. Thermal protection is not provided for instantaneous trip circuit breakers. Instantaneous trip circuit breakers will hold about three times its rating on the low setting, five times its next setting, seven times the next, and approximately ten times its rating on the high setting. Certain types allow such settings to be adjusted from 0 to 1700 percent. **(See Figure 18-8)**

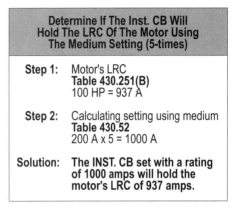

Determine If The Inst. CB Will Hold The LRC Of The Motor Using The Medium Setting (5-times)

Step 1: Motor's LRC
Table 430.251(B)
100 HP = 937 A

Step 2: Calculating setting using medium
Table 430.52
200 A x 5 = 1000 A

Solution: **The INST. CB set with a rating of 1000 amps will hold the motor's LRC of 937 amps.**

NOTE: INSTANTANEOUS CB'S WILL HOLD THIER PRESET OR ADJUSTED RATING UNTIL IT IS EXCEEDED IN AMPS.

INSTANTANEOUS CB IN CONTROLLER
• LOW SETTING
  3 TIMES FRAME
• MEDIUM SETTING
  5 TIMES FRAME
• NEXT TO HIGHEST
  7 TIMES FRAME
• HIGH SETTING
  10 TIMES FRAME
• 3-POLE

FRAME
• 200 A

COMBINATION CB AND CONTROLLER

DESIGN E
• 100 HP
• 460 V
• 3Ø

FLC
• **TABLE 430.250**
• 124 A

LRC
• **TABLE 430.251(B)**
• 937 A

NEC TABLE 430.251(B)
NEC TABLE 430.52

Figure 18-8. An instantaneous trip circuit breaker with its rating set above the locked-rotor current of a motor will allow the motor to start and run.

INVERSE-TIME CIRCUIT BREAKER
TABLE 430.52, COLUMN 5

Inverse-time circuit breakers are designed with instantaneous trip features to detect short-circuits and thermal characteristics to sense slow heat buildup in the circuit due to overloads. If heat should occur in the windings of the motor, the overload values of current will be detected by the thermal action of the circuit breaker and will trip open the circuit, if it is sized properly. The magnetic action of the circuit breaker will clear the circuit if short-circuits or ground-faults should occur on the circuit elements or equipment served.

Motor Tip: Inverse-time circuit breakers will hold about three times their rating for different periods of time based upon their frame size. For example, a motor with a locked-rotor current of 585 amps can be started with a 200 amp circuit breaker.

This can be verified by multiplying the 200 amp circuit breaker by 3 which is equal to 600 amps or 585 amps divided by 3 is equal to 195 amps. By rounding up to the next size circuit breaker per **240.6(A)**, the size circuit breaker is 200 amps. Note that this size circuit breaker allows the motor to start and run. **(See Figure 18-9)**

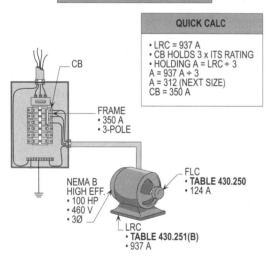

Determine If The CB Will Hold The LRC Of The Motor

Step 1: Motor's LRC
Table 430.251(B)
100 HP = 937 A

Step 2: Calculating size CB
350 A x 3 = 1050 A
Note: CB size is 350 A

Solution: **The circuit breaker with a holding power of 1050 amps will hold the motor's LRC of 937 amps.**

NOTE: CB'S WILL HOLD THREE TIMES THEIR RATING FOR PERIODS OF TIME BASED UPON THEIR FRAME SIZE.

QUICK CALC

• LRC = 937 A
• CB HOLDS 3 x ITS RATING
• HOLDING A = LRC ÷ 3
A = 937 A ÷ 3
A = 312 (NEXT SIZE)
CB = 350 A

CB

FRAME
• 350 A
• 3-POLE

NEMA B
HIGH EFF.
• 100 HP
• 460 V
• 3Ø

FLC
• **TABLE 430.250**
• 124 A

LRC
• **TABLE 430.251(B)**
• 937 A

NEC TABLE 430.251(B)
NEC TABLE 430.52

Figure 18-9. CB's sized at least three times their rating, and having an amp rating above the locked-rotor current of the motor are capable of holding the LRC.

OBTAINING FLC RATINGS
TABLES 430.247 THRU 430.250

The FLC ratings for single-phase and three-phase, DC and AC motors are obtained from **Tables 430.247** through **430.250**. The starting currents (LRC) are obtained from **Tables 430.251(A) and (B)**.

FLC RATING FOR DC MOTORS
TABLE 430.247

The FLC rating for DC motors is determined from the values listed in **Table 430.247** for motors running at base speed. For example, the FLC rating for a 120 volt, 10 HP, DC motor is 76 amps.

FLC RATING FOR
SINGLE-PHASE MOTORS
TABLE 430.248

The FLC rating for a single-phase motor is determined from the values listed in **Table 430.248** for motors running at usual speeds and motors having normal torque characteristics. For example, the FLC rating for a 208 volt, 5 HP, single-phase motor is 30.8 amps.

FLC RATING FOR
THREE-PHASE MOTORS
TABLE 430.250

The FLC rating for a three-phase motor is determined from the values listed in **Table 430.250** for motors running at speeds used for belted motors and motors with normal torque characteristics. For example, the FLC rating for 460 volt, 30 HP, three-phase motor is 40 amps.

STARTING CURRENTS
FOR SINGLE-PHASE MOTORS
TABLE 430.251(A)

The starting current (LRC) for a single-phase motor is determined from the values listed in **Table 430.251(A)**. For example, the starting current (LRC) for a 230 volt, 7 1/2 HP, single-phase motor is 240 amps.

STARTING CURRENT
FOR THREE-PHASE MOTORS
TABLE 430.251(B)

The starting current (LRC) for a three-phase motor is determined from the values listed in **Table 430.251(B)**. For example, the starting current (LRC) for a 208 volt, 50 HP, three-phase, Design B motor is 802 amps.

FLC FOR UNLISTED MOTORS TABLES 430.247 THRU 430.250

The following methods can be used to determine the full-load current rating in amps for motors that are not listed in **Tables 430.247** through **430.250**.

(1) The horsepower rating of a listed motor shall be selected which is below the unlisted motor.

(2) The motor's full-load current rating shall be divided by its horsepower rating to obtain the multiplier.

(3) The multiplier times the HP of the unlisted motor derives FLC in amps for the unlisted motor.

> **Motor Tip:** The full-load current rating of the motor is determined by multiplying these values by the horsepower rating of the unlisted motor. **(See Figure 18-10)**

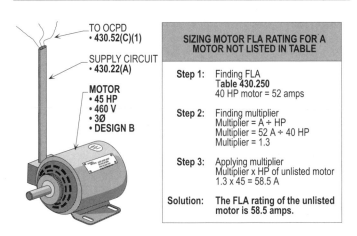

NEC TABLES 430.248 THROUGH 430.250

Figure 18-10. The above illustrates the procedure for calculating the FLA of a motor not listed in **Tables 430.248** through **430.250**.

FLC RATINGS USING RULE-OF-THUMB METHOD

The full-load current of a motor may be found by applying the rule-of-thumb method to HP values in **Table 430.248** for single-phase and **Table 430.250** for three-phase. The Table current will not always be exactly the same as the rule-of-thumb amps.

Overcurrent protection devices, conductors, and other elements can be sized with the full-load current ratings when using the rule-of-thumb method. The full-load current ratings are within a usable range when applying the rule-of-thumb method to determine the full-load current rating in amps. These amperage ratings will provide values to compute elements for a complete and safe electrical motor system. One of the following percentages (multiplier) can be applied when using a rule-of-thumb method to derive full-load amps for a particular size motor:

(1) When installing 550, 575, or 600 volt, three-phase motors, the horsepower rating of the motor shall be multiplied by 1.00 to obtain the full-load current in amps.

(2) When installing 440, 460, or 480 volt, three-phase motors, the horsepower rating of the motor shall be multiplied by 1.25 to obtain the full-load current in amps.

(3) When installing 220, 230, or 240 volt, three-phase motors, the horsepower rating of the motor shall be multiplied by 2.50 to obtain the full-load current in amps.

(4) When installing 220, 230, or 240 volt, single-phase motors, the horsepower rating of the motor shall be multiplied by 5.00 to obtain the full-load current in amps.

(5) When installing 110, 115, or 120 volt, single-phase motors, the horsepower rating of the motor shall be multiplied by 10.00 to obtain the full-load current in amps.

See Figure 18-11 for a detailed illustration of computing full-load current in amps for motors using the rule-of-thumb method.

SIZING MAXIMUM OCPD 430.52(C)(1), Ex.'s 2(a) THRU (c)

Where the rating specified in **Table 430.52** is not sufficient for the starting current of the motor, the following ratings (percentages) shall be applied:

(1) Nontime-delay fuses (400 percent)
(2) Time-delay fuses (225 percent)
(3) Inverse time circuit breakers (400 and 300 percent)
(4) Instantaneous trip circuit breakers (0-1700 percent based on Design letter or NEMA B High Eff.)

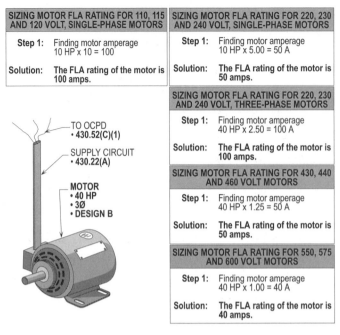

SIZING MOTOR FLA RATING FOR 110, 115 AND 120 VOLT, SINGLE-PHASE MOTORS		SIZING MOTOR FLA RATING FOR 220, 230 AND 240 VOLT, SINGLE-PHASE MOTORS	
Step 1:	Finding motor amperage 10 HP x 10 = 100	Step 1:	Finding motor amperage 10 HP x 5.00 = 50 A
Solution:	The FLA rating of the motor is 100 amps.	Solution:	The FLA rating of the motor is 50 amps.

SIZING MOTOR FLA RATING FOR 220, 230 AND 240 VOLT, THREE-PHASE MOTORS	
Step 1:	Finding motor amperage 40 HP x 2.50 = 100 A
Solution:	The FLA rating of the motor is 100 amps.

SIZING MOTOR FLA RATING FOR 430, 440 AND 460 VOLT MOTORS	
Step 1:	Finding motor amperage 40 HP x 1.25 = 50 A
Solution:	The FLA rating of the motor is 50 amps.

SIZING MOTOR FLA RATING FOR 550, 575 AND 600 VOLT MOTORS	
Step 1:	Finding motor amperage 40 HP x 1.00 = 40 A
Solution:	The FLA rating of the motor is 40 amps.

NEC TABLES 430.148 THRU 430.250

Figure 18-11. The above illustration is a method used by the electrical industry to determine the FLA of motors found in **Tables 430.248** through **430.250**.

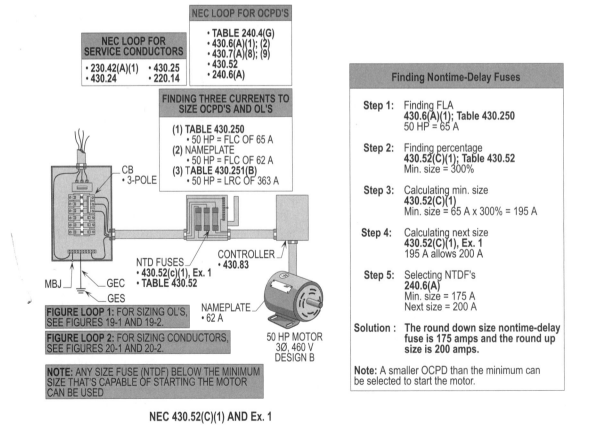

NEC 430.52(C)(1) AND Ex. 1

Figure 18-12(a). Determining the minimum and next size nontime-delay fuses per **Table 430.52** to start and run a motor.

SIZING OCPD'S TO ALLOW MOTORS TO START AND RUN 430.52(C)(1); TABLE 430.52

The branch-circuit protection for a motor may be a fuse or circuit breaker located in the line at the point where the branch-circuit originates. The fuse or circuit breaker is located either at a service cabinet or distribution panel or in the motor control center. When there is only one motor on a branch-circuit, the fuse or circuit breaker is sized according to **Table 430.52** and **430.52(C)(1)**.

Proper use of the Table will require an explanation of the "Design letters." A Design letter provides certain electrical characteristics of a particular motor that are needed to size the OCPD and to permit the motor to start and accelerate its load. To apply **Table 430.52**, it is necessary to take the following four steps:

(1) Select the phase of the motor
  • Single-phase
  • Three-phase (polyphase)

(2) Select the type of motor
  • Squirrel-cage induction
  • Wound-rotor
  • DC
  • Synchronous
  Note that motors can be single-phase or three-phase types.

(3) Select the Design letter of the motor
  • Design B
  • Design C
  • Design D

(4) Select the type OCPD
  • Column 2 is for NTDF's
  • Column 3 is for TDF's
  • Column 4 is for CB's with instantaneous trip settings or adjustments
  • Column 5 is for CB's with both instantaneous trip settings and thermal trip characteristics

See Figures 18-12(a) through (d) for sizing and selecting the size OCPD's per **able 430.52** to allow a motor to start and run its driven load. Note that either the minimum (round up) or next size (round up) OCPD can be used for a particular size motor.

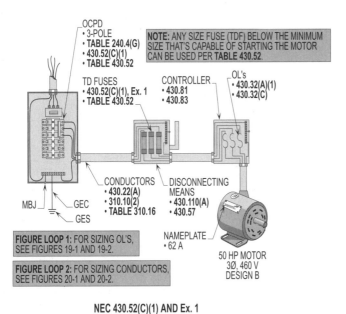

NOTE: ANY SIZE FUSE (TDF) BELOW THE MINIMUM SIZE THAT'S CAPABLE OF STARTING THE MOTOR CAN BE USED PER TABLE 430.52.

OCPD
• 3-POLE
• **TABLE 240.4(G)**
• 430.52(C)(1)
• **TABLE 430.52**

TD FUSES
• 430.52(C)(1), Ex. 1
• **TABLE 430.52**

CONTROLLER
• 430.81
• 430.83

OL's
• 430.32(A)(1)
• 430.32(C)

CONDUCTORS
• 430.22(A)
• 310.10(2)
• **TABLE 310.16**

DISCONNECTING MEANS
• 430.110(A)
• 430.57

MBJ — GEC — GES

NAMEPLATE
• 62 A

50 HP MOTOR
3Ø, 460 V
DESIGN B

FIGURE LOOP 1: FOR SIZING OL'S, SEE FIGURES 19-1 AND 19-2.

FIGURE LOOP 2: FOR SIZING CONDUCTORS, SEE FIGURES 20-1 AND 20-2.

NEC 430.52(C)(1) AND Ex. 1

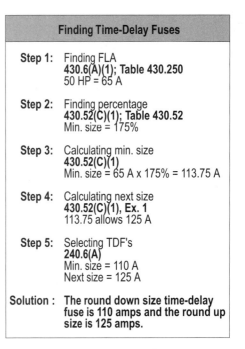

Finding Time-Delay Fuses

Step 1:	Finding FLA **430.6(A)(1); Table 430.250** 50 HP = 65 A
Step 2:	Finding percentage **430.52(C)(1); Table 430.52** Min. size = 175%
Step 3:	Calculating min. size **430.52(C)(1)** Min. size = 65 A x 175% = 113.75 A
Step 4:	Calculating next size **430.52(C)(1), Ex. 1** 113.75 allows 125 A
Step 5:	Selecting TDF's **240.6(A)** Min. size = 110 A Next size = 125 A
Solution :	**The round down size time-delay fuse is 110 amps and the round up size is 125 amps.**

Figure 18-12(b). Determining the minimum and next size time-delay fuses per **Table 430.52** to start and run a motor.

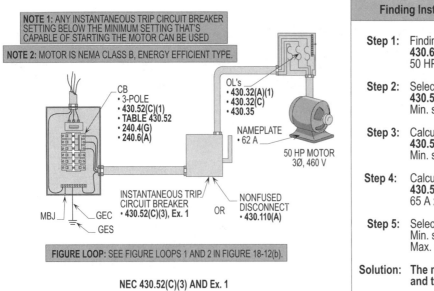

Finding Instantaneous Trip Circuit Breaker

Step 1: Finding FLA
430.6(A)(1); Table 430.250
50 HP = 65 A

Step 2: Selecting the min. setting
430.52(C)(3); Table 430.52
Min. setting = 1100%

Step 3: Calculating min. setting
430.52(C)(3)
Min. setting = 65 A x 1100% = 715 A

Step 4: Calculating max. setting
430.52(C)(3), Ex. 1
65 A x 1700% = 1105 A

Step 5: Selecting inst. trip circuit breaker
Min. setting = 715 A
Max. setting = 1105 A

Solution: **The minimum setting is 715 amps
and the maximum setting is 1105 amps.**

Figure 18-12(c). Determining the minimum and maximum setting for an instantaneous trip circuit breaker to start and run a Class B energy efficient motor.

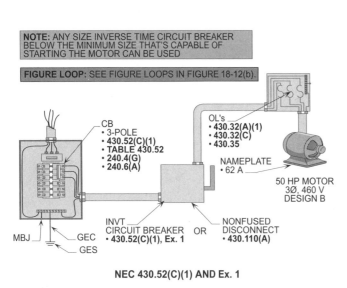

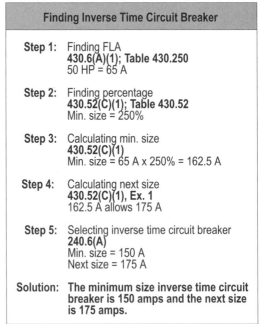

Finding Inverse Time Circuit Breaker

Step 1: Finding FLA
430.6(A)(1); Table 430.250
50 HP = 65 A

Step 2: Finding percentage
430.52(C)(1); Table 430.52
Min. size = 250%

Step 3: Calculating min. size
430.52(C)(1)
Min. size = 65 A x 250% = 162.5 A

Step 4: Calculating next size
430.52(C)(1), Ex. 1
162.5 A allows 175 A

Step 5: Selecting inverse time circuit breaker
240.6(A)
Min. size = 150 A
Next size = 175 A

Solution: **The minimum size inverse time circuit
breaker is 150 amps and the next size
is 175 amps.**

Figure 18-12(d). Determining the minimum and (round up) next size (round up) inverse time circuit breaker to start and run a motor.

NONTIME-DELAY FUSES USING THE MAXIMUM SIZE 430.52(C)(1) Ex. 2(a)

If the minimum or next size OCPD does not allow the motor start and run, the maximum size rating of a nontime-delay fuse, not exceeding 600 amps, can be increased but must not exceed 400 percent of the full-load current of the motor. **(See Figure 18-13)**

TIME-DELAY FUSES USING MAXIMUM SIZE 430.52(C)(1), Ex. 2(b)

To allow a motor to start and run, the rating of a time-delay fuse shall be permitted to be increased but must not exceed 225 percent of the full-load current of the motor. **(See Figure 18-14)**

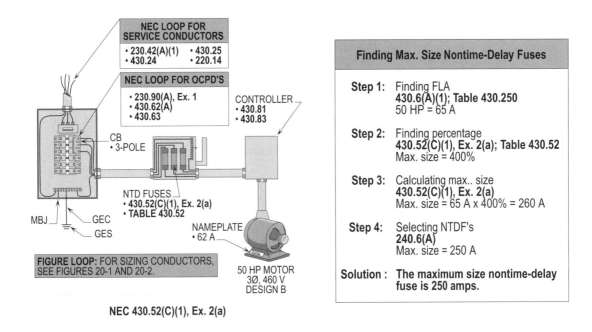

NEC 430.52(C)(1), Ex. 2(a)

Figure 18-13. Nontime-delay fuses can be increased to a maximum size of 400 percent of the motor's full-load current rating in amps.

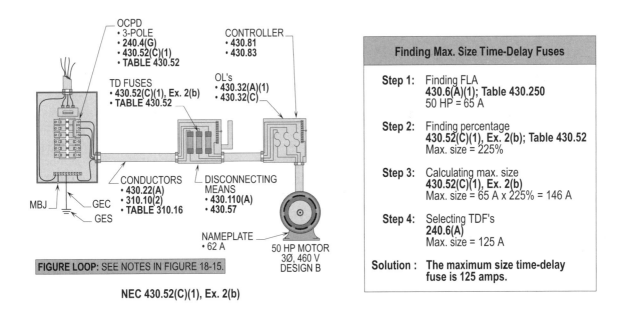

NEC 430.52(C)(1), Ex. 2(b)

Figure 18-14. Determining the maximum size time-delay fuse to start and run a motor.

INVERSE TIME CIRCUIT BREAKERS
430.52(C)(1), Ex. 2(c)

The rating for inverse time circuit breakers can be increased but must not exceed 400 percent for a full-load current of 100 amps or less. Full-load currents greater than 100 amps may be increased up to 300 percent. **(See Figure 18-15)**

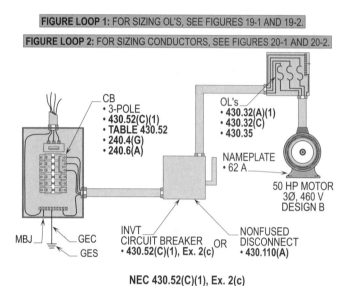

FIGURE LOOP 1: FOR SIZING OL'S, SEE FIGURES 19-1 AND 19-2.

FIGURE LOOP 2: FOR SIZING CONDUCTORS, SEE FIGURES 20-1 AND 20-2.

NEC 430.52(C)(1), Ex. 2(c)

Finding Max. Size Inverse Time CB
Step 1: Finding FLA 430.6(A)(1); Table 430.250 50 HP = 65 A
Step 2: Finding percentage 430.52(C)(1), Ex. 2(c); Table 430.52 Max. size = 400%
Step 3: Calculating max. size 430.52(C)(1), Ex. 2(c) Max. size = 65 A x 400% = 260 A
Step 5: Selecting inverse time circuit breaker 240.6(A) Max. size = 250 A
Solution: The maximum size inverse time circuit breaker is 250 amps.

Figure 18-15. Inverse time circuit breakers can be increased to a maximum size of 400 percent of the motor's full-load current rating in amps.

MOTORS CONNECTED TO INDIVIDUAL BRANCH-CIRCUITS
430.53

Two or more motors or one or more motors and other loads shall be permitted to be connected to an individual branch-circuit under the following conditions:

(1) Motor not over 1 HP,

(2) Smallest rated motor protected, and

(3) Listed for group installations.

MOTOR NOT OVER 1 HP
430.53(A)

Two or more motors may be installed without individual overcurrent protection devices, if rated less than 1 horsepower each and the full-load current rating of each motor does not exceed 6 amps. Motors not rated over 1 horsepower must be within sight, manually started, and portable. Section **430.32** must be applied for running overload protection for each motor, if these conditions are to be met. The overcurrent protection device rated at 20 amps or less can protect 120 volt or less branch-circuits supplying these motors. Branch-circuits of 600 volts or less can also be protected by a 15 amp or less OCPD. **(See Figure 18-16)**

SMALLEST RATED MOTOR PROTECTED
430.53(B)

The branch-circuit OCPD must protect the smallest rated motor of the group for two or more motors of different ratings if the largest motor is allowed to start. The smallest rated motor of the group shall have the overcurrent protection device set at no higher value than allowed per **Table 430.52**. The smallest rated motor and other motors of the group shall be provided with overload protection, if necessary per **430.32**. **(See Figure 18-17)**

OTHER GROUP INSTALLATIONS
430.53(C)

Two or more motors of any size may be installed and connected to an individual branch-circuit. However, the

largest of the group must be protected by the percentages listed in **Table 430.52** for sizing and selecting fuses or circuit breakers. Each motor controller and component installed in the group must be approved for such use. The following are elements that must be sized and selected properly:

(1) Overcurrent protection devices,

(2) Controllers, and

(3) Running overload protection devices.

The elements may be installed as a listed factory assembly or field installed as separate assemblies listed for such conditions of use.

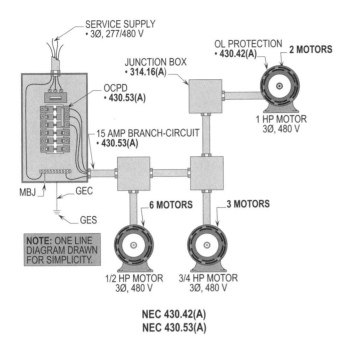

Finding OCPD For Several Motors On A Branch-Circuit

Step 1: Finding amps
430.6(A)(1); Table 430.250
1/2 HP   = 1.1 A
3/4 HP   = 1.6 A
1   HP   = 2.1 A

Step 2: Calculating FLA
430.53(A)
1.1 A x 6   = 6.6 A
1.6 A x 3   = 4.8 A
2.1 A x 2   = 4.2 A
Total load = 15.6 A

Step 3: Calculating permitted load
430.22(A)
15 A CB x 80% = 12 A

Step 4: Verifying loading
430.53(A)
15.6 A is greater than 12 A

Solution: The eleven motors are not allowed to be connected to the 15 amp branch-circuit.

Note: All motors are not started at the same time.

Figure 18-16. Determining the number of motors allowed on a 15 amp branch-circuit.

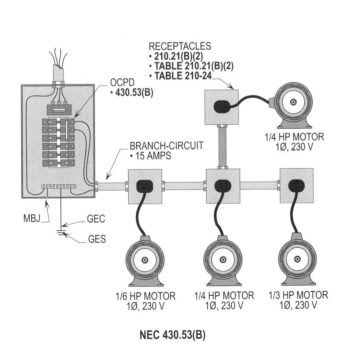

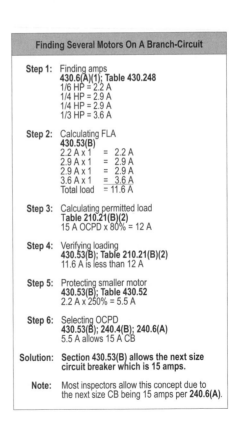

Finding Several Motors On A Branch-Circuit

Step 1: Finding amps
430.6(A)(1); Table 430.248
1/6 HP = 2.2 A
1/4 HP = 2.9 A
1/4 HP = 2.9 A
1/3 HP = 3.6 A

Step 2: Calculating FLA
430.53(B)
2.2 A x 1   = 2.2 A
2.9 A x 1   = 2.9 A
2.9 A x 1   = 2.9 A
3.6 A x 1   = 3.6 A
Total load = 11.6 A

Step 3: Calculating permitted load
Table 210.21(B)(2)
15 A OCPD x 80% = 12 A

Step 4: Verifying loading
430.53(B); Table 210.21(B)(2)
11.6 A is less than 12 A

Step 5: Protecting smaller motor
430.53(B); Table 430.52
2.2 A x 250% = 5.5 A

Step 6: Selecting OCPD
430.53(B); 240.4(B); 240.6(A)
5.5 A allows 15 A CB

Solution: Section 430.53(B) allows the next size circuit breaker which is 15 amps.

Note: Most inspectors allow this concept due to the next size CB being 15 amps per **240.6(A)**.

Figure 18-17. Determining the number of motors allowed on a 15 amp branch-circuit.

SINGLE MOTOR TAPS
430.53(D)

Any number of motor taps may be installed where a fuse or circuit breaker is installed at the point where each motor is tapped to the line. This type of installation made from a feeder is a branch-circuit per **430.28** and **430.53(D)**. **(See Figure 18-18)**

(1) Apply **Table 430.52** to select the largest motor.
(2) Size the largest OCPD for any one motor of the group.
(3) Add FLA of remaining motors.
(4) Do not exceed this value with OCPD rating.

See Figure 18-19 for a detailed procedure for sizing OCPD for a feeder motor circuit.

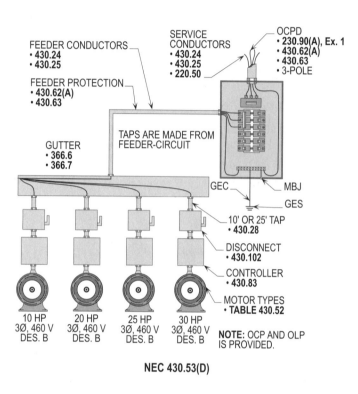

NEC 430.53(D)

Figure 18-18. Taps can be made from a feeder-circuit with the proper size conductors and OCPD with each tap.

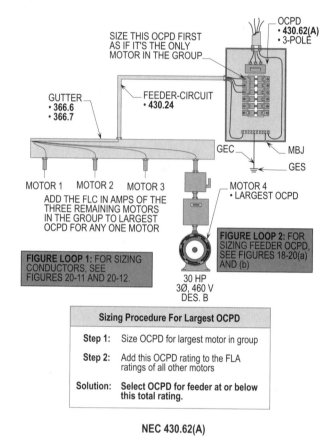

NEC 430.62(A)

Figure 18-19. Determining the OCPD for a feeder-circuit with several motors being protected from short-circuit and ground-fault conditions.

AUTOMATIC RESTARTING
430.43

A motor that can automatically restart after overloading (tripping) must not be installed unless the motor is approved for such use. Automatically restarting of a motor after shutdown must not be installed if the automatic restarting of the motor can cause injury to personnel.

SIZING OCPD FOR FEEDER LOADS
430.62(A)

To determine the size overcurrent protection device to be installed for a feeder supplying two or motors, the following procedures must be applied:

OCPD FOR MOTORS ON A FEEDER-CIRCUIT
430.62(A); (B)

The procedure for sizing the OCPD or a feeder-circuit with two or more motors can be sized by the following procedure:

(1) Sizing OCPD based upon motor's FLC,
(2) Sizing OCPD based upon conductor's ampacity, or
(3) Conductor with ampacities greater than motor's FLC rating.

SIZING OCPD BASED UPON MOTORS FLC 430.62(A)

The OCPD for a feeder-circuit supplying two or more motors is based upon the largest OCPD for any motor of the group plus the FLC in amps of the remaining motors. The procedure requires a selection of OCPD's by rounding down to find the sum of the calculated value if it does not correspond to a standard device. Note that the size OCPD for a feeder-circuit supplying two or more motors is determined by selecting the next lower rating, if the calculation based upon the percentages per **Table 430.52** times the FLA per **Table 430.250** does not correspond to a standard OCPD per **240.6(A)**. **(See Figures 18-20(a) through (d))**

SIZING OCPD BASED UPON CONDUCTORS AMPACITY 430.62(A)

The OCPD for a feeder-circuit supplying two or more motors may be selected based upon the ampacity of the feeder-conductors, if this procedure provides the greater size device. However, the OCPD sized per **Table 430.52** usually provides the largest device. **(See Figure 18-21)**

CONDUCTORS WITH AMPACITIES GREATER THAN MOTOR'S FLC 430.63

Feeder-circuits may be utilized in supplying one or more motors plus other loads. The size of the OCPD is calculated per **Articles 430 and 220** of the NEC. Note that if the ampacity of the conductors produce the largest OCPD, this size must be used. **(See Figure 18-22)**

FUTURE ADDITIONS 430.62(B)

For large industrial plants or large capacity installations, feeders with greater capacity are usually installed to provide for a future addition of loads or changes that might be made. The ratings or settings of the feeder overcurrent protective devices are based on the rated ampacity of the feeder conductors.

Where a feeder carries a motor load in addition to lighting and/or the appliance loads, the capacity of the feeder must be calculated per **Articles 210 and 220**. To this calculation is added the capacity of the motor or motors as in **430.62**. These totals are combined to find the ampacity of the feeder conductors and the overcurrent protection devices for such feeders.

> **Motor Tip:** If two motors of the same horsepower are used, only one is considered as the larger. This motor is then used to calculate the OCPD rating and the sum is added to the other motors at 100 percent of their FLC rating.

Where two or more motors are started simultaneously, the feeder sizes and overcurrent protection devices must be computed accordingly and requires higher ratings.

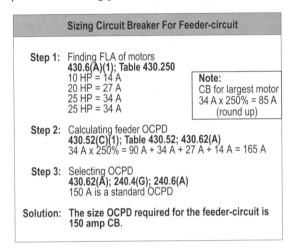

Sizing Circuit Breaker For Feeder-circuit

Step 1: Finding FLA of motors
430.6(A)(1); Table 430.250
10 HP = 14 A
20 HP = 27 A
25 HP = 34 A
25 HP = 34 A

Note:
CB for largest motor
34 A x 250% = 85 A
(round up)

Step 2: Calculating feeder OCPD
430.52(C)(1); Table 430.52; 430.62(A)
34 A x 250% = 90 A + 34 A + 27 A + 14 A = 165 A

Step 3: Selecting OCPD
430.62(A); 240.4(G); 240.6(A)
150 A is a standard OCPD

Solution: **The size OCPD required for the feeder-circuit is 150 amp CB.**

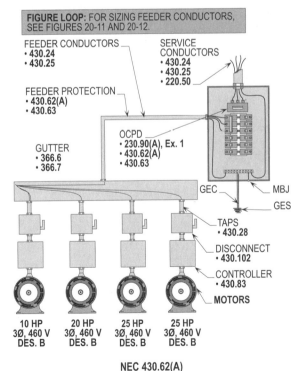

FIGURE LOOP: FOR SIZING FEEDER CONDUCTORS, SEE FIGURES 20-11 AND 20-12.
FEEDER CONDUCTORS • 430.24 • 430.25
SERVICE CONDUCTORS • 430.24 • 430.25 • 220.50
FEEDER PROTECTION • 430.62(A) • 430.63
OCPD • 230.90(A), Ex. 1 • 430.62(A) • 430.63
GUTTER • 366.6 • 366.7
GEC
MBJ
GES
TAPS • 430.28
DISCONNECT • 430.102
CONTROLLER • 430.83
MOTORS
10 HP 3Ø, 460 V DES. B
20 HP 3Ø, 460 V DES. B
25 HP 3Ø, 460 V DES. B
25 HP 3Ø, 460 V DES. B

NEC 430.62(A)

Figure 18-20(a). Sizing the circuit breaker for a feeder-circuit supplying two or more motors.

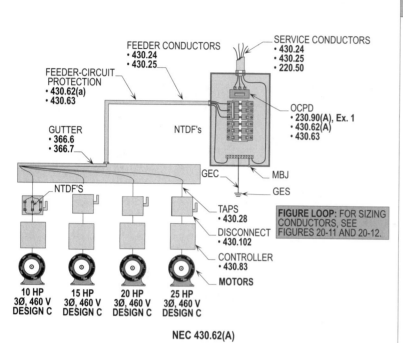

Sizing Nontime-Delay Fuses For Feeder-Circuit

Step 1: Finding FLA of motors
430.6(A)(1); Table 430.250
10 HP = 14 A
15 HP = 21 A
20 HP = 27 A
25 HP = 34 A

Step 2: Calculating largest OCPD
430.52(C)(1), Ex. 1; Table 430.52; 430.62(A); 240.6(A)
34 A x 300% = 102 A requires 110 A NTDF's
27 A x 300% = 81 A requires 90 A NTDF's
21 A x 300% = 63 A requires 70 A NTDF's
14 A x 300% = 42 A requires 45 A NTDF's

Step 3: Calculating OCPD for feeder
430.62(A); 430.52(C)(1), Ex. 1
Largest OCPD = 110 A
Plus remaining motors = 27 A
= 21 A
= 14 A
Total amps = 172 A

Step 4: Selecting OCPD for feeder
430.62(A); 240.4(G); 240.6(A)
172 A requires 150 A NTDF's
(172 A is not a standard OCPD)

Solution: The size nontime-delay fuses for the feeder-circuit are 150 amps based upon the motor's code letters.

Note: There is no Ex. in 430.62(A) to allow the next higher size OCPD above 172 amps.

Figure 18-20(b). Sizing nontime-delay fuses for a feeder-circuit supplying two or more motors.

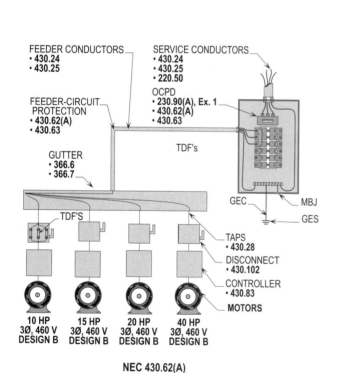

Sizing Time-Delay Fuses For Feeder-Circuit

Step 1: Finding FLA of motors
430.6(A)(1); Table 430.250
10 HP = 14 A
15 HP = 21 A
20 HP = 27 A
40 HP = 52 A

Step 2: Calculating feeder OCPD
430.52(C)(1); Table 430.52; 430.62(A); 240.6(A)
52 A x 175% = 91 A
requires 100 A OCPD = 100 A
Plus other motors
in group = 27 A
= 21 A
= 14 A
Total amps = 162 A

Step 3: Selecting OCPD for feeder
430.62(A); 240.4(G); 240.6(A)
162 A requires 150 A
(162 A is not a standard OCPD)

Solution: The size OCPD required for the feeder-circuit is 150 amps.

Note: Ex. 1 to 430.52(C)(1) is applied in Step 2 above.

Matching Values using 6 x FLC
Table 430.251(A) & (B); use 6 x FLC's
• 14 A + 21 A + 27 A + 52 A = 114 A
• 114 A LCR x 6 = 684 A
• 684 A LRC
150 A TDF x 5 = 750 A
(150 A x 5 = 750 A)
• 750 A allows all 4 motors to start

Figure 18-20(c). Sizing time-delay fuses for a feeder-circuit supplying two or more motors.

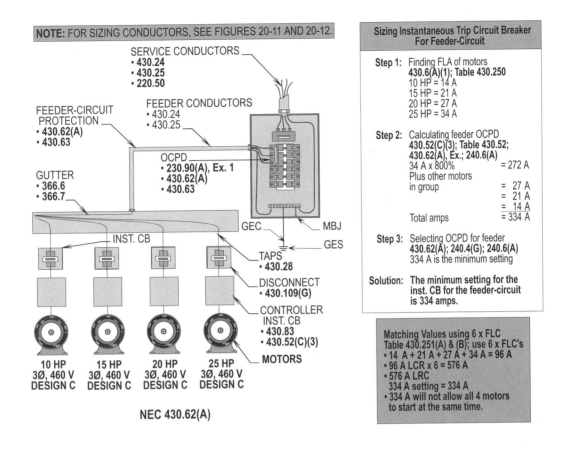

NOTE: FOR SIZING CONDUCTORS, SEE FIGURES 20-11 AND 20-12.

SERVICE CONDUCTORS
• 430.24
• 430.25
• 220.50

FEEDER-CIRCUIT
PROTECTION
• 430.62(A)
• 430.63

FEEDER CONDUCTORS
• 430.24
• 430.25

OCPD
• 230.90(A), Ex. 1
• 430.62(A)
• 430.63

GUTTER
• 366.6
• 366.7

INST. CB

GEC   MBJ
GES

TAPS
• 430.28

DISCONNECT
• 430.109(G)

CONTROLLER
INST. CB
• 430.83
• 430.52(C)(3)

MOTORS

10 HP	15 HP	20 HP	25 HP
3Ø, 460 V	3Ø, 460 V	3Ø, 460 V	3Ø, 460 V
DESIGN C	DESIGN C	DESIGN C	DESIGN C

NEC 430.62(A)

Sizing Instantaneous Trip Circuit Breaker For Feeder-Circuit

Step 1: Finding FLA of motors
430.6(A)(1); Table 430.250
10 HP = 14 A
15 HP = 21 A
20 HP = 27 A
25 HP = 34 A

Step 2: Calculating feeder OCPD
430.52(C)(3); Table 430.52;
430.62(A), Ex.; 240.6(A)
34 A x 800%   = 272 A
Plus other motors
in group   = 27 A
          = 21 A
          = 14 A
Total amps   = 334 A

Step 3: Selecting OCPD for feeder
430.62(A); 240.4(G); 240.6(A)
334 A is the minimum setting

Solution: The minimum setting for the inst. CB for the feeder-circuit is 334 amps.

Matching Values using 6 x FLC
Table 430.251(A) & (B); use 6 x FLC's
• 14 A + 21 A + 27 A + 34 A = 96 A
• 96 A LCR x 6 = 576 A
• 576 A LRC
   334 A setting = 334 A
• 334 A will not allow all 4 motors to start at the same time.

Figure 18-20(d). Sizing an instantaneous trip circuit breaker for a feeder-circuit supplying two or more motors.

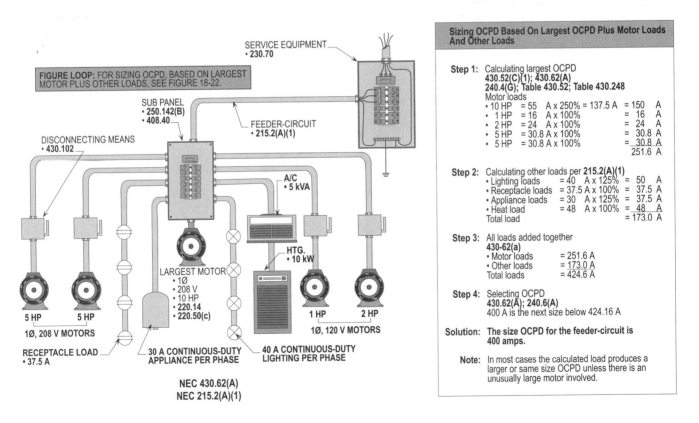

FIGURE LOOP: FOR SIZING OCPD, BASED ON LARGEST MOTOR PLUS OTHER LOADS, SEE FIGURE 18-22.

SERVICE EQUIPMENT
• 230.70

SUB PANEL
• 250.142(B)
• 408.40

FEEDER-CIRCUIT
• 215.2(A)(1)

DISCONNECTING MEANS
• 430.102

A/C
• 5 kVA

HTG.
• 10 kW

LARGEST MOTOR
• 1Ø
• 208 V
• 10 HP
• 220.14
• 220.50(c)

| 5 HP | 5 HP |
| 1Ø, 208 V MOTORS |

RECEPTACLE LOAD
• 37.5 A

30 A CONTINUOUS-DUTY
APPLIANCE PER PHASE

40 A CONTINUOUS-DUTY
LIGHTING PER PHASE

| 1 HP | 2 HP |
| 1Ø, 120 V MOTORS |

NEC 430.62(A)
NEC 215.2(A)(1)

Sizing OCPD Based On Largest OCPD Plus Motor Loads And Other Loads

Step 1: Calculating largest OCPD
430.52(C)(1); 430.62(A)
240.4(G); Table 430.52; Table 430.248
Motor loads
• 10 HP = 55 A x 250% = 137.5 A = 150 A
• 1 HP = 16 A x 100% = 16 A
• 2 HP = 24 A x 100% = 24 A
• 5 HP = 30.8 A x 100% = 30.8 A
• 5 HP = 30.8 A x 100% = 30.8 A
          251.6 A

Step 2: Calculating other loads per 215.2(A)(1)
• Lighting loads = 40 A x 125% = 50 A
• Receptacle loads = 37.5 A x 100% = 37.5 A
• Appliance loads = 30 A x 125% = 37.5 A
• Heat load = 48 A x 100% = 48 A
Total load   = 173.0 A

Step 3: All loads added together
430-62(a)
• Motor loads = 251.6 A
• Other loads = 173.0 A
Total loads = 424.6 A

Step 4: Selecting OCPD
430.62(A); 240.6(A)
400 A is the next size below 424.16 A

Solution: The size OCPD for the feeder-circuit is 400 amps.

Note: In most cases the calculated load produces a larger or same size OCPD unless there is an unusually large motor involved.

Figure 18-21. Sizing a circuit breaker for a feeder-circuit supplying two or more motors plus other loads. Note that the motor's FLC is used to size the OCPD per **Table 430.52 and 430.62(A)**.

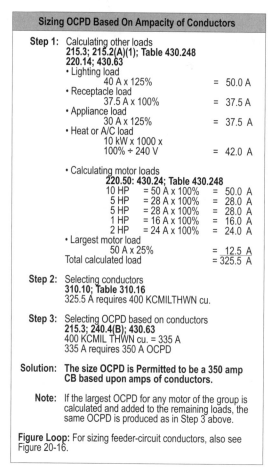

Sizing OCPD Based On Ampacity of Conductors

Step 1: Calculating other loads
215.3; 215.2(A)(1); Table 430.248
220.14; 430.63
• Lighting load
40 A x 125%   = 50.0 A
• Receptacle load
37.5 A x 100%   = 37.5 A
• Appliance load
30 A x 125%   = 37.5 A
• Heat or A/C load
10 kW x 1000 x
100% ÷ 240 V   = 42.0 A

• Calculating motor loads
220.50; 430.24; Table 430.248
10 HP   = 50 A x 100%   = 50.0 A
5 HP   = 28 A x 100%   = 28.0 A
5 HP   = 28 A x 100%   = 28.0 A
1 HP   = 16 A x 100%   = 16.0 A
2 HP   = 24 A x 100%   = 24.0 A
• Largest motor load
50 A x 25%   = 12.5 A
Total calculated load   = 325.5 A

Step 2: Selecting conductors
310.10; Table 310.16
325.5 A requires 400 KCMIL THWN cu.

Step 3: Selecting OCPD based on conductors
215.3; 240.4(B); 430.63
400 KCMIL THWN cu. = 335 A
335 A requires 350 A OCPD

Solution: **The size OCPD is Permitted to be a 350 amp CB based upon amps of conductors.**

Note: If the largest OCPD for any motor of the group is calculated and added to the remaining loads, the same OCPD is produced as in Step 3 above.

Figure Loop: For sizing feeder-circuit conductors, also see Figure 20-16.

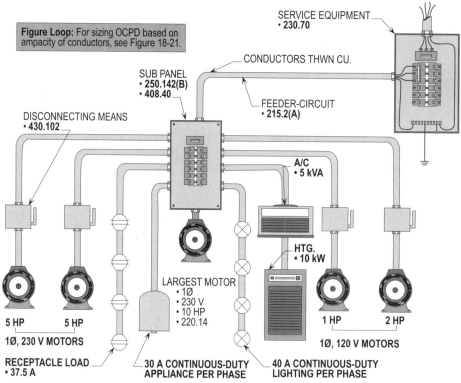

Figure Loop: For sizing OCPD based on ampacity of conductors, see Figure 18-21.

SERVICE EQUIPMENT
• 230.70

CONDUCTORS THWN CU.

SUB PANEL
• 250.142(B)
• 408.40

FEEDER-CIRCUIT
• 215.2(A)

DISCONNECTING MEANS
• 430.102

A/C
• 5 kVA

HTG.
• 10 kW

LARGEST MOTOR
• 1Ø
• 230 V
• 10 HP
• 220.14

5 HP   5 HP

1Ø, 230 V MOTORS

RECEPTACLE LOAD
• 37.5 A

30 A CONTINUOUS-DUTY
APPLIANCE PER PHASE

40 A CONTINUOUS-DUTY
LIGHTING PER PHASE

1 HP   2 HP

1Ø, 120 V MOTORS

NEC 430.63

Figure 18-22. Sizing a circuit breaker for a feeder-circuit supplying two or more motors plus other loads with the OCPD sized on the conductor's ampacity.

Chapter 18: Overcurrent Protection For Individual Motors

Section   **Answer**

T   F   **1.** The motor branch-circuit overcurrent protection device must be capable of carrying the starting current of the motor.

T   F   **2.** Where the percentages of **Table 430.52** times the full-load current of the motor in amps does not correspond to a standard size OCPD, the next higher standard size above this percentage may be used.

T   F   **3.** Time-delay fuses are must not exceed 250 percent of the full-load current rating in amps of the motor.

T   F   **4.** For NEMA B energy - efficiient motors, if needed, the setting on the instantaneous trip circuit breaker can be adjusted up to 1800 percent to allow the motor to start and run.

T   F   **5.** Time-delay fuses will hold 5 times their rating for about 10 seconds.

T   F   **6.** Inverse-time circuit breakers will hold about four times their rating for different periods of times based upon their frame size.

T   F   **7.** The FLC ratings in amps for DC motors is determined from the values listed in **Table 430.248** for motors running at base speed.

T   F   **8.** The starting current (LRC) for a single-phase, Design B motor is determined from the values listed in **Table 430.251(A)**.

T   F   **9.** When installing 480 volt, three-phase motors, the horsepower rating of each motor must be multiplied by 1.25 to obtain the full-load current in amps when applying the rule-of-thumb method.

T   F   **10.** The maximum size rating of a nontime-delay fuse not exceeding 600 amps shall be permitted to be increased but in no case exceed 400 percent of the full-load current of the motor.

11. The rating for an inverse time circuit breaker must be permitted to be increased but shall in no case exceed 300 percent for a full-load current of 100 amps or less.

12. Motors not permanently installed and rated 1 horsepower or less must be within sight from the controller and be manually started.

13. The branch-circuit overcurrent protection device is not permitted to protect the smallest rated motor of the group for two or more motors of different ratings even if the largest motor is allowed to start.

14. The OCPD for a feeder-circuit supplying two or more motors is based upon the largest OCPD for any motor of the group plus the remaining motors.

15. Feeder-circuits may be utilized in supplying one or more motors plus other loads.

16. Inverse time circuit breakers are designed with magnetic trip features to detect and open a circuit due to a _____ or ground-fault condition.

17. The FLC rating for a single-phase motor is determined from the values listed in Table _____.

18. The FLC rating in amps for a three-phase motor is determined from the values listed in Table _____.

19. The starting current (LRC) for a three-phase Design B motor is determined from the values listed in Table _____.

20. When installing 230 volt, single-phase motor, (applying the rule-of-thumb method) the horsepower rating of the motor must be multiplied by _____ to obtain the full-load current (in amps) of the motor.

21. The rating for inverse time circuit breakers shall be permitted to be increased but must in no case exceed _____ percent for a motor's full-load current greater than 100 amps.

22. Two or more motors may be installed without individual overcurrent protection devices if rated less than 1 horsepower each and the full-load current rating of each motor does not exceed _____ amps.

23. An overcurrent protection device rated at _____ amps or less can protect a 120 volt or less branch-circuit supplying motors rated less than 1 horsepower.

24. A motor that can _____ restart after overload tripping must not be installed unless the motor is approved for such use.

25. The size OCPD for a feeder-circuit supplying two or more motors can be determined by selecting the next _____ rating.

26. When nontime-delay fuses are used and they do not exceed 600 amperes in rating, it must be permitted to increase the fuse size up to _____ percent of the motor's full-load current.
    (a) 225   (c) 400
    (b) 300   (d) 1,300

27. Nontime-delay fuses will hold _____ times their rating for approximately 1/4 to 2 seconds based upon the type used.
    (a) 3   (c) 7
    (b) 5   (d) All of the above

28. Instantaneous trip circuit breakers will hold about _____ times its rating, depending on its setting.
    (a) 3   (c) 7
    (b) 5   (d) All of the above

29. What is the FLC rating in amps for a 240 volt, 10 horsepower DC motor?
    (a) 38 amps   (c) 72 amps
    (b) 55 amps   (d) 89 amps

30.   What is the FLC rating in amps for a 208 volt, single-phase, 5 horsepower,   _____   _____
      Design B motor?
         (a) 32.2 amps   **(c)** 28 amps
         (b) 30.8 amps   **(d)** 18.7 amps

31.   What is the FLC rating in amps for a 460 volt, three-phase, 50 horsepower,   _____   _____
      Design B motor?
         (a) 143 amps   **(c)** 65 amps
         (b) 130 amps   **(d)** 52 amps

32.   When installing 600 volt, three-phase motors, (applying the rule-of-thumb   _____   _____
      method) the horsepower rating of the motor must be multiplied by _____
      to obtain the full-load current in amps.
         (a) 1.00   **(c)** 2.50
         (b) 1.25   **(d)** 5.00

33.   When installing 115 volt, single-phase motors, (applying the rule-of-thumb   _____   _____
      method) the horsepower rating of the motor must be multiplied by _____
      to obtain the full-load current in amps.
         (a) 1.25   **(c)** 5.00
         (b) 2.50   **(d)** 10.00

34.   The maximum rating of a time-delay fuse shall be permitted to be increased   _____   _____
      but must, in no case, exceed _____ percent of the full-load current of
      the motor.
         (a) 225   **(c)** 400
         (b) 300   **(d)** 1300

35.   The rating for inverse time circuit breakers shall be permitted to be increased   _____   _____
      but must in no case exceed _____ percent for a full-load current of 100
      amps or less.
         (a) 225   **(c)** 400
         (b) 300   **(d)** 1300

36.   What is the minimum (round down), per **430.52(C)(1)**, and the next size   _____   _____
      nontime-delay fuse, per **430.52(C), Ex. 1**, for allowing a 50 HP, 230 volt, three-
      phase, Design letter B motor to start and run?

37.   What is the next, per **430.52(C)(1)**, maximum (round up) size time-delay fuse,   _____   _____
      per **430.52(C)(1), Ex. 2**, for a 50 HP, 230 volt, three-phase, Design B motor to
      start and run?

38.   What is the minimum and maximum (setting) for an instantaneous trip circuit   _____   _____
      breaker used for starting a 50 HP, 230 volt, three-phase, Design E motor?

39.   What is the minimum (round down) and next size inverse time circuit breaker   _____   _____
      used for starting a 50 HP, 230 volt, three-phase, Design B motor?

40.   What is the maximum (round up) size nontime-delay fuse used for a starting   _____   _____
      50 HP, 230 volt, three-phase, Design letter B motor?

41.   What is the maximum (round up) size time-delay fuse used for starting a 50   _____   _____
      HP, 230 volt, three-phase, Design B motor?

42.   What is the maximum (round down) size inverse time circuit breaker used for   _____   _____
      starting a 50 HP, 230 volt, three-phase, Design B motor?

_____ _____

43. What size OCPD (CB) is required to supply an feeder-circuit load having a group of 10 HP, 15 HP, 20 HP, and 25 HP, 460 volt, three-phase, Design B motors?

_____ _____

44. What size FLA rating is required for a 3 HP, 240 volt, single-phase, Design B motor, when applying the rule-of-thumb method?

_____ _____

45. What size FLA rating is required for a 30 HP, 220 volt, three-phase, Design B motor , when applying the rule-of-thumb method?

_____ _____

46. What size FLA rating is required for a 30 HP, 440 volt, three-phase, Design B motor , when applying the rule-of-thumb method.

_____ _____

47. What size FLA rating is required for a 30 HP, 575 volt, three-phase, Design B motor, when applying the rule-of-thumb method.

Overload Protection for Individual Motors

Devices such as thermal protectors, thermal relays, or fusetrons may be installed to provide running overload protection for motors rated more than 1 horsepower. The service factor or temperature rise of the motor shall be used when sizing and installing the running overload protection for the motor. The running overload protection is normally set to open at 115 or 125 percent of the motor's full-load nameplate current. Under certain operating conditions, the running overload protection can be sized at a greater percentage. Note that the motor's service factor and temperature rise determines the percentages selected to size the overload protection, per **430.32**. Time-delay fuses, selected and sized properly, provide overload or back up overload protection for the conductors and motor windings.

MINIMUM SIZE OVERLOAD PROTECTION
430.32

The amperage for full-load current ratings listed in **Tables 430.247** through **430.250** must be used when sizing the running overload protection. The nameplate's full-load current of the motor shall be used to size the setting of the running overload protection device as outlined in **430.6(A)(2)** and **430.32**.

The running overload protection shall be selected and rated no greater than the following percentages based upon the full-load current rating listed on the motor's nameplate:

(1) Motors with a marked service factor not less than 1.15 use 125 percent times the nameplate's FLA.
(2) Motors with a marked temperature rise not over 40 degrees C use 125 percent times the nameplate's FLA.
(3) All other motors - 115% x FLA.

See Figure 19-1 for a detailed illustration for sizing the minimum running overload protection device.

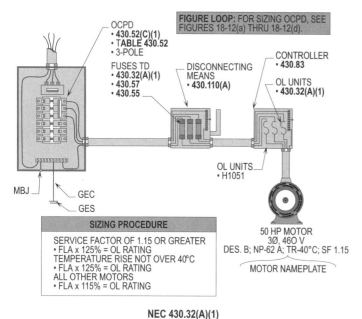

Finding Motor Min. Overload Protection

Sizing OL's Using Fuses

Step 1: Finding FLA
430.6(A)(2)
Nameplate = 62 A

Step 2: Finding percentage
430.32(A)(1)
SF = 125%
TR = 125%

Step 3: Calculating FLA
430.32(A)(1)
62 A x 125% = 77.5 A

Step 4: Selecting TD fuses
430.32(A)(1); 240.6(A)
77.5 A requires 70 A

Solution: **The size time delay fuses are 70 amps.**

Note: 80 amp TDF's provide backup OL protection.

Sizing OL's In Controller

Step 1: Finding FLA
430.6(A)(2); 430.32(A)(1)
62 A x 125% = 77.5 A

Solution: **The size overloads are selected from a manufacturers chart based on 62 amps.**

Note: The OL units are H1051 and can be selected from the chart on the back inside cover of this book.

OCPD
• **430.52(C)(1)**
• **TABLE 430.52**
• 3-POLE

FUSES TD
• **430.32(A)(1)**
• **430.57**
• **430.55**

FIGURE LOOP: FOR SIZING OCPD, SEE FIGURES 18-12(a) THRU 18-12(d).

DISCONNECTING MEANS
• **430.110(A)**

CONTROLLER
• **430.83**

OL UNITS
• **430.32(A)(1)**

OL UNITS
• H1051

MBJ
GEC
GES

SIZING PROCEDURE

SERVICE FACTOR OF 1.15 OR GREATER
• FLA x 125% = OL RATING
TEMPERATURE RISE NOT OVER 40°C
• FLA x 125% = OL RATING
ALL OTHER MOTORS
• FLA x 115% = OL RATING

50 HP MOTOR
3Ø, 460 V
DES. B; NP-62 A; TR-40°C; SF 1.15

MOTOR NAMEPLATE

NEC 430.32(A)(1)

Figure 19-1. Determining the minimum size overloads based upon service factor and temperature rise.

SERVICE FACTOR

For motors marked with a service factor of not less than 1.15 or a service factor less than 1.15, you must use the following percentages of the motor's nameplate amps to ensure protection to the motor's insulation:

(1) Not less than 1.15 - 125 percent of the motor's nameplate rating, in amps.
(2) Less than 1.15 - 115 percent of the motor's nameplate rating, in amps.

TEMPERATURE RISE

The temperature rise in motor windings is affected by the altitude. Less heat is carried from the windings in higher altitudes because thinner air flows through the inlets and outlets of the motor. Elevations of 3300 ft. or less allow thicker air to carry heat away effectively. Motors above elevations of 3300 ft. must be derated. For derating transformers, see pages 7-7 and 7-8 in this book.

Motors must be derated 1 percent for every 330 ft. above 3300 ft. For example, a 460 volt, 50 horsepower, three-phase motor pulling 62 amps and installed at an altitude of 3630 ft. must be derated to a running current of 61.38 amps (62 A x 1% = .62 A) (62 A - .62 = 61.38 A). Motors must be derated 10 percent for every 1000 ft. above 3300 ft. For example, a 460 volt, 50 horsepower, three-phase motor pulling 62 amps and installed at an altitude of 9300 ft. (10% percent for every 1000' above 3300' = 60%) must be derated to a running current of 24.8 amps (62 A x 60% = 37.2 A) (62 A - 37.2 = 24.8 A). Note that 40% of 62 A produces the same results.

For motors marked with a temperature rise of not over 40°C or a temperature rise over 40°C, the following percentages must be used:

(1) Not over 40°C - 125 percent of the motor's nameplate rating, in amps.
(2) Over 40°C - 115 percent of the motor's nameplate rating, in amps.
(3) Motor not marked - 115 percent of the motor's nameplate rating, in amps.

Due to the starting and running current period of the motor, the overload could trip open. A higher percentage can be applied per **430.32(C)** if the overloads should trip open. If the overloads should trip open after applying **430.32(C)(1)**, apply shunting rule per **430.35**.

OTHER CONDITIONS

For 50°C, 55°C, 75°C rise motors, and enclosed motors, having a service factor of 1.0, select one size smaller coil.

When ambient temperature of controller is lower than motor by 26°C (47°F), use one size smaller coil.

When ambient temperature of controller is higher than motor by 26°C (47°F), use one size larger coil.

SIZING OVERLOADS FROM COVER

The motor's nameplate full-load running current in amps is used when sizing the overloads from the cover of a magnetic starter or controller. The motor's full-load current rating is not increased by 125 percent when the overloads are selected using this procedure. **(See Figure 19-1)**

SIZING OVERLOADS FROM CHART

The motor's nameplate full-load current rating in amps is used when sizing the overloads from the chart of a magnetic starter, a motor control center, or the manufacturer's catalog.

MAXIMUM SIZE OVERLOAD PROTECTION
430.32(C)

The running overload protection (overload relay) is permitted to be selected at a higher percentage if the percentages of **430.32(A)(1)** are not sufficient to start and run the motor. The running overload protection must be no larger than the following percentages of the motor's nameplate full-load current rating in amps:

(1) Motors with a marked service factor not less than 1.15 use 140% x motor's nameplate FLA.

(2) Motors with a marked temperature rise not over 40°C use 140% x motor's nameplate FLA.

(3) All other motors - 130% x motor's nameplate FLA.

See Figure 19-2 for a detailed illustration for sizing the maximum running overload protection device.

SINGLE-PHASING

Overloads used in a magnetic starter must be selected at 62 amps or less per **430.6(A)(2)** (using nameplate amps (62 A) of motor) to protect the motor windings from single-phasing or overheating due to the driven load. However, when the percentages per **430.32(A)(1)** are not exceeded, fuses or circuit breakers may be used. For example, overload protection may be provided for a motor by a 70 amp fuse or circuit breaker if the motor will start and run using this size OCPD. A time delay fuse usually holds five times its rating for 10 seconds (70 A x 5 = 350 A).

Single-phasing occurs when one ungrounded (phase) conductor is lost in a three-wire system. Consider that a motor is pulling 62 amps and phase C is lost, the motor will be now be pulling 107.4 amps (62 A x 1.732 = 107.4 A). Note single-phasing has occurred.

Based on the motor's nameplate rating, the 70 amp OCPD does not exceed 125 percent of the motor's nameplate FLC, in amps. The 70 amp OCPD trips open due to an overload of 107.4 amps on phases A and B. **(See Figure 19-3)**

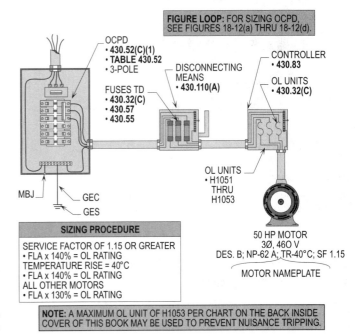

Finding Motor Max. Overload Protection
Sizing OL Units
Step 1: Finding FLA **430.6(A)(2)** Nameplate = 62 A
Step 2: Finding percentage **430.32(C)** SF = 140% TR = 140%
Step 3: Calculating FLA **430.32(C)** 62 A x 140% = 86.8 A
Sizing OL's In Controller
Step 1: Finding FLA **430.6(A)(2); 430.32(C)** 62 A x 140% = 86.8 A
Solution: **The size overloads are selected from a manufacturers chart based on 86.8 amps.**
Note: The OL units are computed at 86.8 amps maximum.

FIGURE LOOP: FOR SIZING OCPD, SEE FIGURES 18-12(a) THRU 18-12(d).

OCPD
• 430.52(C)(1)
• TABLE 430.52
• 3-POLE

FUSES TD
• 430.32(C)
• 430.57
• 430.55

DISCONNECTING MEANS
• 430.110(A)

CONTROLLER
• 430.83

OL UNITS
• 430.32(C)

OL UNITS
• H1051
THRU
H1053

MBJ

GEC

GES

50 HP MOTOR
3Ø, 460 V
DES. B; NP-62 A; TR-40°C; SF 1.15

MOTOR NAMEPLATE

SIZING PROCEDURE
SERVICE FACTOR OF 1.15 OR GREATER
• FLA x 140% = OL RATING
TEMPERATURE RISE = 40°C
• FLA x 140% = OL RATING
ALL OTHER MOTORS
• FLA x 130% = OL RATING

NOTE: A MAXIMUM OL UNIT OF H1053 PER CHART ON THE BACK INSIDE COVER OF THIS BOOK MAY BE USED TO PREVENT NUISANCE TRIPPING.

NEC 430.32(C)

Figure 19-2. Determining the maximum size overloads, based upon the service factor and temperature rise.

The percentage listed in **430.32(A)(1)** may be applied to the following types of devices:

(1) Circuit breakers
(2) Time-delay fuses
(3) Thermal cutouts
(4) Thermal relays
(5) Motor switches with thermal devices
(6) Thermal devices designed into motors

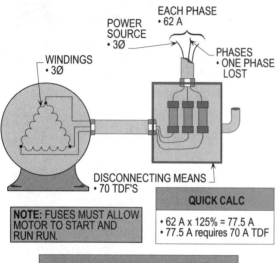

NOTE: FUSES MUST ALLOW MOTOR TO START AND RUN RUN.

QUICK CALC
• 62 A x 125% = 77.5 A
• 77.5 A requires 70 A TDF

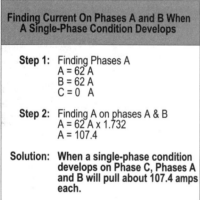

Finding Current On Phases A and B When A Single-Phase Condition Develops

Step 1: Finding Phases A
A = 62 A
B = 62 A
C = 0 A

Step 2: Finding A on phases A & B
A = 62 A x 1.732
A = 107.4

Solution: When a single-phase condition develops on Phase C, Phases A and B will pull about 107.4 amps each.

Figure 19-3. Single-phasing occurs when one phase is lost in a three-wire system. The remaining phases will carry about 1.732 times the original running current of 62 amps.

SHUNTING OL'S DURING STARTING PERIOD 430.35

The following two methods must be used when shunting overloads during the motor's starting period:

(1) Nonautomatically started, and
(2) Automatically started.

NONAUTOMATICALLY STARTED

The overload protection may be shunted for a nonautomatically started motor. However, it is only permitted to shunt or cut out the overload protection during start, provided the shunting device cannot be left in use after starting. Fuses or time-delay circuit breakers are installed and are rated or set so as not to exceed 400 percent of the full-load current rating, in amps, of the motor.

AUTOMATICALLY STARTED

Automatically started motors must not shunt or cut out overload protection. The following exceptions permit overload protection to be shunted or cut out when starting motors automatically:

(1) The starting period of the motor is greater than the time-delay of the available motor overload protective device.

(2) Where a listed means is provided that:
(a) Senses the motor rotation and will automatically prevent the shunting or cutout if the motor fails to start.
(b) The time of shunting is limited for the overload protection or cutout to a point that is less than the locked-rotor current rating of the motor that is being protected.
(c) Causes shutdown, if the running position has not been reached and the motor will have to be restarted manually.

SIZING CONTROLLERS 430.81 AND 430.83

The sizes and types of motor controllers must be installed with a horsepower rating at least equal to the motor being controlled. However, there is an exception to this rule for motors rated at and below a certain horsepower rating. (See Chart on inside back cover for ratings.)

STATIONARY MOTORS 1/8 HORSEPOWER OR LESS 430.81(A)

The branch-circuit protective device may serve as the controller where the motor is rated 1/8 horsepower or less. **(See Figure 19-4)**

For this rule to apply, motors less than 1/8 horsepower, must be stationary and the construction be such that if they fail during operation, the branch-circuit components and/or motor won't be damaged. In other words, the components of the circuit won't be burned out, etc.

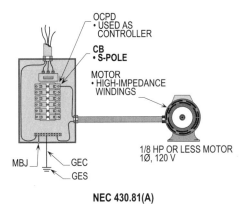

NEC 430.81(A)

Figure 19-4. The branch-circuit OCPD can serve as a controller for a motor of 1/8 HP or less.

PORTABLE MOTOR OF 1/3 HORSEPOWER OR LESS
430.81(B)

The controller may be an attachment plug and receptacle, which is acceptable for use with portable motors rated 1/3 horsepower or less. **(See Figure 19-5)**

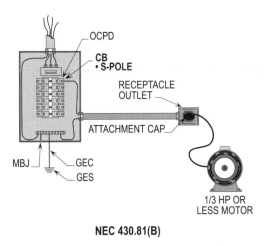

NEC 430.81(B)

Figure 19-5. The controller for 1/3 HP motor or less can be an attachment cap and receptacle.

OTHER THAN HP RATED
430.83(A)

The following conditions, other than horsepower rated controllers, are permitted to be used for energizing and deenergizing circuits supplying motors:

(1) Inverse time circuit breakers.
(2) Stationary motors rated 1/8 horsepower or less and portable motors rated 1/3 horsepower or less.
(3) Stationary motors rated 2 horsepower or less (300 volts or less).
(4) Torque motors.

HORSE POWER RATINGS
430.83(A)(1) THRU (A)(3)

Controllers, other than inverse time circuit breakers and molded case switches, must have horsepower ratings at the application voltage not lower than the horsepower rating of the motor. See the requirements below for CBs and molded case switches:

A branch circuit inverse time circuit breaker rated in amperes is permitted as a controller for all motors. Where this circuit breaker is also used for overload protection, it must conform to the appropriate provisions of this article governing overload protection.

A molded case switch rated in amperes is permitted as a controller for all motors.**(See Figure 19-6)**

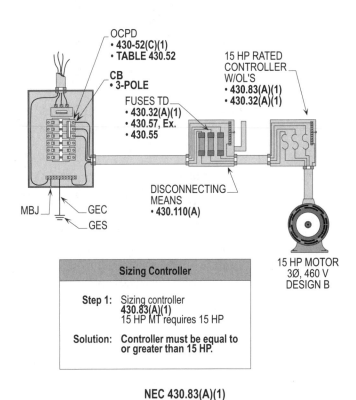

NEC 430.83(A)(1)

Figure 19-6. Controllers other than CBs, and molded case switches must have a horsepower equal to or greater than the horsepower of the motor.

Note that the controller must be sized at least equal to the horsepower of the motor. Care must be exercised when replacing an existing motor with a high efficiency motor and using the existing controller and electrical system.

INVERSE-TIME CIRCUIT BREAKERS
430.83(A)(2)

Inverse-time circuit breakers may be installed as a controller where rated in amps. If such circuit breaker is also used for the overload protection of the motor, it must be sized at 125 percent or less of the motor's nameplate current rating in amps per **430.6(A)(2)** and **430.32(A)(1)**. **(See Figure 19-7)**

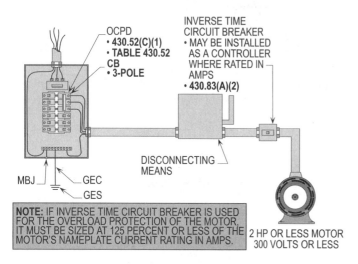

NEC 430.83(A)(2)

Figure 19-7. A branch-circuit breaker rated at 125 percent of the motor's FLA may be used as a controller for the motor and also provide overload protection.

STATIONARY AND PORTABLE MOTORS
430.83(B); 430.81(A); (B)

For stationary motors rated 1/8 horsepower or less and portable motors rated 1/3 horsepower or less, the branch-circuit OCPD's may serve as controllers and are not required to be horsepower rated. These 1/3 or less horsepower rated motors, due to smaller locked-rotor currents, can be disconnected by cord-and-plug connections.

STATIONARY MOTORS
430.83(C)(1); (2)

For a stationary motor rated 2 horsepower or less, the controller may be a general-use switch rated at least twice the motor's full-load current in amps. An AC general-use snap switch may be installed as the controller where the full-load current rating in amps does not exceed 80 percent of the switch. **(See Figure 19-8)**

When the controller is used as a disconnecting means for a motor, the controller must be sized with an interrupting rating of at least 115 percent of the motor's full-load current rating in amps per **430.110(A)**.

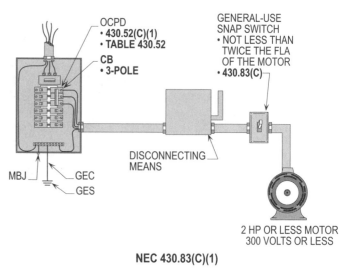

NEC 430.83(C)(1)

Figure 19-8. For stationary motors rated 2 HP or less, a general-use snap switch may be used, if sized not less than twice the motor's full-load current in amps.

TORQUE MOTORS
430.83(D)

The motor controller for a torque motor must have a continuous duty, full-load current rating not less than the nameplate current rating of the motor. **(See Figure 19-9)**

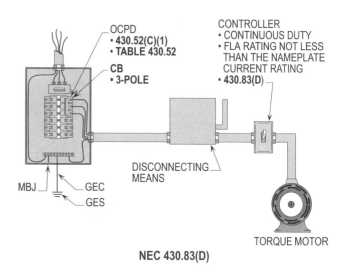

NEC 430.83(D)

Figure 19-9. The controller for a torque motor must be capable of holding the amps indefinitely.

Motor Tip: If the motor controller is rated in horsepower and not marked or rated as above, use **Tables 430.247** through **430.250** to determine the amperage or horsepower rating.

SIZING DISCONNECTING MEANS 430.109(A)(1) THRU (A)(7); 110(A)

Either the disconnecting means for motor circuits must have an ampere rating of at least 115 percent of the full-load current rating of the motor per **430.110(A)** or it must be horsepower rated and capable of deenergizing locked-rotor currents per **Tables 430.251(A) and (B)** of the NEC. For detailed rules, see **430.109**. **(See Figure 19-10)**

OTHER THAN HP RATED 430.109(A) THRU (G)

The following permit other than a horsepower rated disconnecting means to be used to deenergize the power circuit to certain types of motors:

(1) Motor Circuit Switch. A listed motor-circuit switch rated in horsepower.

(2) Molded Case Circuit Breaker. A listed molded case circuit breaker.

(3) Molded Case Switch. A listed molded cash switch.

(4) Instantaneous Trip Circuit Breaker. An instantaneous trip circuit breaker that is part of a listed combination motor controller.

(5) Self-Protected Combination Controller. Listed self-protected combination controller.

(6) Manual Motor Controller. Listed manual motor controllers additionally marked "Suitable as Motor Disconnect" must be permitted as a disconnecting means where installed between the final motor branch-circuit short-circuit protective device and the motor. Listed manual motor controllers additionally marked "Suitable as Motor Disconnect" must be permitted as disconnecting means on the line side of the fuses permitted in **430.52(C)(5)**. In this case, the fuses permitted in **430.52(C)(5)** must be considered supplementary fuses, and suitable branch-circuit short-circuit and ground-fault protective devices must be installed on the line side of the manual motor controller additionally marked "Suitable as Motor Disconnect".

(7) System Isolation Equipment. System isolation equipment must be listed for disconnection purposes. System isolation equipment shall be installed on the load side of the overcurrent protection and its disconnecting means. The disconnecting means must be one of the types permitted by **430.119(A)(1)** through **(A)(3)**.

STATIONARY MOTORS 430.109(B)

For a stationary motor rated 1/8 horsepower or less, the branch-circuit protective device may serve as the disconnecting means. This rule is allowed because the windings of such motors do not produce locked-rotor currents

that are high enough to create damage to such motors, circuit conductors, or components. **(See Figure 19-12)**

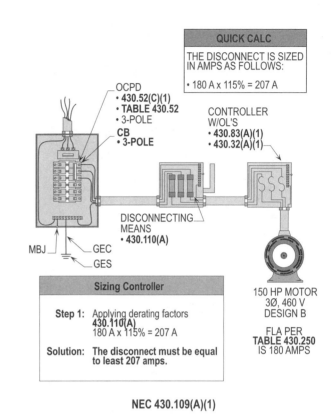

NEC 430.109(A)(1)

Figure 19-10. The disconnecting means must be sized in amps by multiplying the motor's FLA by 115%.

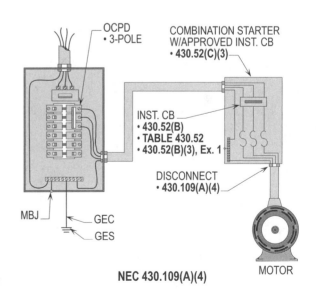

NEC 430.109(A)(4)

Figure 19-11. The disconnecting means for a motor may be an approved instantaneous trip circuit breaker.

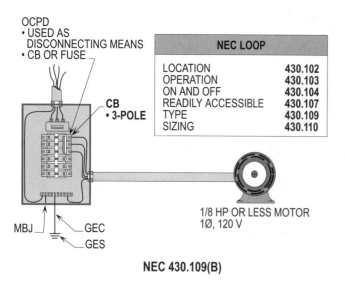

NEC LOOP	
LOCATION	430.102
OPERATION	430.103
ON AND OFF	430.104
READILY ACCESSIBLE	430.107
TYPE	430.109
SIZING	430.110

NEC 430.109(B)

Figure 19-12. Motors rated 1/8 HP or less may be disconnected by the OCPD located in the panelboard supplying power to the circuit.

STATIONARY MOTORS
430.109(C)(1); (2)

For a stationary motor rated 2 horsepower or less, the controller may be a general-use switch rated for at least twice the motor's full-load current. An AC general-use snap switch may be installed as the controller where the motor's full-load current rating does not exceed 80 percent of the switch. **(See Figure 19-13)**

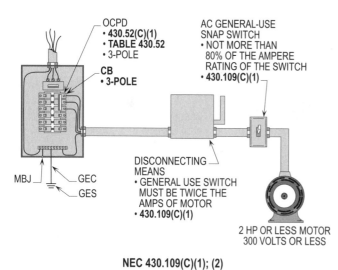

NEC 430.109(C)(1); (2)

Figure 19-13. The above illustration lists the rules pertaining to the disconnecting means for motors rated 2 HP or less.

AUTOTRANSFORMER - TYPE CONTROLLED CONTROLLER
430.109(D)

Motors rated over 2 to 100 horsepower may be installed with a separate disconnecting means (general-use switch), if the motor is equipped with an autotransformer-type controller and all of the following conditions are complied with:

(1) The motor drives a generator that is provided with overload protection.

(2) The controller is capable of interrupting the locked-rotor current of the motor.

(3) The controller is provided with a no-voltage release.

(4) The controller is provided with running overload protection not exceeding 125 percent of the motor's full-load current rating in amps.

(5) Separate fuses or an inverse time circuit breaker is sized at 150 percent or more of the motor's full-load current rating in amps. **(See Figure 19-14)**

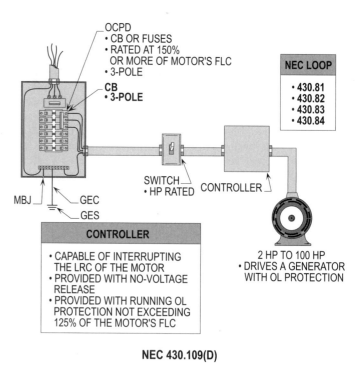

NEC 430.109(D)

Figure 19-14. The above illustration lists the rules for a disconnecting means and controller used to disconnect and control motors rated 2 to 100 HP respectively.

ISOLATING SWITCHES
430.109(E)

The disconnecting means may be a general-use or isolating switch for DC stationary motors rated at 40 horsepower or greater and AC motors rated 100 horsepower or greater. However, such disconnects must be plainly marked "Do not operate under load." **(See Figure 19-15)**

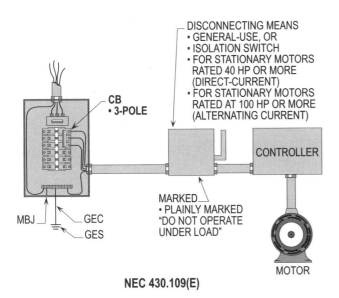

NEC 430.109(E)

Figure 19-15. The above illustration lists the rules for a disconnecting means used to disconnect motors rated at 40 HP or more.

CORD-AND-PLUG CONNECTED MOTORS
430.109(F)

For a cord-and-plug-connected motor, a horsepower-rated attachment plug-and receptacle having ratings no less than the motor ratings shall be permitted to serve as the disconnecting means. A horsepower-rated attachment plug and receptacle shall not be required for a cord-and-plug-connected appliance in accordance with **422.32**, a room air conditioner in accordance with **440.36**, or a portable motor rated 1/3 hp or less.

The OCPD, ahead of the supplying branch-circuit, may be utilized for such purposes. For further information, see **422.33, 422.31(B),** and **440.63**. **(See Figure 19-16)**

TORQUE MOTORS
430.109(G)

The disconnecting means for a torque motor may be installed as a general-use switch. Such switch must be capable of handling the locked-rotor current of the motor indefinitely. **(See Figure 19-17)**

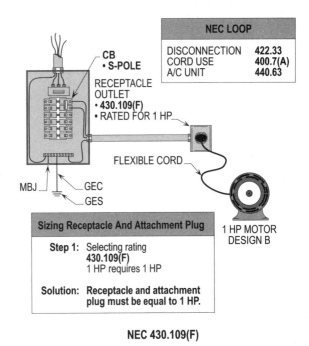

NEC 430.109(F)

Figure 19-16. The HP rating of a receptacle and attachment cap used as a disconnecting means for a motors must be sized to equal the HP of the motor.

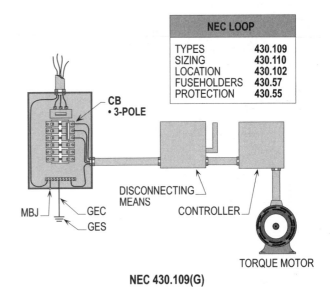

NEC 430.109(G)

Figure 19-17. The disconnecting means for a torque motor may be a general-use switch.

Chapter 19: Overload Protection For Individual Motors

Section   Answer

_____   T   F   **1.** Motors installed below 3300 ft. must be derated due to the altitude.

_____   T   F   **2.** Motors with a temperature rise not over 40°C must have their (minimum) running overload protection (per **430.32(A)(1)**) sized at 115 percent.

_____   T   F   **3.** Motors with a service factor not less than 1.15 must have their (maximum) running overload protection (per **430.32(C)**) sized at 140 percent.

_____   T   F   **4.** Single-phasing occurs when one ungrounded (phase) conductor is lost in a three-phase system.

_____   T   F   **5.** Overload protection is permitted to be shunted for a nonautomatically started motor.

_____   T   F   **6.** Motor controllers are not required to be installed with a horsepower rating equal to the motor that is to be controlled. (General rule)

_____   T   F   **7.** A molded case switch, when properly sized in amps can serve as a controller for a motor.

_____   T   F   **8.** An AC general-use snap switch may be installed as the controller where the full-load current rating (in amps) of the motor does not exceed 80 percent of the switch rating.

_____   T   F   **9.** The disconnecting means for motor circuits, rated at 600 volts or less, must have an ampere rating of at least 125 percent of the full-load current rating (in amps) of the motor.

_____   T   F   **10.** The disconnecting means for a torque motor is permitted per NEC to be a general-use switch.

_____ _____   **11.** Motors with a service factor not less than 1.15 must have the (minimum) running overload protection (per **430.32(A)(1)**) sized at _____ percent.

_____ _____   **12.** Motors with a temperature rise not over 40°C must have the (maximum) running overload protection (per **430.32(C)**) size at _____ percent.

_____ _____   **13.** Overloads used in a magnetic starter must protect the _____ windings.

_____ _____   **14.** The branch-circuit protective device is permitted to serve as the controller where the motor is rated _____ horsepower or less for stationary motors.

_____ _____   **15.** Controllers (general rule) must have a HP rating not _____ than HP rating of the motor.

16. Inverse time circuit breakers used as controllers must be sized at _____ percent or less of the motor's nameplate current rating (in amps) when providing overload protection. (General rule)

17. Motors marked with a service factor of 1.15 or greater may be sized at _____ percent of the motor's FLC in amps, when sizing minimum overload relay protection.

18. Motors rated over _____ horsepower through _____ horsepower are permitted to be installed with a separate disconnecting means but certain conditions must be applied. (autotransformer type controller)

19. The disconnecting means is permitted to be a general-use or isolating switch for DC stationary motors rated at _____ horsepower or greater.

20. A HP rated attachment plug and receptacle with a rating not less than the motor _____ can serve as the disconnecting means.

21. Motors with a service factor other than 1.15 must have the (minimum) running overload protection sized at _____ percent of the FLC (in amps) listed on the nameplate.

      (a) 115   **(c)** 130
      (b) 125   **(d)** 140

22. Motors with a temperature rise over 40°C must have the (maximum) running overload protection sized at _____ percent of the FLC (in amps) listed on the nameplate.

      (a) 115   **(c)** 130
      (b) 125   **(d)** 140

23. The controller is permitted to be an attachment plug and receptacle which is acceptable for use with portable motors rated _____ horsepower or less.

      (a) 1/8   **(c)** 1/2
      (b) 1/3   **(d)** 3/4

24. For a stationary motor rated _____ horsepower or less, the controller is permitted to be a general-use switch rated at least twice the motor's full-load current in amps.

      (a) 1   **(c)** 3
      (b) 2   **(d)** 5

22. System isolation equipment must be _____ for disconnection purposes per NEC.

      (a) approved   **(c)** listed
      (b) identified   **(d)** none of the above

26. What size overload protection (minimum) is required for 20 HP, 460 volt, three-phase, Design B motor with a nameplate rating of 48 amps, temperature rise of 40°C, and a service factor of 1.15?

27. What size overload protection (maximum) is required for 20 HP, 460 volt, three-phase, Design B motor with a nameplate rating of 48 amps, temperature rise of 40°C, and a service factor of 1.15?

28. What is the horsepower rating of the disconnecting means (motor rated switch)   _____ _____
for a 50 HP, 460 volt, three-phase, Design (B) motor?

29. What is the horsepower rating of the controller for a 25 HP, 460 volt, three-   _____ _____
phase, Design B motor?

30. What size horsepower rated receptacle and attachment plug is required for a   _____ _____
20 HP, Design B motor. (Note that installation is not designed to unplug under
load.)

Motor Feeder and Branch-circuit Conductors

Branch-circuit conductors are sized by the percentages that are based on the use of loads that they supply. These loads are rated either continuous or noncontinuous. When sizing branch-circuit conductors, the continuous loads are calculated at 125 percent and the noncontinuous loads are at 100 percent. The duty cycle operation of the driven load may also be used to size conductors.

Capacitors are installed when the power factor is low and the currents are high, thus correcting the power factor and reducing the currents.

SIZING CONDUCTORS FOR SINGLE MOTORS 430.22(A)

Branch-circuit conductors supplying a single-motor must have an ampacity not less than 125 percent of the motor's full-load current rating per T**ables 430.247, 430.248, 430.249**, and **430.250** respectively. For example, a 20 HP, 208 volt, three-phase motor per **Table 430.250** has a full-load current of 59.4 amps. The full-load amps (FLA) for sizing the conductors is determined by multiplying 59.4 A by 125% which equals 74.25 amps.

A motor has a starting current (LRC) of four to six times the full-load current of the motor's FLA for motors marked with code letters A through G and 8 1/2 to 15 times for NEMA, Design B, (high-efficiency motors). Design B, C, and D motors have a starting current of about 4 to 6 times the full-load amps when starting and driving a motor load. Conductors calculated per **430.22(A) and (E)** are adequately sized and normally will not be affected by short starting currents.

Heating effects occur in the conductors when the motor windings develope an overload condition. To reduce this heating effect, the conductor's current-carrying capacity is increased by taking 125 percent of the motor's full-load current rating in amps. For example, a motor with a FLC rating of 42 amps must have conductors with a current-carrying capacity of at least 52.5 amps (42 A x 125% = 52.5 A) to safely carry the load when starting and to also protect the conductor's insulation when overloading occurs.

SIZING CONDUCTORS FOR SINGLE-PHASE MOTORS
430.22(A)

Section **430.6(A)(1)** of the NEC requires the full-load current in amps for single-phase motors to be obtained from **Table 430.248**. This FLC rating in amps is then multiplied by 125 percent per **430.22(A)** to derive the total amps for selecting the conductors from **Table 310.16** to supply power to the motor windings. **(See Figure 20-1)**

SIZING CONDUCTORS FOR THREE-PHASE MOTORS
430.22(A)

Section **430.6(A)(1)** requires the full-load current in amps for three-phase motors to be obtained from **Table 430.250**. This FLC rating in amps is multiplied by 125 percent per **430.22(A)** to derive the total amps to select the conductors for the motor windings. **(See Figure 20-2)**

SIZING CONDUCTORS FOR MULTISPEED MOTORS
430.22(B)

The circuit conductors for multispeed motors must be sized large enough (to the controller) to supply the highest nameplate full-load current rating of the multispeed motor windings involved. A single OCPD is permitted to serve each speed of a multispeed motor per **430.22(B)**. The speed with the greater amps is used to size the OCPD and conductors. Overload protection must be provided for each speed in order to protect each winding from excessive current during an overload condition. **(See Figure 20-3)**

SIZING CONDUCTORS FOR WYE START AND DELTA RUN MOTORS
430.22(C)

The selection of branch-circuit conductors for wye-start and delta run connected motors must be based on the full-load current in amps on the line side of the controller. For continuous use, the selection of conductors between the controller and the motor must be based on 58 percent (1 ÷ 1.732 = .58) of the motor's full-load current times 125 percent. **(See Figure 20-4)**

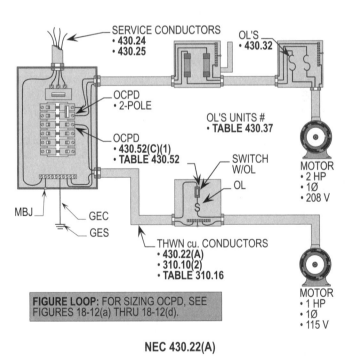

Finding THWN Copper Conductors

Single-phase motors (208 V)

Step 1: Finding FLA
430.6(A)(1); Table 430.248
2 HP = 13.2 A

Step 2: Calculating load
430.22(A)
13.2 A x 125% = 16.5 A

Step 3: Selecting conductors
310.10(2); Table 310.16
16.5 A requires 14 AWG cu.

Solution: **The THWN copper conductors are size 14 AWG.**

Single-phase motors (115 V)

Step 1: Finding FLA
430.6(A)(1); Table 430.248
1 HP = 16 A

Step 2: Calculating load
430.22(A)
16 A x 125% = 20 A

Step 3: Selecting conductors
310.10(2); Table 310.16
20 A requires 14 AWG cu.

Solution: **The THWN copper conductors are size 14 AWG.**

Note: See asterisk to **Table 310.16, Tables 220.3 and 240.4(G)**.

NEC 430.22(A)

Figure 20-1. Determining the size branch-circuit conductors to supply single-phase motors.

Finding THWN Copper Conductors

Three-phase motors (460 V)

Step 1: Finding FLA
430.6(A)(1); Table 430.250
40 HP = 52 A

Step 2: Calculating load
430.22(A)
52 A x 125% = 65 A

Step 3: Selecting conductors
310.10(2); Table 310.16
65 A requires 6 AWG cu.

Solution: **The THWN copper conductors are size 6 AWG.**

Three-phase motors (208 V)

Step 1: Finding FLA
430.6(A)(1); Table 430.250
25 HP = 74.8 A

Step 2: Calculating load
430.22(A)
74.8 A x 125% = 93.5 A

Step 3: Selecting conductors
310.10(2); Table 310.16
93.5 A requires 3 AWG cu.

Solution: **The THWN copper conductors are size 3 AWG.**

Finding THWN Copper Conductors

Sizing branch-circuit conductors

Step 1: Finding FLA
430.22(B)
45 A largest amperage

Step 2: Calculating load
430.22(B)
45 A x 125% = 56.25

Step 3: Selecting conductors
310.10(2); Table 310.16
56.25 A requires 6 AWG cu.

Solution: **The size THWN copper conductors are 6 AWG.**

Sizing multispeed conductors

Step 1: Finding FLA
430.22(B)
3420 RPM = 27 A
2280 RPM = 35 A
1710 RPM = 45 A

Step 2: Calculating load
430.22(B)
27 A x 125% = 33.75 A
35 A x 125% = 43.75 A
45 A x 125% = 56.25 A

Step 3: Selecting conductors
310.10(2); Table 310.16
33.75 A requires 10 AWG cu.
43.75 A requires 8 AWG cu.
56.25 A requires 6 AWG cu.

Solution: **The size THWN copper conductors are 10 AWG cu., 8 AWG cu., and 6 AWG cu. sized for each speed.**

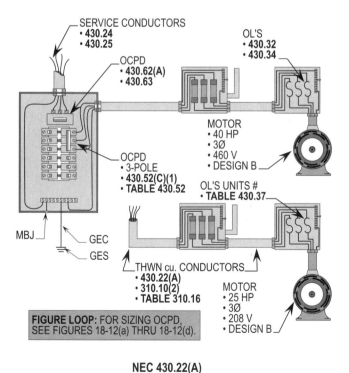

NEC 430.22(A)

Figure 20-2. Determining the size branch-circuit conductors to supply three-phase motors.

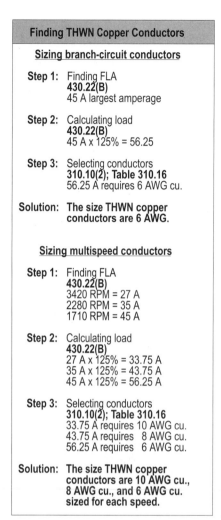

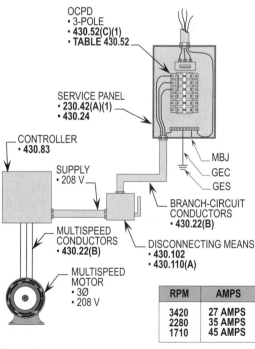

RPM	AMPS
3420	27 AMPS
2280	35 AMPS
1710	45 AMPS

NEC 430.22(B)

Figure 20-3. Determining the size branch-circuit conductors to supply multispeed motors.

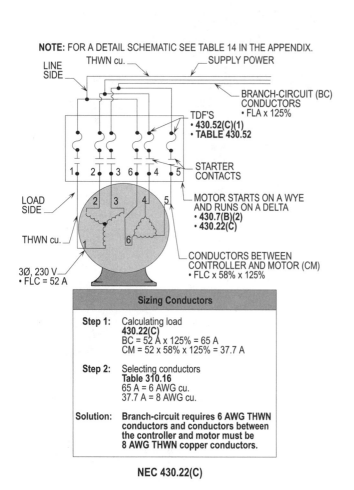

NOTE: FOR A DETAIL SCHEMATIC SEE TABLE 14 IN THE APPENDIX.

LINE SIDE
THWN cu.
SUPPLY POWER
BRANCH-CIRCUIT (BC) CONDUCTORS
• FLA x 125%
TDF'S
• 430.52(C)(1)
• TABLE 430.52
STARTER CONTACTS
LOAD SIDE
MOTOR STARTS ON A WYE AND RUNS ON A DELTA
• 430.7(B)(2)
• 430.22(C)
THWN cu.
CONDUCTORS BETWEEN CONTROLLER AND MOTOR (CM)
• FLC x 58% x 125%
3Ø, 230 V
• FLC = 52 A

Sizing Conductors

Step 1:	Calculating load **430.22(C)** BC = 52 A x 125% = 65 A CM = 52 x 58% x 125% = 37.7 A
Step 2:	Selecting conductors **Table 310.16** 65 A = 6 AWG cu. 37.7 A = 8 AWG cu.
Solution:	**Branch-circuit requires 6 AWG THWN conductors and conductors between the controller and motor must be 8 AWG THWN copper conductors.**

NEC 430.22(C)

Figure 20-4. Determining the size conductors to supply motors starting on a wye and running on a delta. **(See Table 14 in the Appendix)**

SIZING CONDUCTORS FOR DUTY CYCLE MOTORS
430.22(E)

Conductors for a motor used for short-time, intermittent, periodic, or varying duty do not require sizing with a current-carrying capacity of 125 percent of the motor's full-load current in amps. **Table 430.22(E)** permits the conductors to be sized with a percentage times the nameplate current rating based upon the duty cycle classification of the motor.

When sizing conductors to supply individual motors that are used for short-time, intermittent, periodic, or varying duty, the requirements of **Table 430.22(E)** must apply. Varying heat loads are produced on the conductors by the starting and stopping duration of operation cycles which permits conductor sizing changes. In other words, such conductors are never subjected to continuous operation due to on and off periods and therefore conductors are never fully loaded for long intervals of time. For this reason, conductors may

be down sized. **(See Figure 20-5)**

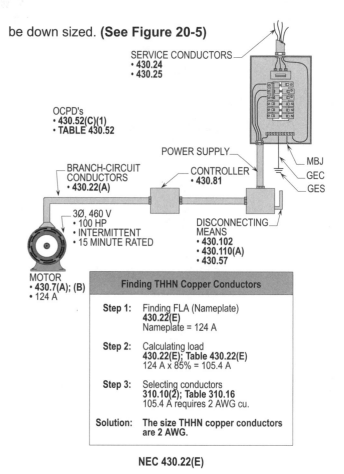

SERVICE CONDUCTORS
• 430.24
• 430.25
OCPD's
• 430.52(C)(1)
• TABLE 430.52
POWER SUPPLY
MBJ
GEC
GES
BRANCH-CIRCUIT CONDUCTORS
• 430.22(A)
CONTROLLER
• 430.81
3Ø, 460 V
• 100 HP
• INTERMITTENT
• 15 MINUTE RATED
DISCONNECTING MEANS
• 430.102
• 430.110(A)
• 430.57
MOTOR
• 430.7(A); (B)
• 124 A

Finding THHN Copper Conductors

Step 1:	Finding FLA (Nameplate) **430.22(E)** Nameplate = 124 A
Step 2:	Calculating load **430.22(E); Table 430.22(E)** 124 A x 85% = 105.4 A
Step 3:	Selecting conductors **310.10(2); Table 310.16** 105.4 A requires 2 AWG cu.
Solution:	**The size THHN copper conductors are 2 AWG.**

NEC 430.22(E)

Figure 20-5. Determining the size conductors to supply duty cycle operated motors.

SIZING CONDUCTORS FOR ADJUSTABLE SPEED DRIVE SYSTEMS
430.122(A); 430.2

Power conversion equipment, when supplied from a branch-circuit, includes all components of the adjustable speed drive system. The rating in amps is used to size the conductors that are based upon the power required by the conversion equipment. When the power conversion equipment provides overcurrent protection for the motor, no additional overload protection is required.

The disconnecting means may be installed in the line supplying the conversion equipment and the rating of the disconnect must not be less than 115 percent of the input current rating in amps of the conversion unit.

Power conversion equipment requires the conductors to be sized at 125 percent of the rated input of such equipment or by the nameplate information per **110.3(B)**.

Motor Tip: Power conversion equipment contains solid state electronics that change the frequency or chops part of the wave form to vary the speed of squirrel-cage motors as needed for the application **(See Figure 20-6)**. Also, see the nameplate on the drive.

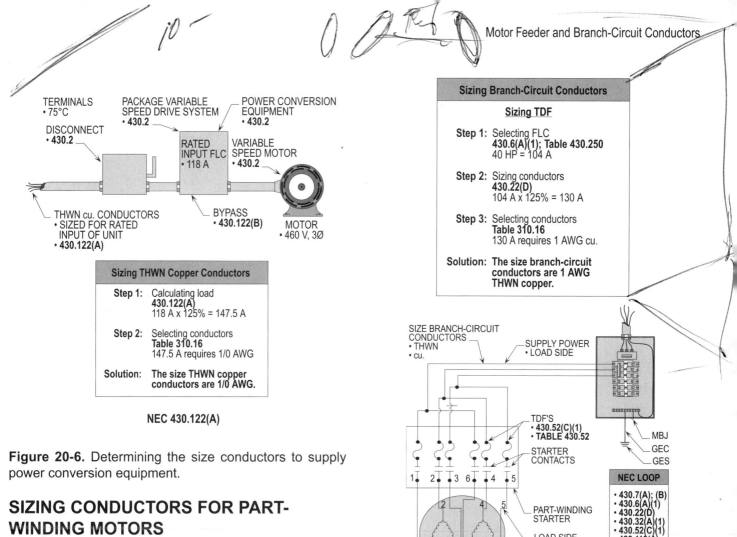

Sizing THWN Copper Conductors

Step 1: Calculating load
430.122(A)
118 A x 125% = 147.5 A

Step 2: Selecting conductors
Table 310.16
147.5 A requires 1/0 AWG

Solution: **The size THWN copper
conductors are 1/0 AWG.**

NEC 430.122(A)

Figure 20-6. Determining the size conductors to supply power conversion equipment.

SIZING CONDUCTORS FOR PART-WINDING MOTORS
430.22(D); 430.4; APPENDIX, TABLE 14

Induction or synchronous motors that have a part-winding start are so designed that when starting, they energize the primary armature winding first. After starting, the remainder of the winding is energized in one or more steps. The purpose of this arrangement is to reduce the initial inrush current until the motor accelerates its speed.

The inrush current, when starting, is the locked-rotor current and this current can be quite high. A standard part-winding-start induction motor is designed so that only half of its winding is energized at the start; then, as it comes up to speed, the other half is energized, so both halves are energized and carry equal current.

Separate overload devices must be used on a standard part-winding-start induction motor to protect the windings from excessive damaging currents. This means that each half of the motor winding must be individually provided with overload protection. These overload requirements are covered in **430.32** and **430.37**. Each half of the windings has a trip current value that is one-half of the specified running current. As required by **430.52**, each of the two motor windings must have branch-circuit, short-circuit, and ground-fault protection that is designed to be not more than one-half the percentages listed in **430.52** and **Table 430.52**. **(See Figure 20-7)**

Sizing Branch-Circuit Conductors

Sizing TDF

Step 1: Selecting FLC
430.6(A)(1); Table 430.250
40 HP = 104 A

Step 2: Sizing conductors
430.22(D)
104 A x 125% = 130 A

Step 3: Selecting conductors
Table 310.16
130 A requires 1 AWG cu.

Solution: **The size branch-circuit
conductors are 1 AWG
THWN copper.**

NEC 430.22(D); APPENDIX, TABLE 14
NEC 430.4

Figure 20-7. Determining the size branch-circuit conductors required to supply part-winding motors. **(See Table 14)**

Motor Tip: Section **430.4, Ex.** allows a single device, with this one-half rating, for both windings, provided that it will permit the motor to start and run. If a time-delay (dual element) fuse is used as a single device for both windings, its rating is permitted, if it does not exceed 150 percent of the motor's full-load current.

WOUND-ROTOR SECONDARY
430.23

Wound-rotor motors are three-phase motors that are installed with two sets of leads. The main leads to the motor windings (field poles) is one set and the secondary leads to the rotor is the other set. The secondary leads connect to the rotor through slip rings, which then connect through a controller to a bank of resistors. The speed of the motor varies when the amount of resistance in the motor's circuit is varied. The rotor will turn slower when the resistance is greater and faster when such resistance is lowered.

SIZING CONDUCTORS FOR CONTINUOUS DUTY
430.23(A)

The conductors must have an ampacity of not less than 125 percent of the full-load secondary current of the motor, where secondary leads are installed between the controller and the rotor. The secondary full-load current rating in amps is obtained from the manufacturer or found on the nameplate of the motor.

SIZING CONDUCTORS FOR OTHER THAN CONTINUOUS DUTY
430.23(B)

When installing a motor to be used for a short-time, intermittent, periodic, or varying duty, the secondary conductors must be sized not less than the full load secondary current in amps per **Table 430.23(B)**. The classification of service determines the correct percentages to select and apply when sizing the conductors, which are based upon the cycles of the motor.

SIZING CONDUCTORS FOR RESISTORS, SEPARATED FROM CONTROLLER
430.23(C)

Where the secondary resistor is separate from the controller, the ampacity of the conductors between the controller and resistor bank must not be less than the resistor's duty classification percentages listed in **Table 430.23(C)**. **(See Figure 20-8)**

DC MOTORS
430.29

Conductors connecting a motor controller to power accelerating and dynamic braking resistors in the armature circuit of DC motors must be sized by the percentages listed in **Table 430.29**. **(See Figure 20-9)**

The conductors supplying power to a DC motor must be sized at 125 percent of the FLC in amps of the motor. OCPD's must be sized to carry the starting current of the motor. **(See Figure 20-10)**

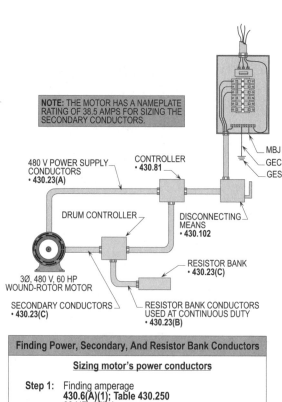

NOTE: THE MOTOR HAS A NAMEPLATE RATING OF 38.5 AMPS FOR SIZING THE SECONDARY CONDUCTORS.

480 V POWER SUPPLY CONDUCTORS • 430.23(A)

CONTROLLER • 430.81

MBJ
GEC
GES

DRUM CONTROLLER

DISCONNECTING MEANS • 430.102

3Ø, 480 V, 60 HP WOUND-ROTOR MOTOR

RESISTOR BANK • 430.23(C)

SECONDARY CONDUCTORS • 430.23(C)

RESISTOR BANK CONDUCTORS USED AT CONTINUOUS DUTY • 430.23(B)

Finding Power, Secondary, And Resistor Bank Conductors

Sizing motor's power conductors

Step 1: Finding amperage
430.6(A)(1); Table 430.250
60 HP = 77 A

Step 2: Calculating amperage
430.23(A); 430.22(A)
77 A x 125% = 96.25 A

Step 3: Selecting conductors
240.4(G); Table 310.16
96.25 A = 3 AWG THWN cu.

Solution: **The size THWN copper conductors are required to be 3 AWG.**

Sizing motors secondary conductors

Step 1: Finding amperage
430.6(A)(1); Table 430.250
FLC = 38.5 A

Step 2: Calculating amperage
430.23(A)
38.5 A x 125% = 48.13 A

Step 3: Selecting conductors
310.10(2); Table 310.16
48.13 A = 8 AWG THWN cu.

Solution: **The size THWN copper conductors are required to be 8 AWG.**

Sizing resistor bank conductors

Step 1: Finding amperage
430.6(A)(1); Table 430.250
FLC = 38.5 A

Step 2: Calculating amperage
430.23(C); Table 430.23(C)
38.5 A x 110% = 42.35 A

Step 3: Selecting conductors
310.10(2); Table 310.16
42.35 A = 8 AWG THWN cu.

Solution: **The size THWN copper conductors are required to be 8 AWG.**

NEC 430.23(A); (B); (C)

Figure 20-8. Determining the size conductors to supply wound-rotor motors.

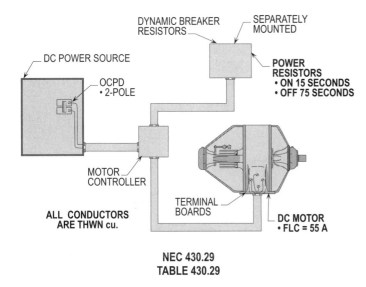

Sizing Conductors For A Dynamic Resistor Bank Mounted Separately From The Controller And DC Motor

Step 1: Calculating load of power resistor bank
430.29; Table 430.29
55 A x 55% = 30.25 A

Step 2: Selecting conductors
430.29; Table 310.16
30.25 A requires 10 AWG cu.

Solution: The size conductors are 10 AWG THWN copper.

NEC 430.29
TABLE 430.29

Figure 20-9. Sizing conductors for a dynamic resistor bank mounted separately from the controller and DC motor.

Sizing OCPD And Conductors For DC Motors

Step 1: Calculating A to size OCPD
430.52(C)(1)
LCR based on arm. (resistance)
= 240 V ÷ .5 (1/2 ohm)
LCR based on arm. = 480 A

Step 2: Calculating OCPD
Table 430.247; 430.52(C)(1); Table 430.52
55 A x 150% = 82.5 A

Step 3: Selecting OCPD
240.4(G); 240.6(A); 430.52(C)(1), Ex. 1
82.5 A requires 90 A

Step 4: Verifying starting of motor
• CB must hold 480 A starting current of motor ÷ 3 (CB holds about 3 times its rating) = 160 A
• 160 A requires 175 A CB
• 175 A CB holds 525 A (175 A x 3 = 525 A)
• 525 A will hold 480 A of LRC

Step 5: Applying max. OCPD
430.52(C)(1), Ex. 2(c)
Max. = 55 A x 400% = 220 A

Solution: A 200 amp OCPD may be used. However, normally a 175 amp OCPD will allow the motor to start and run.

Sizing conductors

Step 1: Finding FLA
430.6(A)(1); Table 430.247
15 HP = 55 A

Step 2: Calculating load
430.22(A)
55 A x 125% = 68.75 A

Step 3: Selecting conductors
310.10(2); Table 310.16
68.75 A requires 4 AWG cu.

Solution: The size THWN copper conductors are 4 AWG.

SIZING CONDUCTORS FOR TWO OR MORE MOTORS
430.24

The full-load current rating of the largest motor must be multiplied by 125 percent when selecting the size conductors for a feeder supplying a group of two or more motors. The remaining motors of the group must have their full-load current ratings added to this value and this total amperage is then used to size the conductors. **(See Figure 20-11)**

SEVERAL MOTORS ON A FEEDER
430.24

Feeder-circuit conductors supplying several motors must be sized to carry 125 percent of the FLC rating of the highest rated current motor plus the sum of the FLC ratings in amps of all remaining motors on the circuit. **(See Figure 20-12)**

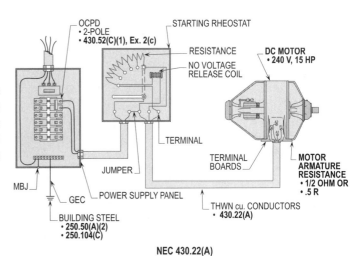

NEC 430.22(A)
TABLE 430.247

Figure 20-10. Sizing OCPD and conductors for a DC motor.

MOTOR IN A HEATING UNIT
430.24, Ex. 2

Section 424.3(B) must be used to calculate the size conductors to supply power to fixed electric space heating units that are equipped with motor operated equipment. **(See Figure 20-14)**

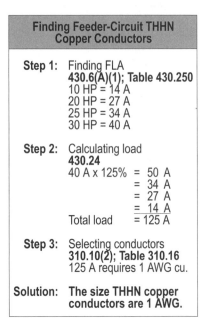

Finding THWN Copper Conductors	
Step 1:	Finding amperage **430.6(A)(1); Table 430.250** 40 HP = 52 A 50 HP = 65 A 60 HP = 77 A

Step 2:	Calculating amperage
	430.24
	77 A x 125% = 96.25 A
	           65 A
	           52 A
	Total load = 213.25 A

Step 3:	Selecting conductors **310.10(2); Table 310.16** 213.25 A requires 4/0 AWG THWN cu.
Solution:	**The size THWN copper conductors are required to be 4/0 AWG.**

Finding Feeder-Circuit THHN Copper Conductors			
Step 1:	Finding FLA **430.6(A)(1); Table 430.250** 10 HP = 14 A 20 HP = 27 A 25 HP = 34 A 30 HP = 40 A		
Step 2:	Calculating load **430.24** 40 A x 125%	=	50 A = 34 A = 27 A = 14 A
	Total load	=	125 A
Step 3:	Selecting conductors **310.10(2); Table 310.16** 125 A requires 1 AWG cu.		
Solution:	**The size THHN copper conductors are 1 AWG.**		

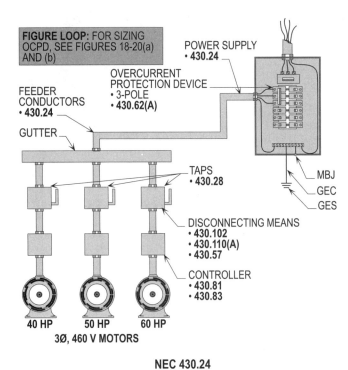

Figure 20-11. Determining the size conductors for a feeder-circuit supplying several motors.

DETERMINING LARGEST MOTOR BASED ON THE DUTY CYCLE
430.24, Ex. 1

Feeder-circuit conductors supplying power to two or more motors that are utilized to serve duty-cycle loads must have the largest motor selected based upon their conditions of use. **(See Figure 20-13)**

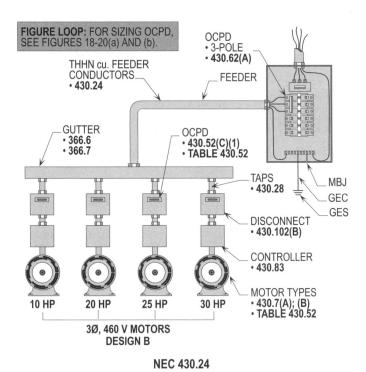

Figure 20-12. Sizing the feeder-circuit conductors to supply power to a group of motors.

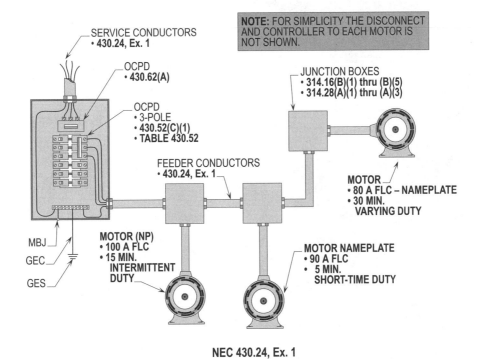

Finding Largest Load And THWN Copper Conductors

Finding largest motor load

Step 1: Applying demand factors
430.24, Ex. 1; 430.22(E); Table 430.22(E)
100 A x 85%   = 85 A
90 A x 110 %   = 99 A
80 A x 150%   = 120 A

Solution: **The largest motor load is 120 amps.**

Sizing THWN cu. conductors

Step 1: Calculating load
430.24, Ex. 1
Largest motor load   = 120 A
Plus others   = 99 A
                     = 85 A
Total load   = 304 A

Step 2: Selecting conductors
310.10(2); Table 310.16
304 A requires 350 KCMIL cu.

Solution: **The size THWN copper conductors
are 350 KCMIL.**

SERVICE CONDUCTORS
• 430.24, Ex. 1

OCPD
• 430.62(A)

OCPD
• 3-POLE
• **430.52(C)(1)**
• **TABLE 430.52**

FEEDER CONDUCTORS
• 430.24, Ex. 1

NOTE: FOR SIMPLICITY THE DISCONNECT
AND CONTROLLER TO EACH MOTOR IS
NOT SHOWN.

JUNCTION BOXES
• **314.16(B)(1) thru (B)(5)**
• **314.28(A)(1) thru (A)(3)**

MOTOR
• 80 A FLC – NAMEPLATE
• 30 MIN.
  VARYING DUTY

MBJ
GEC
GES

MOTOR (NP)
• 100 A FLC
• 15 MIN.
  INTERMITTENT
  DUTY

MOTOR NAMEPLATE
• 90 A FLC
• 5 MIN.
  SHORT-TIME DUTY

NEC 430.24, Ex. 1

Figure 20-13. Sizing feeder-circuit conductors for the largest motor load based on the duty cycle.

MOTORS WITH INTERLOCKS
430.24, Ex. 3

Motors that operate with other loads and are interlocked, so as not to operate at the same time, may have the feeder-circuit conductors based upon the interlocked group producing the greater FLA rating. **(See Figure 20-15)**

Finding THWN Copper Conductors To Supply Power To The Heating Unit

Sizing THWN cu. conductors

Step 1: Finding FLA
430.24, Ex. 2; 424.3(B)
Heating unit
20,000 VA ÷240 V = 83 A
Blower motor = 3 A
Total load = 86 A

Step 2: Calculating load
424.3(B)
86 A x 125% = 107.5 A

Step 3: Selecting THWN cu. conductors
310.10(2); Table 310.16
107.5 A requires 2 AWG cu.

Solution: The size THWN copper conductors are 2 AWG.

Sizing Conductors For The Motors And Other Loads

Finding largest load

Step 1: Finding load for motors 1 and 2
430.24, Ex. 3; 215.2(A)(1)
Motor 1 = 24.2 A x 125% = 30.25 A
Motor 2 = 16.7 A x 100% = 16.7 A
Other loads = 40 A x 125% = 50 A
Total load = 96.95 A

Step 2: Finding load for motors 3 and 4
430.24, Ex. 3; 215.2(A)(1)
Motor 3 = 46.2 A x 125% = 57.75 A
Motor 4 = 30.8 A x 100% = 30.8 A
Other loads = 40 A x 125% = 50 A
Total load = 138.55 A

Step 3: Selecting largest load
430.24, Ex. 3
138.55 A is greater than 96.95 A

Solution: The greater interlocked load is 138.55 amps.

Sizing conductors

Step 1: Selecting conductors
310.10(2); Table 310.16; Step 2 above
138.55 A requires 1/0 AWG cu.

Solution: The size THWN copper conductors are 1/0 AWG.

NOTE 1: BLOWER MOTOR IS 3 AMPS.

NOTE 2: SEE **424.22(B)** FOR RULES CONCERNING SUB DIVIDING THE HEATING ELEMENTS.

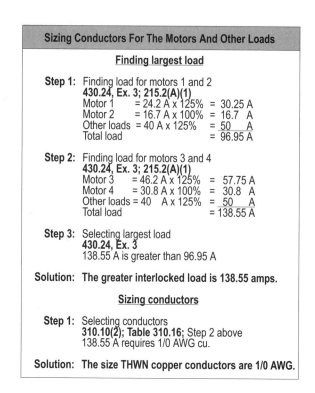

NEC 430.24, Ex. 2

Figure 20-14. Sizing conductors to supply heating unit.

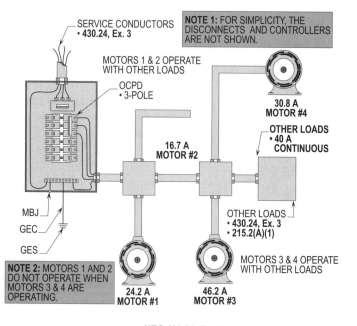

NOTE 1: FOR SIMPLICITY, THE DISCONNECTS AND CONTROLLERS ARE NOT SHOWN.

NOTE 2: MOTORS 1 AND 2 DO NOT OPERATE WHEN MOTORS 3 & 4 ARE OPERATING.

NEC 430.24, Ex. 3

Figure 20-15. Sizing conductors to supply power to the motors and other loads where they do not operate together.

SIZING CONDUCTORS FOR MOTORS AND OTHER LOADS
430.25

The motor load must be calculated per **430.22(A)** or **430.24**, when designing for combination loads that consist of one or more motor loads on the same circuit with lights, receptacles, appliances, or any combination of such loads. For other than motor loads, **Article 220** and other applicable articles of the NEC must be used to calculate such loads. The ampacity required for the feeder-circuit conductors must be equal to the total loads involved. The OCPD's used to protect the conductors from short-circuits and ground-faults are sized per **430.62(A)** or **430-63**. **(See Figure 20-16)**

APPLYING DEMAND FACTORS
430.26

There are specific installations where there may be a special

situation in which a number of motors are connected to a feeder-circuit. And because of their operations, certain motors do not operate together and therefore the feeder-circuit conductors are sized according to the group that has the greater current rating in amps. **(See Figure 20-17)**

SIZING CONDUCTORS FOR CAPACITORS
460.8

The ampacity of capacitor circuit conductors must not be less than 135 percent of the rated current of the capacitor. The leads for a capacitor that supplies a motor must not be less than one-third the ampacity of the motor's circuit conductors. The larger of the two calculations in amps must be used for the capacitor supply conductors. **(See Figure 20-18)**

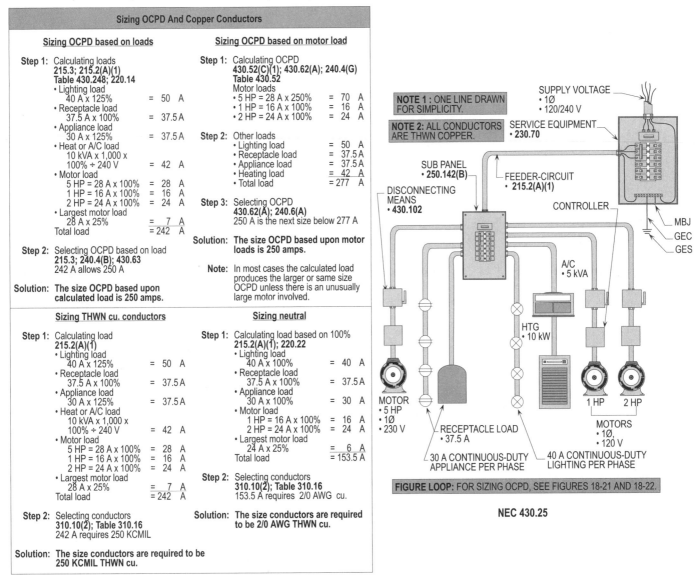

Figure 20-16. Calculating the size conductors for motors and other loads supplied by a feeder-circuit.

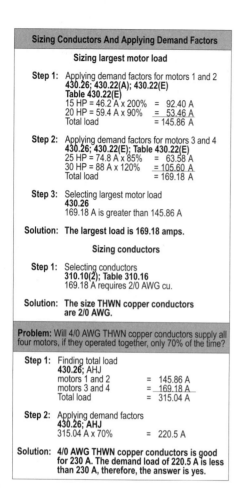

Sizing Conductors And Applying Demand Factors

Sizing largest motor load

Step 1: Applying demand factors for motors 1 and 2
430.26; 430.22(A); 430.22(E)
Table 430.22(E)
15 HP = 46.2 A x 200% = 92.40 A
20 HP = 59.4 A x 90% = 53.46 A
Total load = 145.86 A

Step 2: Applying demand factors for motors 3 and 4
430.26; 430.22(E); Table 430.22(E)
25 HP = 74.8 A x 85% = 63.58 A
30 HP = 88 A x 120% = 105.60 A
Total load = 169.18 A

Step 3: Selecting largest motor load
430.26
169.18 A is greater than 145.86 A

Solution: The largest load is 169.18 amps.

Sizing conductors

Step 1: Selecting conductors
310.10(2); Table 310.16
169.18 A requires 2/0 AWG cu.

Solution: The size THWN copper conductors are 2/0 AWG.

Problem: Will 4/0 AWG THWN copper conductors supply all four motors, if they operated together, only 70% of the time?

Step 1: Finding total load
430.26; AHJ
motors 1 and 2 = 145.86 A
motors 3 and 4 = 169.18 A
Total load = 315.04 A

Step 2: Applying demand factors
430.26; AHJ
315.04 A x 70% = 220.5 A

Solution: 4/0 AWG THWN copper conductors is good for 230 A. The demand load of 220.5 A is less than 230 A, therefore, the answer is yes.

Sizing THWN Conductors To Capacitor

Sizing conductors based on 1/3

Step 1: Finding FLA of motor
Table 430.250
50 HP = 143 A

Step 2: Calculating amps for conductors
430.22(A)
143 A x 125% = 178.8 A

Step 3: Selecting size conductors
310.10(2); Table 310.16
178.8 A requires 3/0 AWG THWN cu.

Step 4: Calculating conductors at 1/3 of 3/0 AWG cu.
460.8(A); Table 310.16
3/0 THWN cu. = 200 A
1/3 of 300 A = 66.7 A

Step 5: Selecting size conductors
Table 310.16
66.7 A requires 4 AWG cu.

Solution: The size THWN copper conductors based on 1/3 of branch-circuit is 4 AWG cu.

Sizing conductors based on 135%

Step 1: Calculating FLA of capacitor
460.8(A)
FLA = kVAR x 1,000 ÷ V x √3
FLA = 25 x 1,000 ÷ 208 V x 1.732
FLA = 69.4 A

Step 2: Calculating conductors
460.8(A)
69.4 A x 135% = 93.7 A

Step 3: Selecting conductors
310.10(2); Table 310.16
93.7 A requires 3 AWG cu.

Solution: The size THWN copper conductors based on 135% of FLA of capacitor are 3 AWG.

Note: Section 460.8(A) requires the largest conductors calculated which are the 3 AWG THWN copper.

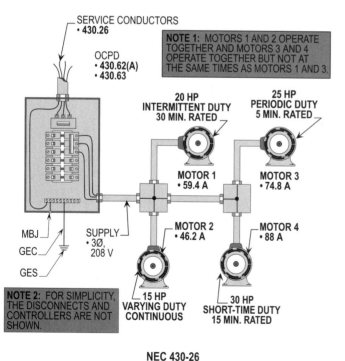

SERVICE CONDUCTORS
• 430.26

OCPD
• 430.62(A)
• 430.63

NOTE 1: MOTORS 1 AND 2 OPERATE TOGETHER AND MOTORS 3 AND 4 OPERATE TOGETHER BUT NOT AT THE SAME TIMES AS MOTORS 1 AND 3.

20 HP
INTERMITTENT DUTY
30 MIN. RATED

25 HP
PERIODIC DUTY
5 MIN. RATED

MOTOR 1
• 59.4 A

MOTOR 3
• 74.8 A

MBJ
GEC
GES

SUPPLY
• 3Ø,
208 V

MOTOR 2
• 46.2 A

MOTOR 4
• 88 A

NOTE 2: FOR SIMPLICITY, THE DISCONNECTS AND CONTROLLERS ARE NOT SHOWN.

15 HP
VARYING DUTY
CONTINUOUS

30 HP
SHORT-TIME DUTY
15 MIN. RATED

NEC 430-26

Figure 20-17. Sizing feeder-circuit conductors when applying demand factors.

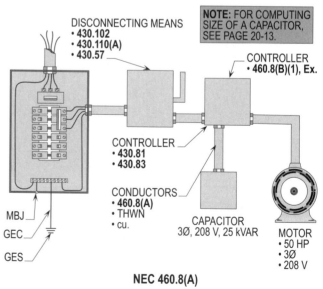

DISCONNECTING MEANS
• 430.102
• 430.110(A)
• 430.57

NOTE: FOR COMPUTING SIZE OF A CAPACITOR, SEE PAGE 20-13.

CONTROLLER
• 460.8(B)(1), Ex.

CONTROLLER
• 430.81
• 430.83

CONDUCTORS
• 460.8(A)
• THWN
• cu.

CAPACITOR
3Ø, 208 V, 25 kVAR

MOTOR
• 50 HP
• 3Ø
• 208 V

MBJ
GEC
GES

NEC 460.8(A)

Figure 20-18. Two calculations have been done and the greater of the two must be selected to size the capacitor's circuit conductors.

FINDING MICROFARADS

The microfarads for a capacitor may be found by applying the following equation:

$$C = \frac{159,300 \times A}{Hz \times V \times 1.732}$$

$$C = \frac{159,300 \times 143\ A}{60 \times 208\ V \times 1.732}$$

$$C = \frac{22,779,900}{21,615.36}$$

$$C = 1,053.9\ MF$$

The above microfarads (MF) were calculated based on the FLC of 143 amps from the 50 HP, three-phase, 208 volt motor in Figure 20-18. Note that the number 159,300 is a constant that is always used when applying the above equation.

MOTOR CONTROL CENTERS
PART VIII OF ARTICLE 430

Motor control centers are used wherever centralized control of a number of motors is feasible and desired. They provide a control location where incoming and outgoing lines to branch circuits can be consolidated.

In addition to the obvious advantages of reduced installation cost, centralized control eliminates the need for time-consuming trips to remote areas in a plant to shut down equipment or restart a motor that has tripped out its circuit due to overload. In the event of a power failure, motor control centers provide a rapid, safe means of restoring power. Orderly and sequential start-up of process motors, fans, pumps, and blowers is critical. In some industrial operations, after a power interruption, automatic restart is dangerous.

OVERCURRENT PROTECTION
430.94

Overcurrent protection shall be provided for these units based upon the current rating of the power bus and the requirements of Article 240. This must be provided either by an overcurrent device located ahead of the motor control center or by a main overcurrent protection device that is within sight of the center.

Motor Control Center Tip: The overcurrent protection cannot exceed the rating of the common power bus of a motor control center. It is permitted to use an overcurrent protective device with a rating less than the common power bus, provided it is of sufficient size to carry the load determined in accordance with **Part II** of **Article 430**.

SERVICE-ENTRANCE EQUIPMENT
430.95

When motor control centers are used as service-entrance equipment, they must have a main disconnect that disconnects all ungrounded conductors. If necessary, a second service disconnect may be used to feed additional equipment. If a grounded conductor is used, a main bonding jumper must be installed. **(See Figure 20-19)**

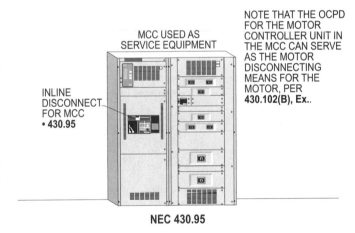

NEC 430.95

Figure 20-19. MCC's used as service equipment must have a main disconnect to disconnect ungrounded conductors.

GROUNDING
430.96

All sections of motor control centers must be bonded together with a conductor or bus sized per **250.122**. All equipment-ground conductors must be connected to this conductor or bus. (A bus is almost always used.)

BUSBARS AND CONDUCTORS
430.97

SUPPORT AND ARRANGEMENT
430.97(A)

Busbars must be protected and held rigidly in place. Conductors shouldn't be installed in vertical sections of the motor control center unless necessary or when protected from the busbars by a barrier.

PHASE ARRANGEMENT
430.97(B)

The phase arrangement for three-phase systems must be A, B, and C from front to back, top to bottom, or left to right. An exception is made for back-to-back units with vertical buswork. **(See Figure 20-20)**

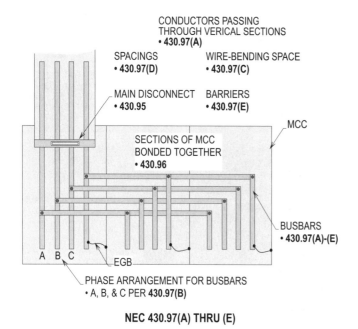

CONDUCTORS PASSING THROUGH VERICAL SECTIONS
• **430.97(A)**

SPACINGS • **430.97(D)**

WIRE-BENDING SPACE • **430.97(C)**

MAIN DISCONNECT • **430.95**

BARRIERS • **430.97(E)**

MCC

SECTIONS OF MCC BONDED TOGETHER • **430.96**

BUSBARS • **430.97(A)-(E)**

A   B   C

EGB

PHASE ARRANGEMENT FOR BUSBARS • A, B, & C PER **430.97(B)**

NEC 430.97(A) THRU (E)

Figure 20-20. Busbars and conductors must be arranged and grounded per **430.97(A)** through **(E)** of the NEC.

MINIMUM WIRE-BENDING SPACE
430.97(C)

The minimum wire-bending space at the motor control center terminals and minimum gutter space must be as required in **Article 312**.

SPACINGS
430.97(D)

Spacings between motor control center bus terminals and other bare metal parts must not be less than specified in **Table 430.97**.

BARRIERS
430.97(E)

Barriers must be placed in all service-entrance motor control centers to isolate service busbars and terminals from the remainder of the motor control center.

MARKING OF MOTOR CONTROL
CENTERS
430.98(A)

Motor control centers must be marked according to **110.21**, and such marking must be plainly visible after installation. Marking must also include common power bus current rating and motor control center short-circuit rating.

Note that motor control units in a motor control center must comply with **430.8** of the NEC.

Motor Control Center Tip: Part VIII, 430.1, FPN 1 refers to installation requirements for motor control centers contained in **110.26(F)**. The requirements of **110.26(F)** specify dedicated space for a motor control center and physical protection from mechanical systems that might leak or otherwise adversely impact a motor control center. **(See Figure 20-21)**

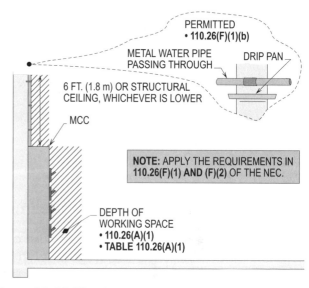

PERMITTED • **110.26(F)(1)(b)**

METAL WATER PIPE PASSING THROUGH

DRIP PAN

6 FT. (1.8 m) OR STRUCTURAL CEILING, WHICHEVER IS LOWER

MCC

NOTE: APPLY THE REQUIREMENTS IN 110.26(F)(1) **AND** (F)(2) OF THE NEC.

DEPTH OF WORKING SPACE • **110.26(A)(1)** • **TABLE 110.26(A)(1)**

Figure 20-21. The dedicated space above a MCC must be provided per **110.26(F)** of the NEC.

Chapter 20: Motor Feeder and Branch-Circuit Conductors

Section	Answer		
_____	T F	**1.**	Branch-circuit conductors supplying three-phase motors must have an ampacity not less than 100 percent of the motor's full-load current rating in amps.
_____	T F	**2.**	The branch-circuit conductors for wye-start and delta-run connected motors must be selected based on the full-load current in amps based on the line side of the controller.
_____	T F	**3.**	Conductors used for periodic duty must always be sized with a current-carrying capacity of 125 percent of the motor's full-load current in amps.
_____	T F	**4.**	When power conversion equipment provides overcurrent protection for the motor, no additional overload protection is required.
_____	T F	**5.**	The full-load current rating (in amps) of the largest motor must be multiplied by 125 percent to select the size of conductors for a feeder supplying a group of two or more motors.
_____	T F	**6.**	Branch-circuit conductors supplying single-phase motors must have an ampacity not less than _____ percent of the motor's full-load current rating in amps.
_____	T F	**7.**	Power conversion equipment requires the conductors to be sized at _____ percent of the rated input of such equipment.
_____	T F	**8.**	The conductors supplying power to a DC motor (continuous duty) must be sized at _____ percent of FLC in amps of the motor.
_____	T F	**9.**	Feeder-circuit conductors supplying several motors must be sized to carry _____ percent of the FLC rating in amps of the highest rated motor plus the sum of the FLC rating in amps of all remaining motors on the circuit.
_____	T F	**10.**	The ampacity of capacitor circuit conductors must not be less than _____ percent of the rated current in amps of the capacitor.
_____ _____		**11.**	What size THWN (branch-circuit) copper conductors are required for a 3 HP, 208 volt, single-phase, Design B motor?
_____ _____		**12.**	What size THWN (branch-circuit) copper conductors are required for a 20 HP, 230 volt, three-phase, Design B motor?
_____ _____		**13.**	What size THWN (branch-circuit) copper conductors are required for a 75 HP, 460 volt, three-phase, 15 minute rated intermittent duty cycle motor?
_____ _____		**14.**	What size THWN (branch-circuit) copper conductors are required to supply power conversion equipment with a rated input of 112 amps?

_____ _____   **15.** What size THWN (branch-circuit) copper conductors (line side) are required to supply a 50 HP, 208 volt, three-phase, Design B part-winding motor?

_____ _____   **16.** What size THWN (feeder) copper conductors are required to supply 30 HP, 40 HP, and 50 HP, 460 volt, three-phase, Design B motors?

_____ _____   **17.** What size THWN (feeder) copper conductors are required to supply a 10 HP, 208 volt, three-phase, 5 minute rated intermittent duty cycle, and 15 HP, 208 volt, three-phase, 15 minute rated intermittent duty cycle motor, and 20 HP, 208 volt, three-phase motor?

_____ _____   **18.** What size THWN (branch-circuit) copper conductors are required to supply a 20 kVAR, 208 volt, three-phase capacitor connected to a 40 HP, 208 volt, three-phase, Design B motor?

Control Circuit Conductors and Components

Control circuit conductors are designed and installed so they may be tapped from the motor power supply circuit or supplied from the service equipment. Overcurrent protection for control circuits that are tapped on the load side of controllers are designed and installed per **430.72**. Overcurrent protection for control circuits supplied from a different source of power other than the motor circuit's source of power is designed and installed per **725.23** and **725.24**. Lower voltage may be provided by control transformers that are installed for controlling motor circuits and related systems.

TYPES OF CONTROL CIRCUITS
430.72 AND 725.23

A motor control circuit tapped on the load side of fuses and circuit breakers, protecting motor branch-circuits, must also protect control conductors. If not protected by these fuses or circuit breakers, supplementary protection devices must be provided for control conductors.

The size of the control circuit conductors and the rating of the motor branch-circuit device will be determined by using one of these methods. Motor control circuits are classified as remote-control circuits where such circuits derive their power from other than the motor branch-circuit conductors. Fuses or circuit breakers may be utilized to protect remote motor control circuits. For further information, see **725.23, 725.24,** and **725.51**.

Motor Control Tip: Remote control circuits must have their disconnecting means located immediately adjacent to the disconnecting means used to disconnect the branch-circuit conductors supplying the controller and motor. Sometimes an interlock in the disconnect for the motor controller is used for this purpose which permits the controller, motor, and remote control circuit to be disconnected simultaneously.

CONDUCTOR PROTECTION
430.72(B)

Conductors, 14 AWG and larger are selected from Tables **310.16** through **310.19** for motor control circuit conductors that are tapped from a motor power circuit. Overcurrent protection for conductors smaller than 14 AWG must not exceed the values listed in **Table 430.72(B), Column A**. Conductors 18 and 16 AWG must be protected at the following amperage ratings:

(1) Conductors 18 AWG must be protected at 7 amps, when used for remote control circuits.

(2) Conductors 16 AWG must be protected at 10 amps, when used for remote control circuits.

Fuses selected at either 1, 3, 6, or 10 amps are normally used to protect these conductors from short-circuits, ground-faults, and overloads. See **240.6(A)** for the selection of these fuse sizes.

ONLY SHORT-CIRCUIT PROTECTION
430.72(B), Ex. 1

Control circuit conductors must have short-circuit and ground-fault protection and must be protected by the motor branch-circuit, short-circuit, and ground-fault protective device, where a hazard is created by the opening of the control circuit.

CONDUCTORS FROM SECONDARY OF CONTROL TRANSFORMERS
430.72(B), Ex. 2

The secondary conductors of the control transformer circuit may be protected by the primary side of the transformer. The transformer must be protected per **450.3(B)** and **Table 450.3(B)**. A two-wire secondary for a transformer installed within the control starter enclosure may be protected per **240.4(F)** and **240.21(C)(1)**.

The secondary conductor ampacity must be multiplied by the secondary-to-primary voltage ratio to provide protection in accordance with **450.3(B)** and **Table 450.3(B)**. Where the rated primary current is 9 amps or greater and 125 percent of this current does not correspond to a standard rating of a fuse or circuit breaker, the next higher standard size may be selected. Where the rated primary current is less than 9 amps, but is 2 amps or greater, an overcurrent protection device rated or set at not more than 167 percent of the primary current may be used. Where the rated primary current is less than 2 amps, an overcurrent protection device rated or set not greater than 300 to 500 percent may be used. **(See Figure 21-1)**

For example, if the primary full-load current of a motor control transformer is less than 2 amps, the OCPD may be calculated and sized at 500 percent times such full-load current in amps. (See **430.72(C)(4)**)

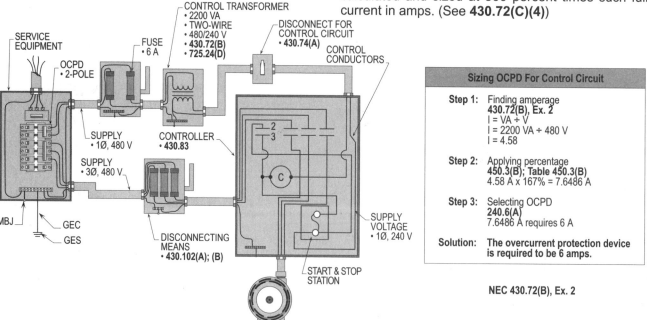

Figure 21-1. Control circuit conductors are supplied by a control transformer and protected by fuses in the primary side.

CONDUCTORS IN ENCLOSURES
430.72(B)(2)

Motor control circuit conductors that do not extend beyond the control equipment enclosure may be protected by the motor branch-circuit fuses or circuit breakers. **Table 430.72(B), Column B** permits this type of installation where the devices do not exceed 400 percent of the ampacity rating of size 14 AWG and larger conductors. Overcurrent protection for conductors smaller than 14 AWG shall not exceed the values listed in **Table 430.72(B), Column B**. Conductors rated 18 through 10 AWG may be protected with the following sized OCPD's.

(1) Conductor 18 AWG = 25 amps
(7 A x 400% = 28 A - requires 25 A OCPD)

(2) Conductor 16 AWG = 40 amps
(10 A x 400% = 40 A - requires 40 A OCPD)

(3) Conductor 14 AWG = 100 amps
(25 A x 400% = 100 A - requires 100 A OCPD)

(4) Conductor 12 AWG = 120 amps
(30 A x 400% = 120 A - requires 110 A OCPD)

(5) Conductor 10 AWG = 160 amps
(40 A x 400% = 160 A - requires 150 A OCPD)

(6) Conductor 8 AWG and larger
(400 percent x ampacity)

Control Tip: The free air ampacities of **Table 310.17** for 60°C wire are used to determine the ampacity ratings for the control circuit conductors. This type of installation has more free space to dissipate the heat where control conductors are installed in the open air space of enclosures instead of enclosed raceways.

See Figure 21-2 for selecting such conductors based upon branch-circuit OCPD rating.

CONDUCTORS RUN REMOTE
430.72(B)(2)

Motor control circuit conductors that extend beyond the control equipment enclosure may be protected by the motor's branch-circuit fuse or circuit breaker. **Table 430.72(B)** for conductors smaller than 14 AWG shall not exceed the values listed in **Table 430.72(B), Column C**. Conductors rated 18 through 10 AWG can be protected with the following sized OCPD's:

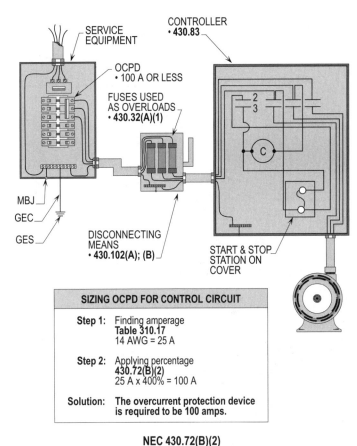

NEC 430.72(B)(2)

Figure 21-2. Control circuit conductors are located in controller and protected by the branch-circuit OCPD.

(1) Conductor 18 AWG = 7 amps
(requires 7 A OCPD)

(2) Conductor 16 AWG = 10 amps
(requires 10 A OCPD)

(3) Conductor 14 AWG = 45 amps
(15 A x 300% = 45 A - requires 45 A OCPD)

(4) Conductor 12 AWG = 60 amps
(20 A x 300% = 60 A - requires 60 A OCPD)

(5) Conductor 10 AWG = 90 amps
(30 A x 300% = 90 A - requires 90 A OCPD)

(6) Conductor 8 AWG and larger
(300 percent x ampacity)

Note that the above protection is required anytime the control circuit is used for remote control of coils in a motor controller enclosure or motor control center.

See Figure 21-3 for selecting such conductors based upon the branch-circuit's OCPD rating.

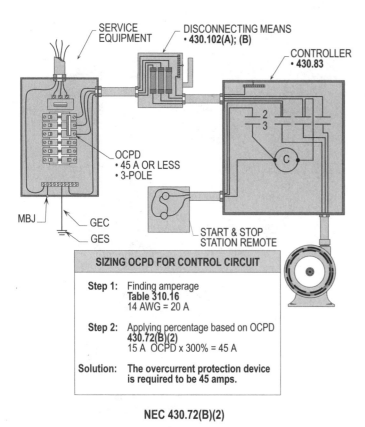

SIZING OCPD FOR CONTROL CIRCUIT

Step 1: Finding amperage
Table 310.16
14 AWG = 20 A

Step 2: Applying percentage based on OCPD
430.72(B)(2)
15 A OCPD x 300% = 45 A

Solution: The overcurrent protection device is required to be 45 amps.

NEC 430.72(B)(2)

Figure 21-3. Control circuit conductors are run remote and protected by the branch-circuit OCPD in the panelboard.

CONTROL CIRCUIT TRANSFORMERS 430.72(C)

Article 450 is used for designing the protection of control circuit transformers. The following control circuit transformers must be sized accordingly:

(1) Transformers with OCP omitted.
(2) Transformers rated Class 1, 2, and 3,
(3) Transformers less than 50 VA,
(4) Transformers less than 2 amps, and
(5) Transformers with other approved means.

TRANSFORMERS WITH OCP OMITTED 430.72(C), Ex.

Overcurrent protection is omitted where the opening of the control circuit would create a hazard. For example, the control circuit for a fire pump motor.

TRANSFORMERS RATED CLASS 1, 2, & 3 430.72(C)(2)

The FLC in amps for a control circuit transformer that is rated 1000 volt-amps or less and 30 volt or less may be increased 167 percent per **725.21(A)(2)**.

Control circuit transformers with limited power sources are required per **725.21(A)(1)** to be designed and protected per **450.3(B), Ex.** Section **430.72(C)(1)** refers to **725.21(A)(1)**, which requires overcurrent protection devices to be designed and placed in the secondary of Class 1 control circuit transformers per **450.3(B), Ex.**

TRANSFORMERS LESS THAN 50 VA 430.72(C)(3)

Protection is not needed for control transformers rated less than 50 VA and are located in the controller enclosure and an integral part of the controller. The motor circuit overcurrent protection device protects the transformer for this type of installation.

TRANSFORMERS LESS THAN 2 AMPS 430.72(C)(4)

Where the rated primary current is less than 2 amps, an overcurrent protection device rated or set not greater than 300 to 500 percent may be used.

For example, if the primary full-load current in amps of a motor control transformer is less than 2 amps, the OCPD may be calculated and sized at 500 percent times such full-load current in amps.

TRANSFORMERS WITH OTHER APPROVED MEANS 430.72(C)(5)

Control circuit transformers may be protected where provided with other approved means.

MECHANICAL PROTECTION OF CONDUCTORS 430.73

Remote motor control circuit conductors that are outside the control device must be installed in a raceway or be suitably protected from physical damage, if damage to the motor control circuit could create a hazard.

Motor control circuits that are grounded on one side must be arranged in such a manner that an accidental ground in control circuits remote from the motor controller will comply with the following:

(1) The motor will not start, and
(2) Will not bypass manually operated shutdown devices or automatic safety shutdown devices.

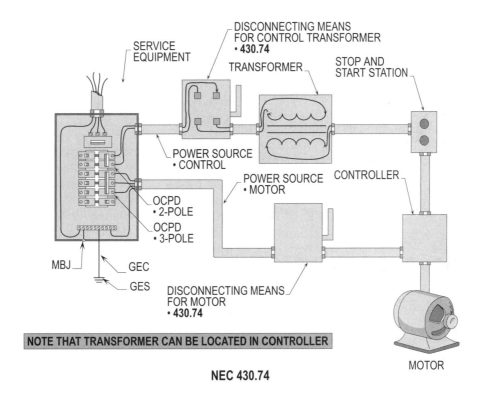

NEC 430.74

Figure 21-4. One disconnecting means or a number of disconnects may be required to disconnect the power supply to a motor and the control circuit located in the controller.

DISCONNECTION OF CONTROL CIRCUIT CONDUCTORS AND POWER
430.74(A)

Motor control circuits must be disconnected from all sources of supply when the disconnecting means is in the open position. The disconnecting means for the starter may be installed to serve as the disconnecting means for the motor circuit conductors if the control circuit conductors are tapped from the line terminal of the magnetic starter. An auxiliary contact must be installed in the disconnecting means of the controller or an additional disconnecting means must be mounted adjacent to the controller to disconnect the motor control circuit conductors if they are fed from another source and not tapped from the starter conductors. **(See Figure 21-4)**

MORE THAN 12 CONDUCTORS
430.74(A), Ex. 1

A disconnecting means is not required to disconnect 12 or more control circuit conductors that are permitted to be located other than adjacent to each other when the following conditions are complied with:

(1) Access to energized parts is limited to qualified persons only.

(2) Access to live parts in motor control circuits is permitted by a warning sign, which is permanently located on the outside of each equipment enclosure door or cover. The sign warns that the disconnecting means for the motor control circuit is located remotely. Such sign must also list the location and identify each disconnect and its use.

CLASS 1 CIRCUITS
725.21

Class 1 circuits are divided into two types. Power-limited and remote-control and signaling circuits. Class 1 - power-limited circuits are limited to 30 volts and 1000 volt-amperes. Class 1 - remote-control and signaling circuits are limited to 600 volts, with no limitations on the power output of the source. Note that the rules pertaining to Class 1 circuits for motor control are reviewed in this Section.

POWER-LIMITED CIRCUITS
725.21(A)

Class 1 power-limited circuits are supplied from a power source that has a rated output of not more than 30 volts and a power limitation of 1000 volt-amps or less. Class 1 power-limited circuits have a current limiter on the power source that supplies them. This limiter is an overcurrent protection device that restricts the amount of supply current to the circuit, in the event of an overload, short-circuit, or ground-fault. These Class 1 circuits may be supplied from a transformer or other type of power supply such as generators or batteries.

REMOTE CONTROL OR SIGNALING CIRCUITS
725.21(B)

Class 1 remote control and signaling circuits are permitted to operate up to 600 volts and have no limitation on the power rating of the source. Class 1 remote control systems, generally, must meet most wiring requirements for power and light circuits. Class 1 remote control circuits are commonly used in motor controllers that operate mechanical processes, elevators, conveyors, and equipment that is controlled from one or more remote locations. Class 1 signaling circuits are used in nurse's call systems in hospitals, electric clocks, bank alarm systems, and factory call systems. **See Figure 21-5** for Class 1 circuit requirements.

CLASS 2 AND 3 CIRCUITS
725.41(A); (B)

Class 2 and Class 3 circuits are defined by two Tables 11(A) and (B) in Chapter 9. One table is used for AC current and one for DC current. In general, a Class 2 circuit operating at 24 volts with a power supply durably marked "Class 2" and not exceeding 100 volt-amperes is the type most commonly used.

A Class 2 circuit is defined as that portion of the wiring system between the load side of a Class 2 power source and the connected equipment. Due to its power limitations, a Class 2 circuit is considered safe from a fire initiation standpoint and provides acceptable protection from electric shock.

A Class 3 circuit is defined as that portion of the wiring system between the load side of a Class 3 power source and the connected equipment. Due to its power limitations, a Class 3 circuit is usually considered unsafe from a fire initiation standpoint. Since higher levels of voltage and current for Class 3 circuits are permitted, additional safeguards are specified to provide protection from an electric shock hazard that might be encountered.

Power for Class 2 and Class 3 circuits is limited either inherently (in which no overcurrent protection is required) or by a combination of a power source and overcurrent protection scheme.

The maximum circuit voltage is 150 volts AC or DC for a Class 2 inherently limited power source and 100 volts AC or DC for a Class 3 inherently limited power source. The maximum circuit voltage is 30 volts AC and 60 volts DC for a Class 2 power source limited by overcurrent protection and 150 volts AC or DC for a Class 3 power source limited by overcurrent protection. **(See Figure 21-6)**

For example, heating system thermostats are commonly Class 2 systems and the majority of small bells, buzzers, and annunciator systems are Class 2 circuits. Class 2 also includes small intercommunicating telephone systems in which the voice circuit is supplied by a battery and the ringing circuit by a transformer.

Class 2 and 3 systems do not require the same wiring methods as power, light, and Class 1 systems. However, a 2 in. (50 mm) separation is required between these systems.

CONTROL CIRCUITS IN RACEWAYS, CABLES, AND ENCLOSURES
725.26

Class 1 control circuits may be installed in raceways, cables, and enclosures using the following installation procedures:

(1) Two or more Class 1 circuits, and
(2) Class 1 circuits with power conductors.

TWO OR MORE CLASS 1 CIRCUITS
725.26(A)

Class 1 circuits may occupy the same cable, enclosure, or raceway without regard to whether the individual circuits are AC or DC current, provided all conductors are insulated for the maximum voltage of any conductor in the cable, enclosure, or raceway. **(See Figure 21-7)**

CLASS 1 CIRCUITS WITH POWER CONDUCTORS
725.26(B)(1)

Class 1 circuits and power supply circuits may occupy the same cable, enclosure, or raceway but only in situations where the equipment power system is functionally associated. **(See Figure 21-7)**

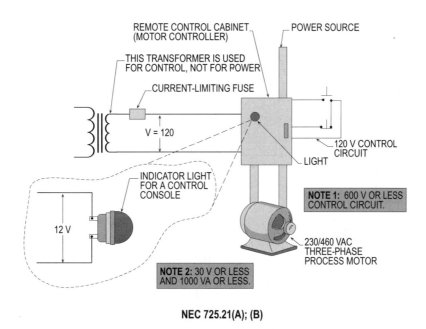

NEC 725.21(A); (B)

Figure 21-5. Class 1 remote control and signaling circuits are permitted to operate up to 600 volts and in some cases have no limitations on the power rating of the source. Control-circuits may also be operated at 30 volts or less and 1000 VA or less. **Note:** Class 1 circuits are either the power limited or non-power limited type.

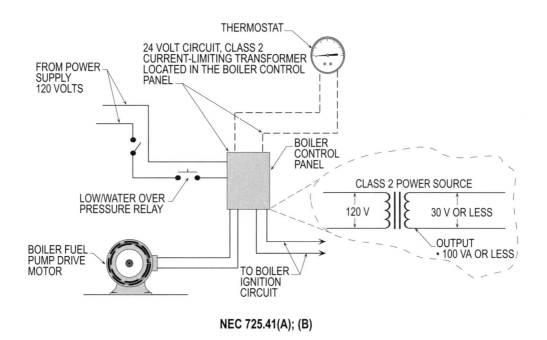

NEC 725.41(A); (B)

Figure 21-6. The maximum voltage is usually 150 volts AC or DC for a Class 2 inherently limited power source and 100 volts AC or DC for a Class 3 inherently limited power source per **Tables 11(A) and 11(B) in Chapter 9**.

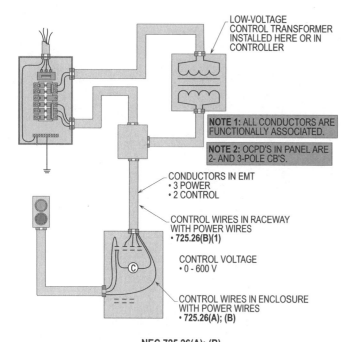

NEC 725.26(A); (B)

Figure 21-7. Class 1 circuits may occupy the same cable, enclosure, or raceway without regard to whether the individual circuits are AC or DC current.

Section **725.26(B)(2)** clarifies that where installed in factory or field-assembled control centers, Class 1 and power supply circuits may be mixed. Section **725.26(B)(3)** also permits mixing for underground (phase) conductors in a manhole, if all of the following conditions are complied with:

(1) The power supply or Class 1 circuit conductors are in a metal enclosed cable or Type UF cable.

(2) The conductors are permanently separated from the power supply conductors by a continuous firmly fixed nonconductor, such as flexible tubing, in addition to the insulation on the wire, and

(3) The conductors are permanently and effectively separated from the power supply conductors and securely fastened to racks, insulators, or other approved supporting means. **(See Figure 21-8)**

WHEN TO DERATE THE AMPACITY 725.28(A); (B)

Where only Class 1 circuit conductors are in a raceway, the number of conductors may be determined by the provisions of **300.17**. The derating factors given in **Table 310.15(B)(2)(a)** apply only, if such conductors carry continuous loads in excess of 10 percent of the ampacity of each control conductor routed through the raceway system.

The number of power supply conductors and Class 1 circuit conductors pulled through a raceway based upon the rules of **725.28** must be determined per **300.17**. The derating factors, given in **Article 310, Table 310.15(B)(2)(a)** to Ampacity Tables of 0 to 2000 volts, apply to the following conditions:

(1) To all conductors where the Class 1 circuit conductors carry continuous loads in excess of 10 percent of the ampacity of each conductor and where the total number of conductors is four or more.

(2) To the power supply conductors only, where the Class 1 circuit conductors do not carry continuous loads in excess of 10 percent of the ampacity of each conductor and where the number of power supply conductors is four or more. **(See Figure 21-9)**

Motor Control Tip: Class 1 circuit conductors installed in cable tray systems must comply with the rules and regulations of **392.9** through **392.11** as well as **725.26(B)(3)**.

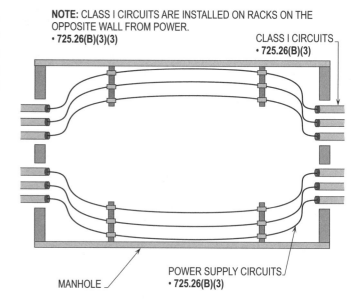

NEC 725.26(B)(3)(1) THRU (3)

Figure 21-8. Class I circuits when properly installed can be installed in manholes with power supply circuits.

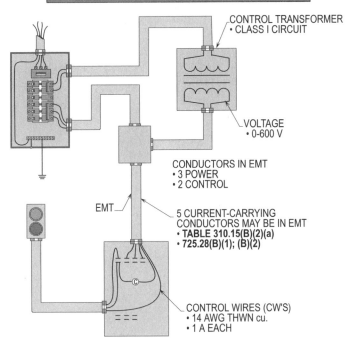

NOTE: TRANSFORMER CAN BE INSTALLED IN CONTROLLER.

CONTROL TRANSFORMER
• CLASS I CIRCUIT

VOLTAGE
• 0-600 V

CONDUCTORS IN EMT
• 3 POWER
• 2 CONTROL

EMT

5 CURRENT-CARRYING
CONDUCTORS MAY BE IN EMT
• TABLE 310.15(B)(2)(a)
• 725.28(B)(1); (B)(2)

CONTROL WIRES (CW'S)
• 14 AWG THWN cu.
• 1 A EACH

NEC 725.28(A) through (C)

Finding Max. Amps Before Control Wires (CW) Are Considered Current-Carrying.	
Step 1:	Finding Amps **Table 310.16** 14 AWG cu. = 20 A
Step 2:	Finding Max. Amps **725.28(B)(1); (B)(2)** 20 A x 10% = 2 A
Solution:	**The 1 amp on each CW is less than 2 amps so no derating of conductor is required.**

Note That Control Voltage Can Be:
• 120 V Circuit   • 208 V Circuit   • 240 V Circuit   • 480 V Circuit   • 550 V Circuit   • 600 V Circuit

Figure 21-9. The derating factors given in **Table 310.15(B)(2)(a)** apply only, if such control conductors carry continuous loads in excess of 10 percent of the ampacity of each conductor routed through the raceway system.

Chapter 21. Control Circuit Conductors and Components

Section	Answer			
_____	T	F	1.	18 AWG copper conductors must be protected at 10 amps when used for remote control circuits.
_____	T	F	2.	Motor control circuit conductors that do not extend beyond the control equipment enclosure are permitted to be protected by the motor's branch-circuit fuses.
_____	T	F	3.	Motor control circuit conductors that extend between the control equipment enclosure cannot be protected by the motor branch-circuit fuse.
_____	T	F	4.	The secondary conductors of a control transformer may be protected by the primary side of the transformer.
_____	T	F	5.	Protection is not needed for control transformers rated less than 50 VA that are located in the controller enclosure and an integral part of the controller.
_____	T	F	6.	Overcurrent protection is omitted where the opening of the control circuit could create a hazard.
_____	T	F	7.	The maximum voltage is usually 100 volts or less AC or DC for a Class 2 inherently limited power source.
_____	T	F	8.	Class 1 circuits and power supply circuits are not permitted to occupy the same raceway.
_____	T	F	9.	The maximum voltage for a Class 3 inherently limited power source may be 150 volts AC or DC.
_____	T	F	10.	Class 1 remote control or signaling circuits are permitted to operate up to 600 volts and have no limitation on the power rating of the source.
_____ _____			11.	16 AWG copper conductors must be protected at _____ amps when used for remote control circuits.
_____ _____			12.	Where the rated primary current is less than _____ amps, an overcurrent protection device rated or set not greater than 300 percent to 500 percent may be used.
_____ _____			13.	Power-limited Class 1 circuits are limited to _____ volts and _____ volt-amperes.
_____ _____			14.	The maximum circuit voltage is _____ volts AC and _____ volts DC for a Class 2 power source limited by overcurrent protection at 100 VA.

_____ _____

15. Class 2 systems require a _____ inch separation between power, light, and Class 1 systems.

_____ _____

16. What size OCPD is permitted for motor control circuit conductors that are located in the controller and supplied by 12 AWG copper conductors?

_____ _____

17. What size OCPD is permitted for motor control circuit conductors that are run remote and supplied by 12 AWG copper conductors?

_____ _____

18. What size OCPD is required for motor control circuit conductors that are supplied by a 2,400 VA, 480 volt, two-wire control transformer?

Connecting Controls for Operation

A means for starting and stopping must be provided for all electric motors and their driven load. Either manual magnetic starters or motor control centers are used as a controlling means for commercial and industrial motors.

The power supply is connected to the manual starter in series through the contacts to the motor leads. Magnetic starters are controlled by pressure, temperature, light, start-and-stop buttons, etc., which provide automatic starting and stopping of motors.

Note: Review Chapter 23 and Tables 1 through 14 of Appendix A for troubleshooting techniques on control circuits and components.

MAGNETIC STARTERS

Magnetic starters are the most common type of controllers in the electrical industry. They are equipped with normally open (NO) power contacts that can be closed by applying voltage to their closing coils. Coil voltages may range in values from 24 to 480 volts and control voltage is used to close the contacts and provide power to the motor. The device used to control the voltage to the coil may be manually or automatically controlled. Different wiring procedures are required for each method. The motor windings are protected from overload conditions by an overload relay unit that is provided in the magnetic starter circuitry. **(See Figure 22-1)**

COMPONENTS

The terminals L_1, L_2, and L_3 are the terminals used to connect the branch-circuit conductors from the power line to the magnetic starter. The branch-circuit conductors for the magnetic starter are sized and selected per **430.22(A)** and **Table 310.16** of the NEC.

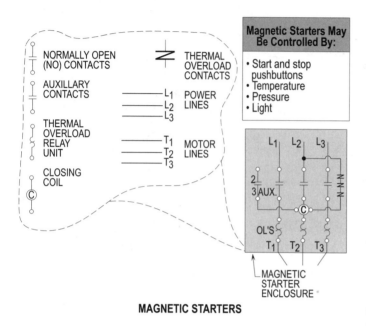

Magnetic Starters May Be Controlled By:
- Start and stop pushbuttons
- Temperature
- Pressure
- Light

MAGNETIC STARTERS

Figure 22-1. The above illustrates a magnetic starter that is designed to be controlled automatically or nonautomatically.

The terminals T_1, T_2, and T_3 are used to connect the magnetic starter to the motor leads. The minimum wire bending space at the terminals within the enclosure housing for the magnetic starter must comply with **Table 430.10(B)**. Where terminal housings are provided on motors, the minimum space required must comply with **430.12**.

The branch-circuit conductors are connected from the power supply to the motor leads by stationary contacts. Contacts will eventually become tarnished from the making and breaking when starting and stopping motors with their driven loads. The contacts are energized and deenergized by the closing coil. The control circuit operating this coil is designed and selected per **430.72** and **725.23**. The connection of the power circuit conductors to the motor leads are bridged from auxiliary contact points 2 and 3. The temperature rise in the motor windings is sensed by the thermal overload relay unit. If the setting of the overload relay is exceeded by the temperature rise in the motor windings, the coil circuit is opened by the overload contacts, dropping out the power circuit conductors to the motor. Overload contacts are designed to be connected in series from L_2 to the coil and from that point to the controlling devices supplied by L_1. The coil control circuit is opened by the overload contacts due to the heat of an overload condition.

For example: What is the minimum wire bending space required for 1 - 2 AWG conductor per terminal within the enclosure housing for a magnetic starter?

Step 1: Finding space
Table 430.10(B)
2 AWG conductor = 2 1/2"

Solution: **The minimum wire bending space required is 2 1/2 in.**

TWO-WIRE CONTROL SYSTEMS

Two-wire control circuits are designed and installed to eliminate a voltage release during a power failure. This type of installation (no voltage release) means that the coil circuit is maintained through the contacts of the pilot device until it is disconnected. The contacts to the pilot device controlling the circuit to the coil usually remains closed and connects power immediately to the coil when the power to the circuit is restored. Two-wire devices are designed and installed to be used for two-wire control circuits. These type of devices are single-pole switches, pressure switches, float switches, thermostats, limit switches, etc. **(See Figure 22-2)**

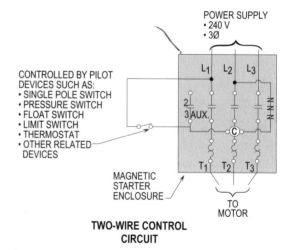

TWO-WIRE CONTROL CIRCUIT

Figure 22-2. The above illustrates a magnetic starter that is controlled by a two-wire control circuit.

THREE-WIRE CONTROL SYSTEMS

Three-wire control circuits are designed and installed to provide a voltage release during a power failure. Power is energized to the coil by pushing the starting button, which is normally open (NO), causing the contacts to close and energize power to the motor.

A three-wire control circuit consists of a start button with normally open (NO) contacts and a stop button with normally closed (NC) contacts. The auxiliary contacts are connected in parallel for the start button and in series for the stop button. No voltage protection means that the coil circuit is maintained through the normally closed contacts of the stop button. The control circuit is completed with an extra set of contacts (auxiliary contacts 2 and 3) through the stop button and holds power to the circuit until the stop button is pressed, which deenergizes the circuit to the coil and drops out the control circuit. The extra set of contacts (auxiliaries 2 and 3) will not close again until the coil has been energized by the start button. **(See Figure 22-3)**

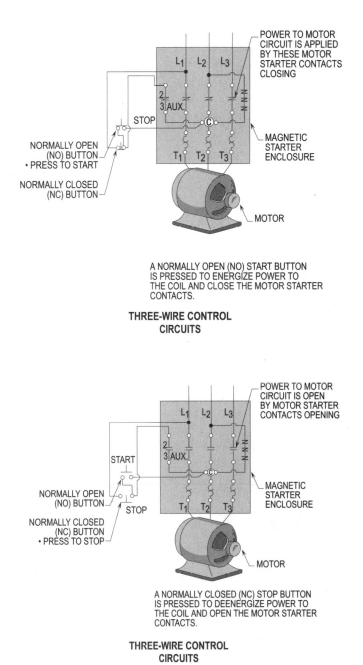

A NORMALLY OPEN (NO) START BUTTON IS PRESSED TO ENERGIZE POWER TO THE COIL AND CLOSE THE MOTOR STARTER CONTACTS.

THREE-WIRE CONTROL CIRCUITS

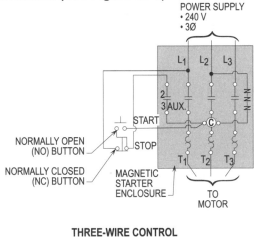

THREE-WIRE CONTROL CIRCUITS

Figure 22-3. A three-wire circuit installed to control voltage to the coil of a magnetic starter.

By pressing the normally open start button in a three-wire control circuit, power energizes the coil and closes the motor starter contacts. By pressing the normally closed stop button in a three-wire control circuit, power is deenergized to the holding circuit to the coil, opening the motor starter contacts disconnecting voltage to the motor. **(See Figure 22-4)**

CONTROL DEVICES

Motor control circuits may be equipped with control devices to perform a variety of operations. The following types of control devices may be added to the control circuit to regulate the starting and stopping of the motor.

(1) Start buttons,
(2) Stop buttons,
(3) Jog buttons,
(4) Auxiliary contacts,
(5) Emergency or extra motor stop buttons,
(6) Hand-off automatic switches,
(7) Forward-reverse stop stations,
(8) Float switches, and
(9) Pressure switches, etc.

A NORMALLY CLOSED (NC) STOP BUTTON IS PRESSED TO DEENERGIZE POWER TO THE COIL AND OPEN THE MOTOR STARTER CONTACTS.

THREE-WIRE CONTROL CIRCUITS

Figure 22-4. The normally open start button connects power to the coil and the normally closed stop button disconnects power to the holding circuit to the coil in a three-wire control circuit.

START STATIONS

Extra start buttons may be added as needed for control purposes but they must be connected in parallel with the start button to energize the control circuit to the coil. These contacts close and start the motor. **(See Figure 22-5)**

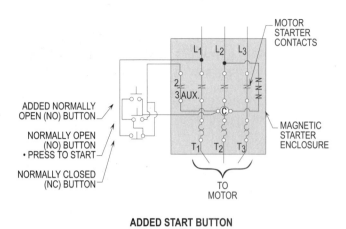

ADDED START BUTTON

Figure 22-5. Extra start buttons may be added for the control of motor control circuits and power circuits.

STOP STATIONS

Extra stop buttons can be added as needed for control purposes but must be connected in series to deenergize power to the coil of the magnetic starter. These extra stop buttons may be located at various locations to stop the motor. **(See Figure 22-6)**

Control Tip: A motor may be stopped by any type of switch that is connected in series with the holding circuit (auxiliary contacts 2 and 3) to the coil. Note that contacts connected in series interrupts the power source and disconnects the circuit.

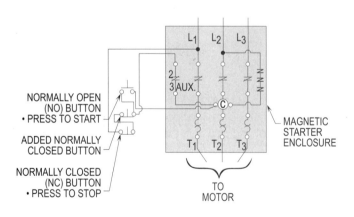

ADDED STOP BUTTON

Figure 22-6. Extra stop buttons may be added for the motor control and power circuits.

JOG STATIONS

Jog buttons are installed for jogging and lets the motor run as long as the jog button is depresssed. When jogging the motor, the magnetic starter must be wired with a job relay in the control circuit so that there is no chance of locking-in. The jog button has normally open contacts that are connected in parallel with the start button and two normally closed contacts (CR&M) that are connected in series with the stop button and auxiliary terminal No. 3. The jog button is held down to connect the power to the main coil and job the motor. The jog button's normally open contacts will prevent the holding coil from locking-in and running the motor. **(See Figure 22-7)**

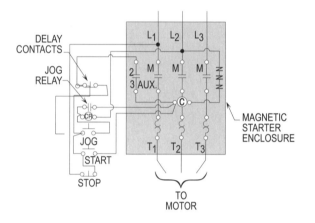

ADDED JOG BUTTON

Figure 22.7. When jogging the motor, the magnetic starter must be wired with a jog relay in the control circuit so that there is no chance of locking in the circuit.

AUXILIARY CONTACTS

An extra auxiliary contact may be added to one side of a magnetic starter to control a circuit to another coil or device. The extra auxiliary contact may be installed either normally open or normally closed. If the auxiliary contact is normally open, its function is to close the control circuit. Auxiliary contacts are installed with one side being connected to L_1 and the other side being routed and connected to the coil to be controlled. **(See Figure 22-8)**

MOTOR STOP STATIONS

A master stop button may be installed for safety when it is connected in series with the wire from L_1 to the first stop button in the control circuit. The contacts remain in the open position when the master stop button is turned off manually. The coil of the magnetic starter cannot be energized since the contacts of the stop button are in the open position which disconnects the power of L_1 from the components of the control circuit. **(See Figure 22-9)**

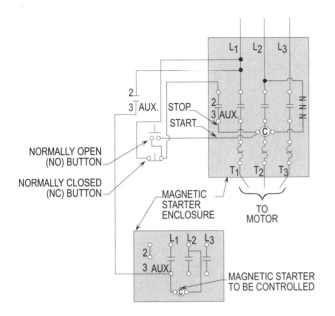

ADDING AUXILLARY CONTACTS

Figure 22-8. An extra auxiliary contact may be added to one side of a magnetic starter to control a circuit to another coil or device.

HAND-OFF AUTOMATIC SWITCHES

A motor may be started manually or automatically by installing a hand-off automatic switch. When starting the motor automatically, a remote control device may be installed. The coil may be energized to start the motor by installing pilot devices such as float switches, limit switches, and pressure switches etc. **(See Figure 22-10)**

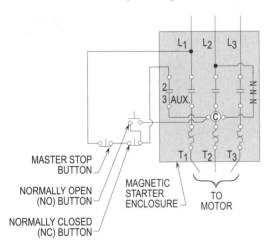

ADDED MASTER STOP BUTTON

Figure 22-9. A master stop button may be installed for safety when it is connected in series with the wire from L_1 to the first stop button in the control circuit.

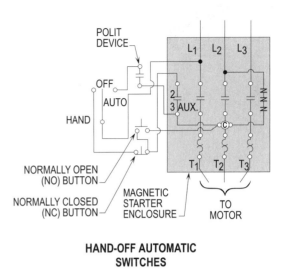

HAND-OFF AUTOMATIC
SWITCHES

Figure 22-10. Installing a hand-off automatic switch in the control circuit to start a motor.

FORWARD-REVERSE STOP STATIONS

The motor may be wired to rotate in the forward or reverse direction by installing a forward-reverse stop pushbutton station. The rotation of the motor is stopped by pressing the stop button. By pressing the forward button, terminal 3 and terminals 3 through 6, are connected to one side of the forward control coil. The power circuit conductors are connected to the motor by the contacts of the forward magnetic starter being closed and energizing the coil. The reverse button for reverse rotation must not be pressed until the stop button is pressed and the motor has stopped. By pressing the reverse button, terminals 5 through7 are energized to one side of the reverse control coil. The power circuit conductors are connected to the motor by the contacts of the reverse magnetic starter being closed and energizing the coil. **(See Figure 22-11**

FLOAT SWITCHES

Float switches (liquid level switches) may be installed and controlled by hardware that floats on a liquid. For example, a sump pump in a basement is controlled by the use of a float switch. The float permits the switch to turn the pump motor on when the water reaches a level in the sump. The float of the switch will turn the pump motor off when the water reaches a lower level.

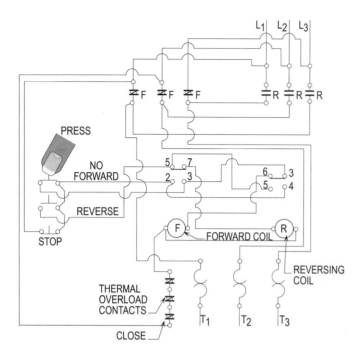

FORWARD - REVERSE STOP BUTTONS

Figure 22-11. The coil is energized by pressing the forward button for forward rotation and the reverse button for reversal rotation.

PRESSURE SWITCHES

Pressure switches (pressure operated) are vacuum switches which are defined as the absence of pressure. Pressure switches are designed and installed in applications where fluids (gases or liquids) are placed under pressure and turns equipment on or off at some preset operating pressure or at a pressure level beyond which it would be unsafe to operate. For example, pressure switches are installed in steam generators and electrically operated air compressors to maintain the correct operating pressures and to detect the pressure level beyond which it would be unsafe to operate.

Chapter 22. Connecting Controls For Operation

Section	Answer				
T	T	F		1.	Magnetic starters are the most commonly used types of controllers to connect and disconnect power to motors.
T	T	F		2.	The device used to control the voltage to the coil may be a manual or automatic type.
F	T	F		3.	Coil voltage may range in values from 24 volts to 240 volts.
T	T	(F)		4.	Terminals L_1, L_2, and L_3 are the terminals on the magnetic starter used to connect the branch-circuit conductors to the motor leads.
F	(T)	F		5.	Overload contacts are designed to be connected in series from L_2 to the coil and from that point to the controlling devices supplied by L_1.
F	T	F	✓	6.	Two-wire devices are designed and installed to be used for three-wire control circuits.
T	T	F	✓	7.	A three-wire control circuit consists of a start button with normally open contacts and a stop button with normally closed contacts.
F	(T)	F		8.	Extra stop buttons may be added as needed for control purposes but must be connected in series to disconnect the power to the coil of the magnetic starter.
F	(T)	F		9.	A extra closing coil may be added to one side of a magnetic starter to control a circuit to another closing coil or device.
F	T	F	✓	10.	A master stop button can be installed for safety when it is connected in parallel with the wire from L_1 to the first stop button in the control circuit.
	Parallel ✓			11.	Start buttons are connected in _____ with the auxiliary contacts.
	SERIES ✓			12.	Stop buttons are connected in _____ with the auxiliary contacts and coil.
	Series ✓			13.	A motor may be stopped by any type of switch that is connected in _____ because power is disconnected to the holding circuit of the coil.
	Jogging			14.	Jog buttons are installed for _____ operation and lets the motor run as long as the jog button is depressed.
	open			15.	The jog button's normally _____ contacts will prevent the holding coil from locking-in and running the motor when the job button is released.
Manually	Automatically			16.	A motor may be started _____ or _____ by installing a hand-off automatic switch.

17. The motor may be wired to run in _____ or _____ rotation by installing a forward-reverse stop pushbutton station.

18. Float switches are installed and controlled by hardware that __Float__ on liquid.

19. Pressure switches are designed so that when a __Fluid__ is placed under pressure, the switch contacts will turn the equipment on or off.

20. By pressing the normally open __Start__ button in a three-wire control circuit, power is energized to the coil and closes the motor starter contacts.

21. What are the components marked A through I?

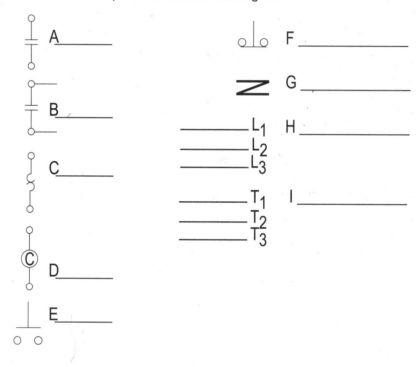

A _____   F _____

B _____   G _____

C _____   H _____

D _____   I _____

E _____

22. Connect the magnetic starter for two-wire operation?

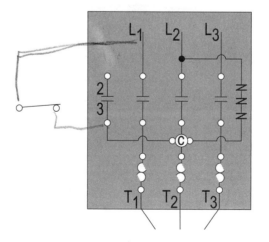

23. Connect the three-wire circuit to control a magnetic starter?   _____ _____ 22-9

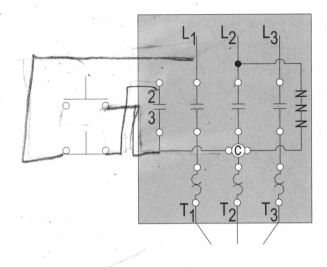

24. Connect the additional stop button for the magnetic starter?   _____ _____

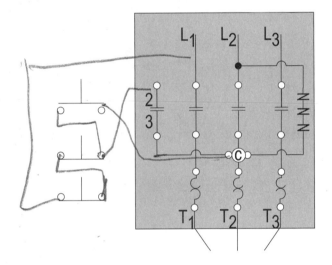

25. Connect the additional start button for the magnetic starter?   _____ _____

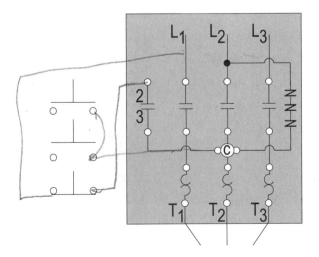

_____ _____

26. Connect the jog button to the magnetic starter for the jogging of the motor?

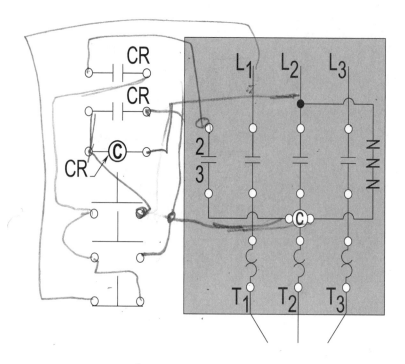

_____ _____

27. Connect the additional auxiliary contact to the magnetic starter to control a circuit to another coil or device?

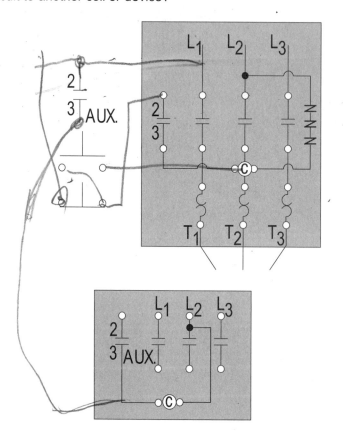

28. Connect the master stop button for disconnecting the control circuit?   _____ _____ 22-11

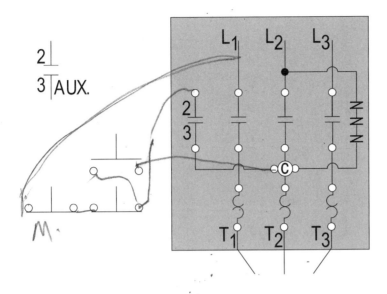

29. Connect the hand-off automatic switch in the control circuit to start a motor?   _____ _____

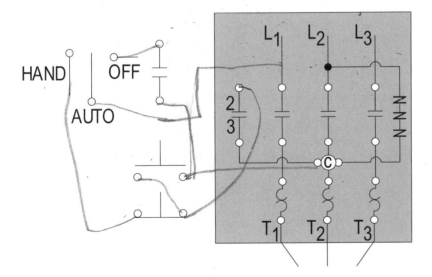

Troubleshooting Motor Windings and Components

Most commercial and industrial motor applications require the controller to be capable of being operated from remote locations. This scheme of automatic operation occurs in response to signals from such pilot devices as thermostats, pressure of float switches, limit switches, etc. These devices allow the magnetic starter to be controlled as necessary from any location. However, manual control can also be used with the starter mounted so that the operator has easy access to the controls.

A motor that fails to run must be checked to determine the problem. These problems can be defective windings in the motor or an electrical apparatus associated with the operation of the motor.

This chapter covers the techniques necessary to troubleshoot electrical motors and components pertaining to their operation and control.

Note: Review Chapter 22 for control connections. For troubleshooting tips, see Tables 1 through 14 of Appendix A in the back of this book.

TROUBLESHOOTING SPLIT-PHASE MOTORS

When troubleshooting single-phase, split-phase motors there are various components with different electrical characteristics which have to be considered before attempting to test and determine operating problems.

CONNECTING LEADS AND WINDINGS

There are basically two procedures for connecting the leads of motors to the power supply line. The first step is to identify the motor winding conductors. The NEMA tagging methods for new motors identifies the running winding leads as T_1 and T_2. The starting winding leads are tagged T_3 and T_4 respectively. For older motors, the identification method of tagging winding leads is M_1 and M_2 for the running windings and S_1 and S_2 for the starting windings. However, some leads in older motors are tagged as follows:

(1) S_1 and S_2 for starting, and
(2) R_3 and R_4 for running.

Note that for some motors, the leads may be color coded as follows:

(1) Red is T_1,
(2) Blue is T_2,
(3) Yellow is T_3, and
(4) Black is T_4.

The above color coding is typical. However, color coding can vary greatly from manufacturer to manufacturer.

See Figure 23-1 for a detailed illustration of connecting the leads of a motor based upon its tagging method.

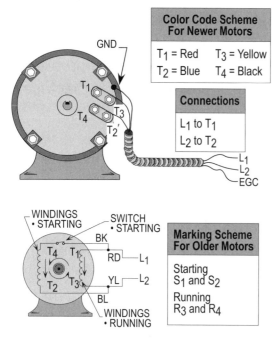

**CONNECTING WINDINGS PER
COLOR CODE AND MARKINGS**

Figure 23-1. The above illustration shows the windings of a single-phase motor being connected by markings and color coding of leads.

SINGLE-PHASE, SQUIRREL-CAGE INDUCTION MOTORS

Single-phase, split-phase induction motors consist of a housing and a laminated iron core stator with embedded windings located inside the motor housing. A rotor made of copper bars is set in slots in an iron core and connected by copper rings around both ends of the core with plates that are bolted to the housing. The motor enclosure also supports the bearings, the rotor shaft, and centrifugal switch. The centrifugal switch opens the circuit to the starting winding when the motor reaches its running speed.

> **Motor Tip:** The type of rotor mentioned above is often called a squirrel-cage rotor since the configuration of the copper bars resembles a cage.

TESTING WINDINGS

To detect defects in a single-phase, split-phase induction motor, both the running and starting windings must be tested for :

(1) Grounds,
(2) Open-circuits, and
(3) Short-circuits.

TESTING FOR GROUNDS

A winding is grounded if it makes an electrical contact with the metal of the motor housing, etc. To determine whether the winding is grounded, a continuity tester may be used. One test lead of the tester must be connected to the winding and the other lead to the motor frame. If a reading is taken, the winding is grounded to the motor's enclosure in some way. In like manner, one lead of the ohmmeter is connected to the winding and the other to the motor's frame and if a reading can be made, the winding is grounded.

TESTING FOR OPEN-CIRCUITS

The cause of an open-circuit in a split-phase motor can be a loose or dirty connection or broken conductor, which may be in either the running or the starting winding or centrifugal switch.

To determine whether the running winding is open or not, the leads of the tester or ohmmeter are connected to the ends of the winding. If the tester or the ohmmeter has a reading, the circuit is complete. If there is no continuity on the tester or no reading on the ohmmeter, there is an open-circuit. **(See Figure 23-2)**

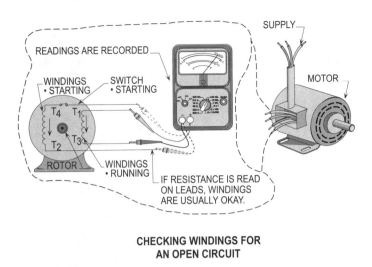

**CHECKING WINDINGS FOR
AN OPEN CIRCUIT**

Figure 23-2. If the windings to the run and start show continuity using a ohmmeter, the windings are usually okay.

TESTING FOR SHORT-CIRCUITS

Two or more turns of wire that make contact with each other electrically are the cause of short-circuits. There are other cases where excessive heat develops from overloads and makes the insulation defective and causes shorts to occur. A short-circuit is easy to spot because smoke comes from the winding while the motor is running or drawing excessive current at no load conditions. Any one of the following can be utilized to find a short-circuit:

(1) Running the motor and locating the hot winding. This winding is normally the one that is short-circuited when tested.
(2) Placing a growler on the core of the stator and moving it from slot to slot until a rapid vibration occurs. This coil is short-circuited.
(3) Connecting the winding to a low DC voltage and a voltage measurement taken. The winding with the least voltage drop is the one that is short-circuited.

See Figure 23-3 for a detailed illustration of how to locate short-circuits, using the growler method.

TESTING CAPACITORS

The reasons for a capacitor to suddenly become defective can be caused by any one of the following:

(1) Overheating, and
(2) Excessive voltage.

Note that a defective capacitor must be replaced with one having about the same value of capacitance. If one with a different capacitance value is used, the motor may not have the necessary starting torque to start and run the motor.

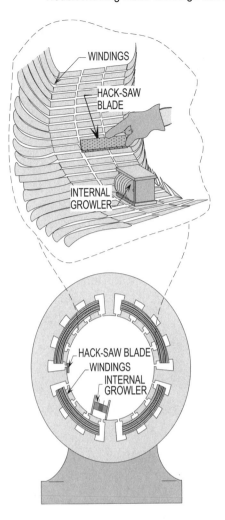

USING A GROWLER TO TEST FOR GROUNDS

Figure 23-3. The above illustration shows a growler and hack saw blade being utilized to test for short-circuits.

When checking capacitors, they must be removed from the circuit and their capacitance measured. Such measurement will detect either open or shorted capacitors. For best results, measure the resistance of a capacitor with an ohmmeter and if the capacitor is shorted, the meter will read less than 10 ohms. However, if the capacitor is good, the resistance reading will be about 50 ohms or greater.

An open-circuit on a capacitor can be checked by using a neon voltage tester. If the neon tester does not glow during the test, the capacitor can be considered defective. Note that this tester is not to be used on mica grid capacitors.

Capacitors can be checked by placing the leads of an ohmmeter to the capacitor bridge. If the needle pegs and then falls to zero, the capacitor is good. An in-line fuse can be connected to one side of the capacitor's terminal. If the fuse does not blow, the capacitor is good. **See Figure 23-4** for a detailed illustration of how to test capacitors for an open or short-circuit.

> **Motor Testing Tip:** Check the type of capacitor used before selecting the testing method to test the capacitor.

> **Motor Testing Tip:** If the fiber insulating washers do not correct the problem, the switch must be changed

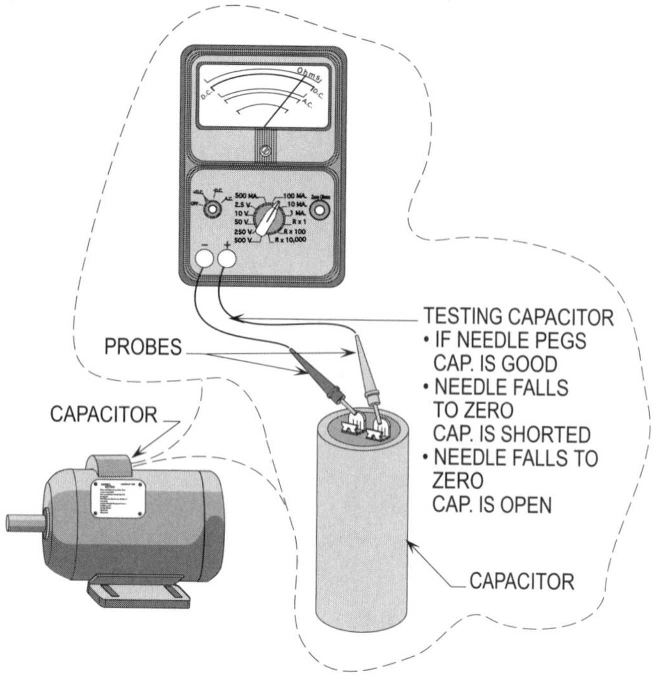

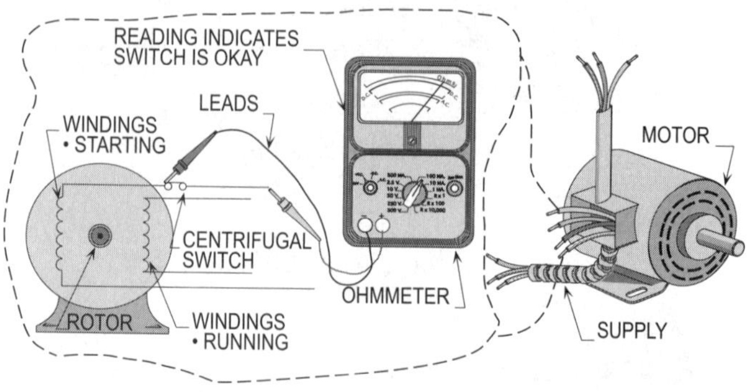

TESTING A CAPACITOR WITH AN OHMMETER

Figure 23-4. The above illustration shows a capacitor being tested with an ohmmeter.

TESTING CENTRIFUGAL SWITCHES

A centrifugal switch that is defective causes considerable trouble that is difficult to find unless the troubleshooter is familiar with the operating characteristics of such switches. If the switch fails to close when the rotor stops, the motor will not start again when supplied with line power. The switch's failure to close normally is caused by dirt, grit, or some other foreign material getting into the contacts of the switch.

If the centrifugal switch and the starting windings are to be tested for an open-circuit, connect the test leads to the starting winding circuit. If there is no reading, the contacts of the centrifugal switch may not be closed.

To verify this condition, the rotor can be pushed lengthwise toward the front end. If this causes the contacts to close the tester will show a reading. Such trouble can be corrected by adding several insulating washers to the pulley end of the motor shaft to push the rotor forward. If a reading of the tester cannot be taken, the trouble is in the centrifugal switch (starting switch). **See Figure 23-5** for the testing procedure to verify if a centrifugal switch is good or bad.

TESTING CENTRIFUGAL SWITCH WITH OHMMETER

Figure 23-5. The above illustration shows the testing of a centrifugal switch using an ohmmeter.

SINGLE-PHASE, SHADED POLE MOTORS

Shaded pole motors are single-phase induction motors equipped with a short-circuited auxiliary winding that is displaced in a magnetic position from the main winding. The auxiliary winding is called the shading coil and surrounds a portion of the pole. The main winding surrounds the entire pole and may consist of one or more coils per pole to provide the proper running power.

TESTING WINDINGS

The windings can be tested by using an ohmmeter. If the windings are not defective, the ohmmeter will have a reading which proves that the windings have continuity. A battery test light or lamp can also be used to make such test. For example, the test light will glow if the windings have continuity and will not glow if they are broken. **(See Figure 23-6)**

TESTING REVERSE SWITCHES

Reverse switches can be tested by placing one of the tester leads to the line side of the switch and the other lead to the load side. If a reading is taken, the switch contacts are good. Note that a multiposition switch with two or more speeds is checked using the same procedure. **(See Figure 23-7)**

UNIVERSAL MOTORS

Universal motors are an adaptation of series connected DC motors and they are so named universal from the fact that they can be connected on either AC or DC and operate in the same manner.

Basically, the universal motor contains field windings on the stator within the frame, an armature with the ends of its windings brought out to a commutator at one end. Carbon brushes are held in place by the motor's end plate, which allows them to have contact with the commutator.

When an AC or DC current is applied to a universal motor, such current flows through the field coils and the armature windings, which are in series. The magnetic field set up by the field coils in the stator reacts with the current-carrying wires on the armature and produces the desired rotation of the motor and equipment served.

TESTING WINDINGS

To ensure winding continuity, the field windings of a universal motor must be measured using an ohmmeter or light tester. If a reading cannot be measured, an open-circuit is present and the following test must be performed:

    (1) Test motor leads for an open-circuit,
    (2) Check brushes for the right setting,
    (3) Check the cleanness of the commutator,
    (4) Check the spring tension of the brushes riding on the commutator, and
    (5) Test the windings for grounds.

See Figure 23-8 for a detailed illustration of troubleshooting a universal motor.

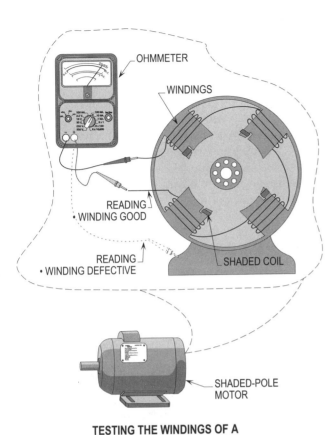

TESTING THE WINDINGS OF A
SHADED-POLE MOTOR

Figure 23-6. The above illustration shows the windings of a shaded-pole motor being tested for continuity.

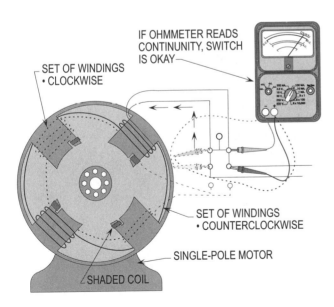

TESTING THE WINDINGS OF A
SHADED-POLE MOTOR

Figure 23-7. The above illustrates the procedure for testing the reversing switch to a shaded-pole motor.

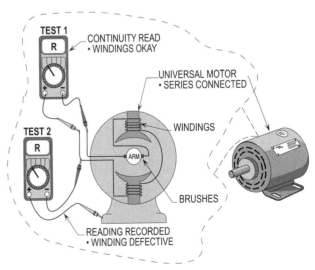

TESTING THE WINDINGS OF A UNIVERSAL MOTOR

Figure 23-8. The above illustration shows the procedure for testing the windings of a universal motor.

TESTING BRUSHES

The continuity of the brushes through the armature winding can be tested by placing one lead of the ohmmeter to one side of the brushes and other lead to the other side of the brushes. If a measurement can be read, the brushes are setting properly on the commutator. Therefore, good continuity should be made. If a reading is not obtained, the setting of the brushes must be checked as follows:

(1) Check for the wrong brush position,
(2) Check for brushes off-neutral plane, and
(3) Check setting of brushes riding on the commutator.

See Figure 23-9 for a detailed illustration on how to check the continuity of brushes.

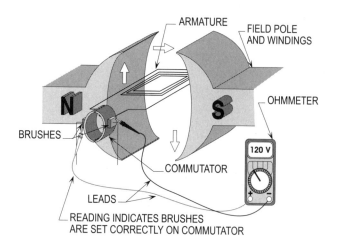

TESTING BRUSHES ON A UNIVERSAL MOTOR

Figure 23-9. The above illustration shows one of the methods used to test brushes on a universal motor.

SINGLE-PHASE REPULSION MOTORS

Single-phase repulsion motors are divided into the following types:

(1) Repulsion,
(2) Repulsion induction-run,
(3) Repulsion induction start, and
(4) Induction run.

Even though there are several motor types, there are specific construction characteristics that are definitely common to each type. Such characteristics are as follows:

(1) Each has a stator with a running winding similar to that of a split-phase motor.
(2) The rotor has a slotted core with embedded windings.
(3) Bearings are mounted in the end plates to support the rotor shaft.
(4) Carbon brushes are fitted in holders and ride on the commutator.
(5) Either the front end plate or the rotor shaft supports the brush holders.

Troubleshooting techniques are based on the type of repulsion motor under test.

TESTING WINDINGS

The windings of repulsion motors can be tested phase to ground by the use of an ohmmeter. If there is a reading from one lead of the motor windings to the frame of the motor, there is a short-circuit of some kind.

If the motor fails to start and run when the switch is energized, the trouble may be any one of the following:

(1) Burned out fuse or tripped CB.
(2) Worn bearings.
(3) Brushes stuck in the holder.
(4) Worn brushes.
(5) Open-circuit in the stator or armature.
(6) Wrong brush-holder position.
(7) Shorted armature.

See Figure 23-10 for a detailed illustration of how to test the windings of repulsion motors for continuity.

TESTING BRUSHES

The brushes can be tested by placing one lead of the ohmmeter to one side of the brushes and the other lead to the other side of the brushes. If a reading is measured, the continuity of the brushes to the commutator and armature windings are usually good.

See Figure 23-11 for the procedures to use when testing the conductivity of the brushes and their relationship to the commutator and armature windings.

TROUBLESHOOTING THREE-PHASE INDUCTION MOTORS

A typical three-phase induction motor has three main parts and they are as follows:

(1) Stator,
(2) Rotor, and
(3) End plates.

The stator consists of a steel frame and a laminated iron core and winding formed of individual coils placed in slots. The rotor may be constructed of a squirrel-cage or wound rotor type.

Three-phase induction motors have relatively constant speed characteristics and are available in designs that provide a variety of torque values. Some are designed to have a high starting torque and others with a low starting torque. Some draw a normal starting current while others are designed with a high starting current.

The end plates or brackets are bolted to each side of the stator enclosure and contains the bearings in which the shaft rotates freely.

CONNECTING LEADS AND WINDINGS

There are two methods by which to connect the stator windings of a three-phase induction motor to a three-phase power supply and they are as follows:

(1) Wye or star (人)
(2) Delta or triangle (Δ)

When using either method, the windings are so connected that only three leads come from the windings in the stator, which make the line connections a very simple task.

See Figures 23-12 (a) and (b) for a detailed illustration of how to connect the leads from the windings of three-phase motors to the power supply leads.

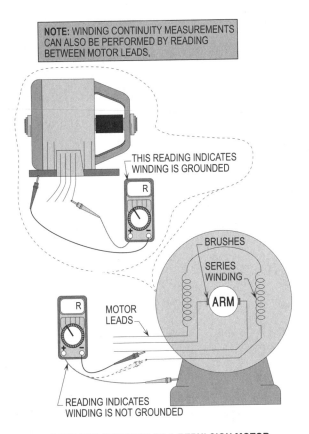

Figure 23-10. The above illustration shows the procedure for testing the windings of a repulsion motor.

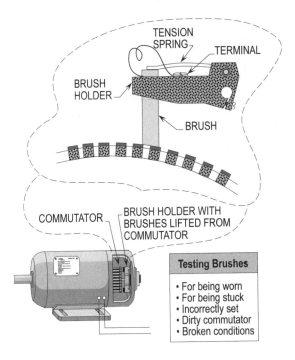

TESTING BRUSHES FOR A REPULSION MOTOR

Figure 23-11. The above illustrates the procedures for testing brushes in a repulsion motor.

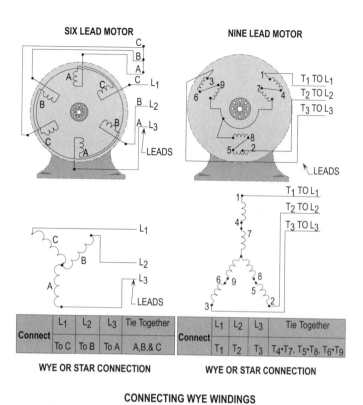

CONNECTING WYE WINDINGS

Figure 23-12(a). The above illustration shows the procedure for connecting six and nine lead motors in a wye configuration.

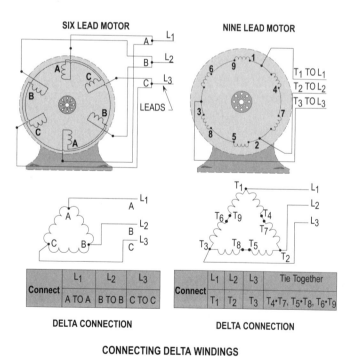

CONNECTING DELTA WINDINGS

Figure 23-12(b). The above illustration shows the procedure for connecting six and nine lead motors in a delta configuration.

TESTING AND FINDING THE LEADS OF WYE MOTORS

Wye-connected motors have four individual circuits to find and mark and they are as follows:

(1) Three circuits with two leads each, and
(2) One circuit with three leads.

TAKING MEASUREMENTS

The leads and internal connections can easily be identified by drawing the windings to resemble a wye and then in a right handed motion spirally decrease and number each winding end as shown in **Figure 23-13**.

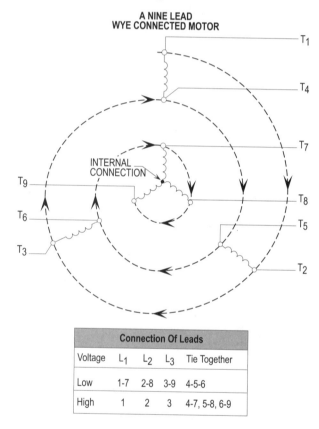

Connection Of Leads			
Voltage	L_1	L_2	L_3	Tie Together
Low	1-7	2-8	3-9	4-5-6
High	1	2	3	4-7, 5-8, 6-9

MARKING THE LEADS OF A WYE MOTOR

Figure 23-13. The above illustrates the procedure for marking the leads of a wye motor by drawing and following a decreasing spiral (circle) and numbering each lead as shown.

Each circuit winding can be determined by using an ohmmeter or continuity tester as follows:

(1) Connect one lead of the tester to any selected lead of the circuit and read for continuity between each of the other eight leads.

(2) When a continuity measurement between two other leads is found, the three-wire circuits have been found that makes up the internal winding of the wye.

(3) Readings must be continually made until all four circuits have been found.

(4) After finding and isolating the leads, mark the three-wire leads T_7, T_8, and T_9.

(5) Temporarily mark the other leads as follows:
  (a) T_1 and T_4 for one-circuit winding.
  (b) T_2 and T_5 for the second-circuit winding.
  (c) T_3 and T_6 for the third-circuit winding.

> **Testing Tip:** The circuit windings are now marked and ready for voltage testing.

See Figure 23-14(a) for a detailed illustration of applying this procedure.

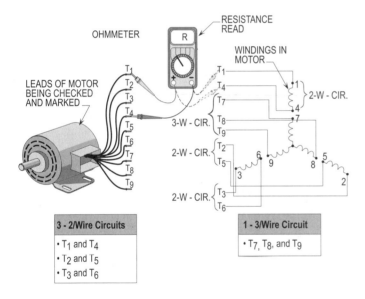

3 - 2/Wire Circuits
• T_1 and T_4
• T_2 and T_5
• T_3 and T_6

1 - 3/Wire Circuit
• T_7, T_8, and T_9

FINDING NUMBER OF CIRCUITS

Figure 23-14(a). The above illustration shows the procedure for determining the number of circuits in a wye motor and temporary marking them for testing procedures.

VOLTAGE TESTING WINDING CIRCUITS

The windings of wye-connected motors can be tested for correct markings by applying 240 volts to leads T_7, T_8, and T_9 respectively. Windings T_1 and T_4, T_2 and T_5, and T_3 and T_6 will act like the secondary of a transformer with T_7, T_8, and T_9 serving as the primary. In other words, a voltage is set up in the secondary by the applied voltage in the primary. There is a transformer relationship taking place in the

windings of the motor. After starting the motor and it is running, take readings on the induced windings (T_1 and T_4, T_2 and T_5, and T_3 and T_6) by using a voltmeter as follows:

(1) Read each circuit,

(2) Voltage reading should be about 125 to 130 volts,

(3) Voltage readings could be less than 125 to 130 volts, and

(4) Readings may be 75 to 85 volts which is okay as long as they are about equal.

See Figure 23-14(b) for a detailed illustration of applying this procedure.

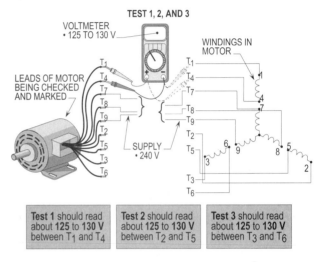

Test 1 should read about **125 to 130 V** between T_1 and T_4	Test 2 should read about **125 to 130 V** between T_2 and T_5	Test 3 should read about **125 to 130 V** between T_3 and T_6

TESTING LEADS MARKED T_1 - T_4, T_2 - T_5, AND T_3 - T_6

Figure 23-14(b). The above illustration shows leads T_7, T_8, and T_9 being supplied by 240 volts and leads T_1 and T_4, T_2 and T_5, and T_3 and T_6 being read to check if such circuits measure about 125 to 130 volts each. Motor is ready to test other leads. **(See Figure 23-14(c)).**

TESTING OTHER LEADS

With the motor running, the other lead that is temporarily identified as T_4 must be connected to T_7 and tested as follows:

(1) Read the voltage between T_1 and T_8, and

(2) Read the voltage between T_1 and T_9.

If both readings are measured with values of about 330 to 340 volts respectively, leads T_1 and T_4 can be permanently marked T_1 and T_4. However, if such readings are about 125 to 130 volts, reverse T_1 to T_4 and test remaining leads as follows:

(1) If readings between T_1 and T_8, and T_1 and T_9 are not equal,

(2) Disconnect T_4 from T_7 and connect T_4 to T_8 and the supply line.

(3) Read the voltage between T_1 and T_7 and T_1 and T_9.

(4) If voltage readings are equal and about 330 to 340 volts.

(5) Mark T_1 as T_2 permanently.

(6) Mark T_4 as T_5 permanently.

However, if readings are about equal between 125 to 130 volts, mark the leads as follows:

(1) Disconnect leads T_1 and T_4.

(2) Interchange and mark T_2 and T_5.

(3) Change T_1 to T_5.

(4) Change T_4 to T_2.

If the voltage readings are different, disconnect and reconnect as follows:

(1) Disconnect T_4 from T_8.

(2) Read between T_1 and T_7 and T_1 and T_8.

(3) If readings are about 330 to 340 volts are measured.

(4) Permanently identify T_3 and T_6.

See Figures 23-14(c) and (d) for a detailed illustration of applying the above procedures.

Note that the same methods of identification is used for the other two circuits which are temporarily marked T_2 and T_5 and T_3 and T_6. A position must be determined where both circuits have readings that are about equal and measure 330 to 340 volts. **(See Figure 23-14(e))**

After all the leads of the circuits have been marked, leads T_4, T_5, and T_6 must be connected together and voltage readings taken between T_1, T_2, and T_3. Such readings should have voltages measuring about 230 volts.

With the motor power off, disconnect leads T_7, T_8, and T_9 then connect leads T_1, T_2, and T_3 to the power supply line. T_1 must be connected to the line that T_7 was previously connected to and T_2 to the same line as T_8 was connected. T_3 has to be connected to the same line as T_9 and T_4, T_5, and T_6 is still connected together to make up the wye connection.

Start the motor unloaded and if all lead markings are right, the motor rotation in relationship with T_1, T_2, and T_3 connected in the same manner as when T_7, T_8, and T_9 were connected.

Motor Testing Tip: The above voltage readings are based upon a three-phase, 230/460 volt, induction motor. Voltage readings will vary for a motor having a different voltage.

Test 4: Connect T_4 to T_7 and test as follows:
- Read V between T_1 and T_8
- Read V between T_1 and 9

Note: If readings are about 330 to 340 V, mark leads T_1 and T_4 permanently.

Test 5: If readings are about 125 to 130 V, reverse T_1 and T_4 and test as follows:
- Readings between T_1 and T_8, and T_1 and T_9 are not equal.
- Disconnect T_4 from T_7 and connect T_4 to T_8 and line

Note: If readings are about 330 to 340 V, mark T_1 as T_2 and T_4 as T_5 permanently.

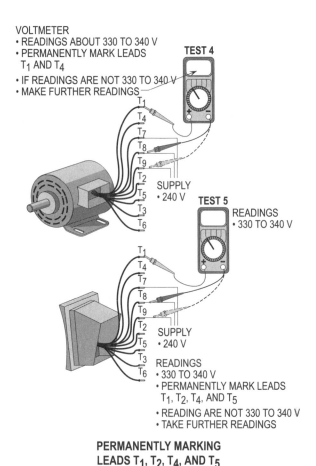

VOLTMETER
- READINGS ABOUT 330 TO 340 V
- PERMANENTLY MARK LEADS T_1 AND T_4
- IF READINGS ARE NOT 330 TO 340 V
- MAKE FURTHER READINGS

TEST 4

T_1
T_4
T_7
T_8
T_9
T_2
T_5
T_3
T_6

SUPPLY
- 240 V

TEST 5

READINGS
- 330 TO 340 V

T_1
T_4
T_7
T_8
T_9
T_2
T_5
T_3
T_6

SUPPLY
- 240 V

READINGS
- 330 TO 340 V
- PERMANENTLY MARK LEADS T_1, T_2, T_4, AND T_5
- READING ARE NOT 330 TO 340 V
- TAKE FURTHER READINGS

PERMANENTLY MARKING LEADS T_1, T_2, T_4, AND T_5

Figure 23-14(c). The above illustrates the procedure for identifying the leads of a wye motor.

Test 6: If readings from Test 5 are about 125 to 130 V, mark leads as follows and perform Test 7.
- Disconnect T_1 and T_4
- Interchange and mark T_2 and T_5
- Change T_1 to T_5 and T_4 to T_2

Note: If readings are the same as above, disconnect and reconnect as follows:
- Disconnect T_4 from T_8 and reconnect to T_9 and perform test.
- Take readings between T_1 and T_7, and T_1 and T_8.
- If readings are 330 to 340 V, permanently mark T_3 and T_6.
- Leads T_7, T_8, and T_9 should be correctly marked.

Final Test: If the other two circuits T2 and T5, and T3 and T6 read about 330 to 340 V, permanently mark and identify as such.

Note 1: If the V read is different, reconnect leads until such reading is measured.

Note 2: Rotation of motor can now be checked, and put into service.

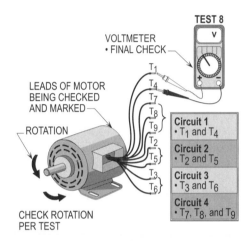

FINAL PROCEDURE FOR CHECKING AND MARKING LEADS

Figure 23-14(e). The above is the final procedure for testing, checking, and marking leads of a wye-connected motor.

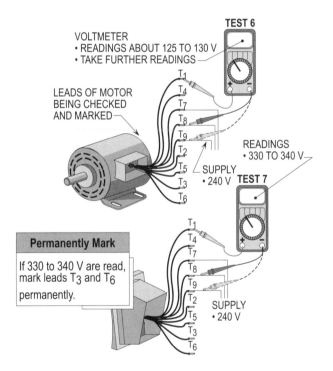

PERMANENTLY MARKING LEADS T_3 and T_6

Figure 23-14(d). The above illustrates the procedure for identifying the leads of a wye motor.

TESTING AND FINDING THE LEADS OF DELTA MOTORS

Delta-connected motors with nine leads have only three circuits with three leads each to find and mark.

TAKING MEASUREMENTS

The leads can be easily found by drawing the windings to resemble a delta triangle and then in a right hand motion spirally decrease and number each coil end as shown in **Figure 23-15**.

Each circuit winding can be determined by using a ohmmeter or continuity tester as follows:

(1) Read the resistance between one lead and the others until two match up to one winding of leads.

(2) The first lead is used to find the other two common windings.

(3) The common lead is marked T_1 and the other leads temporarily marked T_4 and T_9.

(4) The common lead of the next group is determined and marked T_2 with the other leads temporarily being marked T_5 and T_7.

(5) The common lead of the final group is found and marked T_3 with the other leads being temporarily marked T_6 and T_8.

After properly marking the leads, T_1, T_4, and T_9 are connected to a 240 volt power supply. With the motor subjected to no load conditions, lead T_7 is then connected to the power supply line. **(See Figure 23-16(a))**

Voltage readings are taken as follows to find and mark the leads of the motor:

(1) Read the voltage between T_1 and T_2.
(2) If the voltage reads about 460 volts, the markings are right and can be permanently identified.
(3) Readings of 400 volts or less interchange T_5 and T_7 or T_4 and T_9.
(4) Read the voltage again, if readings of about 230 volts are measured, then read 5.
(5) Interchange leads T_5 with T_7 and T_4 with T_9.
(6) The readings should now measure about 460 volts between T_1 and T_2.
(7) The leads connected together as T_4 and T_7 can be permanently identified.
(8) The remaining leads in each group can be marked as T_9 and T_5.

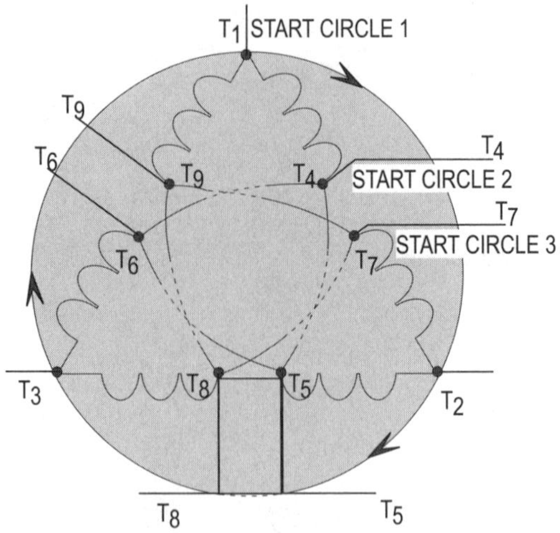

FINAL READINGS

Connect one of the leads to the last winding of the group and measure T_9, identify as follows:

(1) If about 460 volts is read between T_1 and T_3.
(2) The lead can be permanently marked as T_6.
(3) If a reading of 400 volts or less is measured, interchange T_6 and T_8.
(4) If 460 volts is read between T_1 and T_3.
(5) T_6 is changed to T_8 and permanently identified. Note that T_8 is changed to T_6.
(6) If about 460 volts can be read between leads T_1, T_2, and T_3 the leads are considered correctly marked.

See **Figures 23-16(b)** through **(g)** for testing procedures pertaining to Test 1 through Test 6.

Connection Of Leads				
Voltage	L_1	L_2	L_3	Tie Together
Low	1-6-7	2-4-8	3-5-9	———
High	1	2	3	4-7, 5-8, 6-9

MARKING THE LEADS OF A DELTA MOTOR

Figure 23-15. The above illustrates the procedure for marking the leads of a delta motor by drawing a decreasing spiral (circle) and numbering each lead as shown.

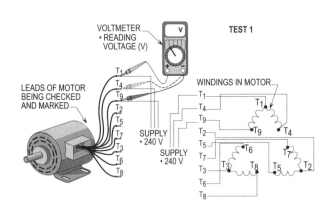

CONNECTING 240 V TO MOTOR LEADS

Figure 23-16(b). The above illustration shows 240 volt supply being connected to leads T_1, T_4, and T_9 so that the windings can be tested and permanently marked

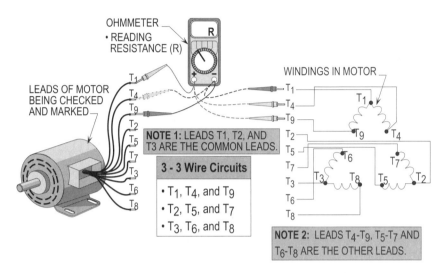

FINDING NUMBER OF CIRCUITS

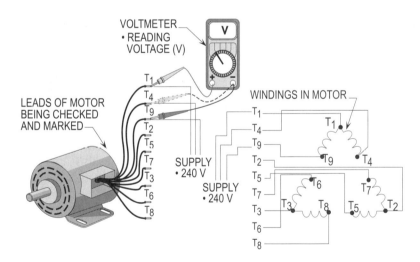

CONNECTING 240 V TO MOTOR LEADS

Figure 23-16(a). The above illustration shows the procedure for determining the number of circuits in a delta motor and temporarily marking them for testing purposes.

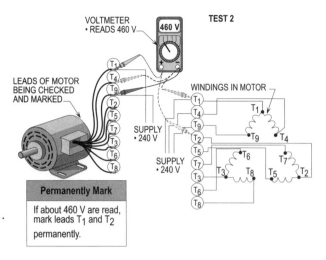

READING LEADS AND MARKING THEM

Figure 23-16(c). The above illustration shows the procedure for reading voltage on leads and marking them based upon voltage measurements.

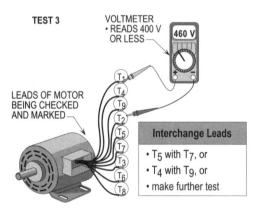

READING VOLTAGE AND MARKING LEADS

Figure 23-16(d). The above illustration shows a voltage reading on leads and either permanently marking them or making further tests to determine the permanent markings.

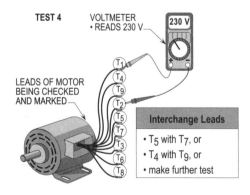

READING VOLTAGE AND MARKING LEADS

Figure 23-16(e). The above illustration shows a voltage reading on leads and either permanently marking them, or making further test.

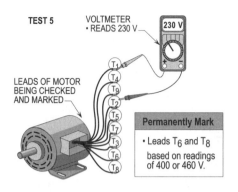

READING VOLTAGE AND MARKING LEADS

Figure 23-16(f). The above illustrations shows a voltage reading on leads and either permanently marking them, or making further test.

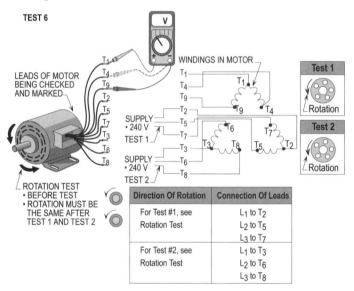

CHECKING LEAD IDENTIFICATIONS BY ROTATION OF MOTOR

Figure 23-16(g). Two tests are performed to identify all leads of a delta-connected motor. If the previous connections of T_1, T_4, and T_9 do not produce proper voltage measurements, connect leads T_5 with T_7 or T_4 with T_9 and then apply test 2 and the rotor of the motor should rotate in the same direction as in test 1. (See rotation test above)

DOUBLE CHECK MARKINGS

With the motor supply off, reconnect T_2, T_5, and T_7 to the supply line. Lead T_2 is connected to line as T_1. Lead T_5 is connected to where T_4 was previously connected. In like manner, line T_7 is connected where T_9 was connected. The motor should rotate in the same direction as before.

After stopping the motor and disconnecting the power source, connect leads T_3, T_6, and T_8 to the power supply line where T_2, T_5, and T_7 were previously connected. The motor should rotate in the same direction as before. **(See Figure 23-16(g))**

Motor Testing Tip: The above voltage readings are based upon a three-phase, 230/460 volt, induction motor. Voltage readings will vary for motors having a different voltage.

TROUBLESHOOTING WOUND-ROTOR MOTORS

Wound-rotor motors are induction motors that are equipped with stator windings called primary windings that are connected in a three-phase wye or delta configuration.

The rotors of such motors are wound with insulated windings that are connected with slip rings mounted on the rotors. Rotor windings are called secondary windings and they have the same number of poles as the stator windings.
Rotor windings are connected to external resistors by brushes and slip rings. The resistors can be used to reduce the starting current of the motor and also regulate the speed of the motor. Reduce the resistance in the rotor, the motor will speed up. Increase the resistance in the rotor, the motor will slow down.

TESTING SLIP RINGS

The continuity between the slip rings and brushes to the rotor can be interrupted by any of the following conditions:

    (1) Slip rings are dirty.
    (2) Slip rings are not set against the rotor.
    (3) Slip rings are broken.

Any condition above can be corrected by cleaning slip rings or refitting brushes to make good electrical contact. If slip rings are broken, they must be replaced. **(See Figure 23-17)**

TESTING BRUSHES

The continuity of the brushes to the rotor can be interrupted by any one of the following conditions:

    (1) Brushes are chipped or broken.
    (2) Brushes not making proper contact.
    (3) Variac setting not correct.

Any condition above can be corrected by replacing brushes or readjusting spring tension or completely replacing such. Reset variac as necessary to make proper continuity. **(See Figure 23-18)**

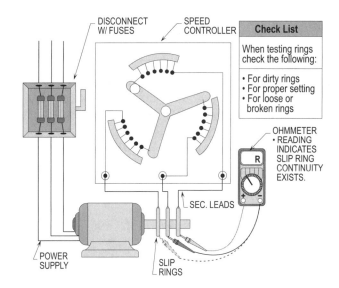

TESTING SLIP RINGS FOR CONTINUITY

Figure 23-17. The above illustration shows the main items to check when slip rings are considered the source of trouble.

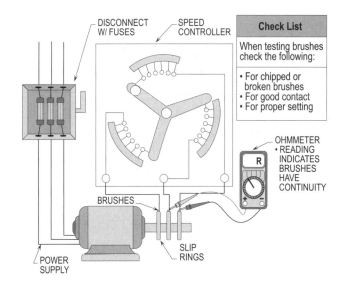

TESTING BRUSHES FOR GOOD CONTINUITY

Figure 23-18. The above illustration shows the procedure for checking the continuity of the brushes to the commutator.

TESTING CONTROLLER AND RESISTORS

The circuitry from the resistor bank or drum controller may be open. Such a condition won't allow the resistance to be increased or decreased to the rotor for starting or speed control. Any one of the following conditions can cause the above problem:

(1) Fuse may be blown.

(2) CB may be open.

(3) Windings of rheostat may be open.

To solve such a problem, check fuses and replace the one that is blown or verify if a CB is tripped open. Replace rheostat if defective or bridge across the resistor windings to complete circuit. **(See Figure 23-19)**

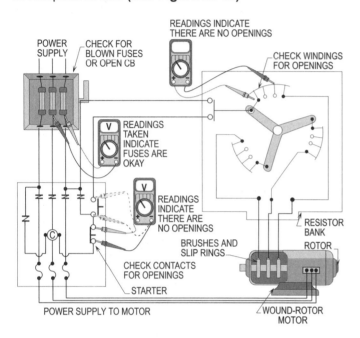

TESTING CONTROLLER AND RESISTORS

Figure 23-19. When checking the starting and running problems of wound-rotor motors, test for open contacts of start and stop buttons and resistor (rheostat) windings. Also test for blown fuses and CB's in primary windings that may be tripped open.

TESTING WINDINGS

To test windings, connect one lead of the ohmmeter to the frame of the motor and the other lead to one of the motor leads. If a reading is measured, then the winding is grounded. To ensure the test is adequately performed, move the test lead to each lead of the motor. However, if a measurement is not read to the frame of the motor, read the leads to the windings and a reading should be obtained. **(See Figure 23-20)**

TROUBLESHOOTING SYNCHRONOUS MOTORS

Synchronous motors have stators which are constructed in the same manner as regular squirrel-cage induction motors.

In addition, the rotors of synchronous motors have coils (damper windings) wound on laminated poles and connected to slip rings on shafts. A squirrel-cage winding is usually embedded in the pole faces to start the motor. For control, a small DC generator called an exciter is utilized to energize the rotor coils. Note that synchronous motors operate at synchronous speed. They can be controlled to produce leading current and thus improve on power factor.

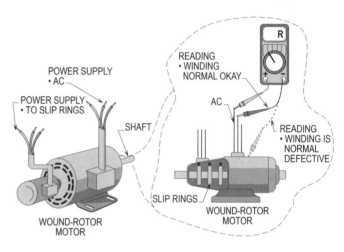

TESTING THE WINDINGS OF A WOUND-ROTOR MOTOR

Figure 23-20. The above illustration shows the procedure for checking the continuity of windings between phases and the ground. **Note:** Metal frame of the motor is grounded to the metal conduit or EGC in the circuit.

TESTING EXCITER

Another motor or damper windings on rotor can be used to start a synchronous motor. At some point, slightly below the motor's synchronous speed, a DC source of power is fed into the rotor through slip rings and the motor will run at its synchronous speed. The following test can be made to determine if the exciter is delivering DC current to the rotor:

(1) Test for a defective exciter,

(2) Test for DC power (voltage),

(3) Test for an open exciter circuit, and

(4) Test for a low exciter output.

If any of the above problems exist, they can be corrected by applying one of the following troubleshooting techniques:

(1) Turn the rotor by hand and check the exciter output.

(2) Check for a DC open-circuit.

(3) Check for a blown fuse.

(4) Check for an open CB.

(5) Check the variac windings for an open-circuit.
(6) Check the exciter variac for proper setting.
(7) Check the exciter variac for short.

If any of the defects above are found, fix or replace as necessary. **(See Figure 23-21)**

TESTING SLIP RINGS

DC voltage is applied through brushes to slip rings and then to the rotor. If the slip rings fail to conduct this voltage to the rotor, the motor will not function properly. Any one of the following problems can cause the slip rings to interrupt the DC supply to the rotor:

(1) Dirt on slip rings.
(2) Slip rings are broken.
(3) Slip rings are open.
(4) Open-circuit from the DC source.

Anyone of the conditions above can be corrected by cleaning or resetting the slip rings and by checking the circuitry to see if an open-circuit exists. A blown fuse must be replaced and a tripped CB must be reset to restore the DC power voltage to the rotor windings.

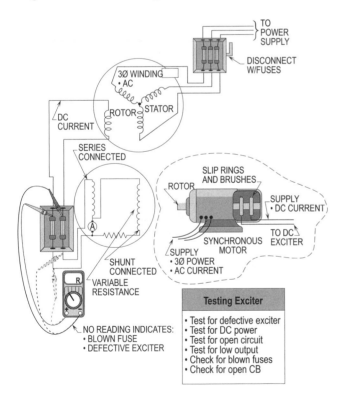

TESTING EXCITER OUTPUT

Figure 23-21. The above illustration shows the testing procedure for determining the output of the exciter power for a synchronous motor.

TROUBLESHOOTING DC MOTORS

The major parts of DC motors are the armature, field poles, frames, end brackets, and brush riggings. Armatures are the rotating parts of such motors and they consist of laminated iron cores with slots. Coils of wire are placed in these slots. Each core is pressed on a steel shaft which holds the commutator of a DC motor and current is conducted from the brushes to the coils in the slots.

TESTING THE OUTPUT

To check the output for an open-circuit, there are three complete circuits to be tested. (See **Figures 23-22(a) and (b)**) The testing can be performed as follows:

(1) Test for an opening in the armature circuit.
(2) Test for a problem in the brushes.
(3) Test for connection to the brushes.
(4) Test for openings in the series or shunt field.
(5) Test for a circuit reading from one winding to another.

Testing Procedure
• For opening of shunt windings, test between F_1 and F_2.
• For opening of series windings, test between S_1 and S_2.
• For opening of armature winding, test between A_1 and A_2.
Note: Three tests must be performed.

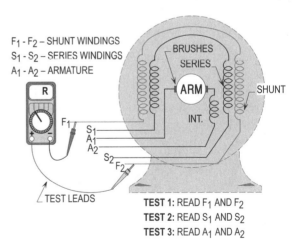

F_1 - F_2 – SHUNT WINDINGS
S_1 - S_2 – SERIES WINDINGS
A_1 - A_2 – ARMATURE

TEST 1: READ F_1 AND F_2
TEST 2: READ S_1 AND S_2
TEST 3: READ A_1 AND A_2

TESTING FOR OPEN CIRCUITS

Figure 23-22(a). The above illustrates the procedure for testing the armature, shunt windings, and series windings for open-circuits.

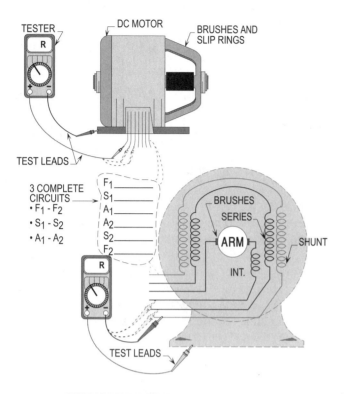

TESTING WINDINGS FOR OPEN OR GROUND

Figure 23-22(b). The above illustration shows the procedure for testing the windings of a DC motor.

Anyone of the above problems can be corrected by replacing or resetting the brushes as needed. If there is an opening in the series or shunt field, correct by fixing the loose connection on the open winding.

TESTING WINDINGS

The leads to DC motors can be identified by using an ohmmeter or continuity tester as follows:

(1) The first step is to find the three circuits of the armature, the series field, and the shunt field.
(2) By performing the test above, three pairs of leads are obtained.
(3) The ohmmeter will read a higher resistance for one pair of leads (shunt winding).
(4) The ohmmeter will read a lower resistance for the remaining two sets of leads.

MARKING LEADS

By removing the brushes, readings can be taken and the ohmmeter should not record a measurement. These leads are to be connected to the armature and must be marked A_1 and A_2.

The remaining pair connect to the series field leads. Such leads should be marked as follows:

(1) The shunt field leads must be marked F_1 and F_2.
(2) The series field leads must be marked S_1 and S_2.

After final checks, the motor is ready to be connected to the power supply and put into service. **(See Figures 23-22(a) and (b))**

TROUBLESHOOTING CONTROL CIRCUITS

Before attempting to troubleshoot a control circuit, it is necessary to understand the basic operation and construction of magnetic starters.

A typical magnetic starter consists of a magnet assembly, a coil, an armature, and contacts. The armature is controlled by current through such coil. The contacts are mechanically connected to the armature so that when the armature is in the closed position, the contacts are closed. This action of the starter connects power to the motor. When the coil is energized and the armature and contacts are in the closed position, the starter is in the picked-up position and the armature is in the sealed-in position.

Note that the coil has a fair amount of inrush current when energized by the control device. Such inrush current can be as high as 5 to 10 times the sealed-in current.

DETERMINING INRUSH CURRENT

Information on magnetic coils are normally listed in units of volt-amperes (VA) per manufacturers specifications. For example, for a magnetic starter rated 500 VA inrush and 50 VA sealed-in, the inrush current of a 120 volt coil is 500/120 volt, or 4.2 amps. A starter with a 480 volt coil pulls only 500/480 volt, or 1.04 amps inrush, and 50/480 volt, or .104 amps sealed-in current. (50 VA x 10 = 500 VA)

See Figure 23-23 for a detailed illustration of how to compute such inrush current.

TESTING FUSES

When testing fuses you must first test the incoming power supply line to verify if there is a voltage. If a reading between the ungrounded phases cannot be measured, check the following:

(1) Blown fuse,
(2) Open CB,
(3) Poor connections, and
(4) Broken switch blades.

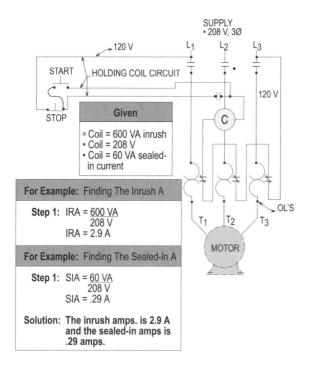

CALCULATING THE INRUSH AND SEALED-IN
AMPS OF A MAGNETIC STARTER

Figure 23-23. The above illustrates the procedure for computing the inrush and seal-in amps for a coil in a magnetic starter.

A voltage tester rated for the correct voltage can be used as follows:

(1) Test for voltage between L_1 and L_2.
(2) Test for voltage between L_1 and L_3.
(3) Test for voltage between L_2 and L_3.

If voltage readings are measured and they are the same as the supply voltage to the starter, there is not a problem with the circuit supplying the voltage to the line side of the magnetic starter. **(See Figure 23-24)**

Motor Testing Tip: The above test must never be taken for L_1, L_2, or L_3 to ground because a back-feed through the motor windings can be read with one blown fuse and the other two fuses can still be energized.

The fuses in the disconnect switch should be tested for defects as follows where the three-phase supply circuit is rated at 480 volts:

(1) A measurement of 480 volts to the load side of fuses L_1 and L_2 indicates L_1 and L_2 are not defective.
As shown in **Figure 23-25,** measurements between (L_1 and L_2), (L_1 and L_3) and (L_2 and L_3) indicates that fuse B is defective.

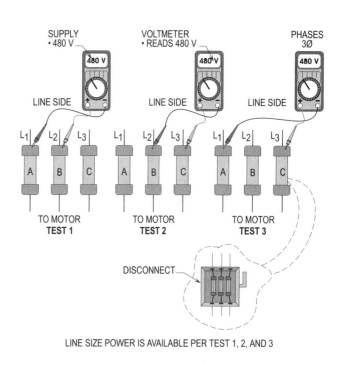

LINE SIZE POWER IS AVAILABLE PER TEST 1, 2, AND 3

TESTING POWER ON SUPPLY SIDE

Figure 23-24. The above illustrates the procedure for testing fuses. Test 1, 2, and 3 shows fuses are good because they have readings of 480 volts respectfully.

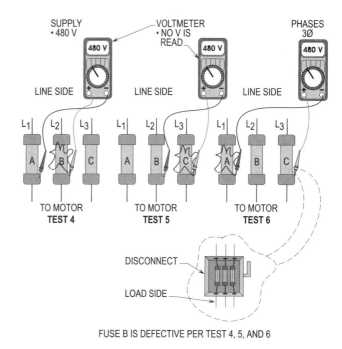

FUSE B IS DEFECTIVE PER TEST 4, 5, AND 6

TESTING FUSES FOR CONTINUITY

Figure 23-25. The above illustrates the procedures for testing fuses. Test 4, 5, and 6 shows fuses B, C, and A are defective because they do not have readings of 480 volt.

TESTING FOR POWER TO MAGNETIC STARTERS

At the line side of the magnetic starter, measure the voltage between L_1, L_2, and L_3. Readings of 480 volts between all three-phases indicates that supply voltage is available at these terminals. **(See Figure 23-26)**

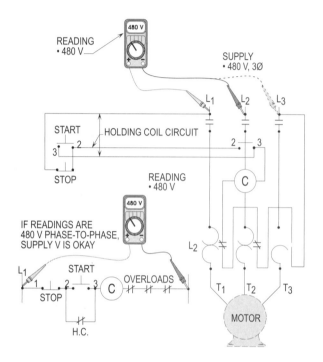

TESTING VOLTAGE TO LINE SIDE OF THE STARTER

Figure 23-26. The above illustrates the procedure for testing the voltage to the line side of the magnetic starter.

TESTING OVERLOADS

The procedure for testing the overloads (test 1) is to read the voltage from L_1 to the line side of L_2, the overload terminals. **(See Figure 23-27)**

If a reading is not measured, there is a broken or loose wire from L_1 to the line side terminal of the overloads previously checked. If a reading is recorded between L_1 and the load side terminal of the overloads, this indicates an open wire or overload contact is present. The following trouble points should be checked for a problem:

(1) Check overloads for open contacts.
(2) Reset overloads for continuity.
(3) Check for loose wires.
(4) Check for burned or discolored elements.

If the above trouble points are checked and measurements are read and they turn out to be correct, the overloads can be eliminated as a source of trouble. However, if readings are not recorded, the overloads can be the source of trouble. The procedure for test 2 is to read the voltage from L_1 to the control side of the coil.

> **Motor Testing Tip:** Do not read from the overload side of the coil to obtain measurement.

If a reading is measured here, the coil is usually good. However, if a reading is not obtained, the coil is normally defective and has an open path. For such a problem, check the following:

(1) Loose wire,
(2) Broken wire, or
(3) Defective coil.

To correct the above problem, fix loose or broken wire or replace coil if necessary. To continue the procedure for test 2, read the voltage between L_1 and the load side terminal of the same overloads. **(See Figure 23-28)**

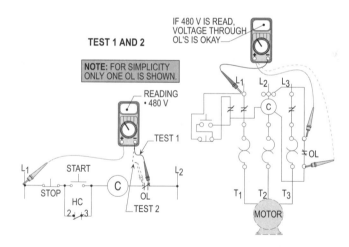

TESTING VOLTAGE TO THE OVERLOADS

Figure 23-27. The above illustrates the procedure for testing the voltage through the overloads and to the coil.

TESTING COIL

To perform test 1 above, read between L_1 and the side of the coil that is connected to the overloads. If a reading can be measured, the wire between the coil and overload is not defective. However, no measurement recorded indicates a loose connection, broken wire or defective coil.

Test 2 is accomplished by reading between L_1 and the control device side of the coil. If a measurement is recorded, the coil should be good. No reading between these points indicates that there is an open path. This open path can be caused by a defective coil, loose connection, or defective overload. **(See Figure 23-29)**

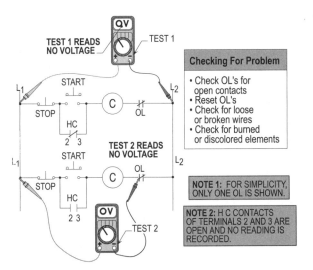

CHECKING PROBLEMS THAT CAUSE
INTERRUPTION OF VOLTAGE TO OL'S

Figure 23-28. The above illustrates common problems that could cause interruption of the voltage through the OL's to the coil terminals.

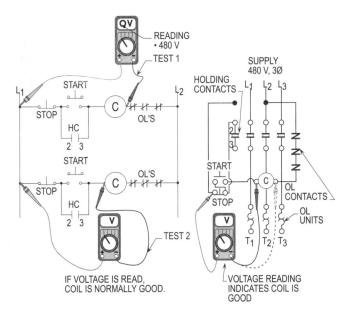

TESTING FOR DEFECTIVE COIL

Figure 23-29. The above illustration shows the procedure for testing the voltage to a coil to determine if it's defective.

TESTING CONTROL DEVICES

There are many types of control devices which are used to energize and deenergize the power to the coil that causes the contacts in the magnetic starter to close. The closing of these contacts provides line voltage to the windings of the motor. The following are procedures utilized to troubleshoot two-wire or three-wire control devices.

TESTING TWO-WIRE DEVICES

First perform a reading between L_1 and the side of the two-wire device connected to the coil. If a reading is measured here, the two-wire switch to the coil is not defective. The two-wire switch can be checked by reading each side of the switch. If a reading between the terminals of the switch can be read, the switch is not defective. If no measurement is recorded, it is an indication that the switch is open or defective and must be replaced. However, many times loose or bad connections between conductors and switch terminals is the source of trouble. Therefore, always check for these problems, by visually looking for arcing, burning, or discolored wire and terminals. **(See Figure 23-30)**

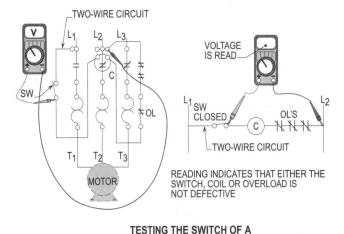

TESTING THE SWITCH OF A
TWO-WIRE CIRCUIT

Figure 23-30. The above illustration shows the procedure for testing the voltage of a two-wire control circuit. If the switch or overload contacts are good, voltage should be read at the coil.

TESTING THREE-WIRE DEVICES

The general procedure for troubleshooting a problem in a three-wire control circuit such as a start and stop push button station can be performed as follows:

(1) Check the circuit OCPD for:
  (a) Power circuit voltage.
  (b) Control circuit voltage.
(2) Open contacts of the stop button.
(3) Closed contacts of the start button.
(4) Open overload contacts.
(5) Defective coil.
(6) Holding contacts of the magnetic starter.

TESTING THE CIRCUIT

Test for voltage between L_1 and L_2 including L_3 if necessary. If there is no voltage measured, one or more fuses are blown or a CB has opened. Don't overlook broken or loose wires, for many times they can be the cause of lost power. However, if voltage readings are present, eliminate problem No. 1 as a source of trouble and move on to problem No. 2. **(See Figure 23-31)**

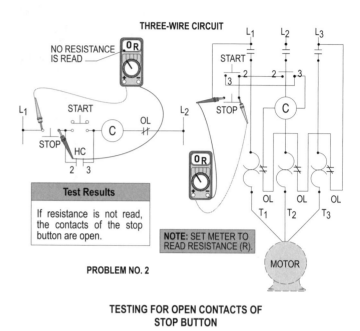

TESTING FOR OPEN CONTACTS OF STOP BUTTON

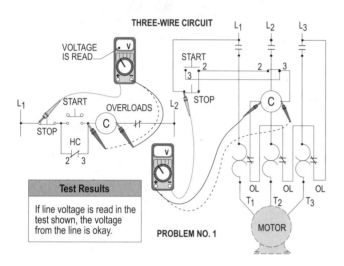

Figure 23-31. The above illustration shows the procedure for testing the line voltage from the line to the contacts of the stop and start button and coil.

Figure 23-32. The above illustration shows the procedure for testing the contacts of a stop button in a three-wire control circuit, using a resistance testing procedure.

TESTING FOR OPEN CONTACTS ON THE STOP BUTTON (NO POWER)

To perform an ohmmeter test, read between L_1 and the side of the stop button that is connected to the start button and terminal 2 of the holding contacts. If a reading is measured, problem No. 2 of open contacts in the stop button can be eliminated and problem No. 3 is now considered. **(See Figure 23-32)**

TESTING FOR CLOSED CONTACTS OF THE START BUTTON (NO POWER)

To perform this test, read from the side of the stop button connected to L_1 and to the side of the start button terminated to the holding coil. If no resistance can be read, the contacts of the start button are open. Note that a recorded measurement indicates that a jammed contact or loose wire is the source of trouble. By eliminating problem No. 3 as the trouble, move on to problem No. 4 and test for voltage through the overloads. **(See Figure 23-33)**

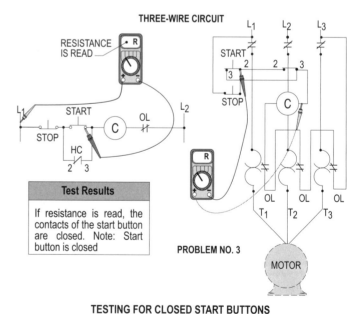

TESTING FOR CLOSED START BUTTONS

Figure 23-33. The above illustration shows the procedure for testing the contacts of a start button in a three-wire control

TESTING FOR OPEN OVERLOAD CONTACTS

Using the leads of the tester, read for voltage between L_1 and the side of the overloads connected to L_2. If no voltage measurement can be read in test 1, there is a loose or broken wire from L_2 through the overloads. If the reading in test 2 is okay, eliminate problem No. 4 and move on to problem area No. 5. **(See Figure 23-34)**

TESTING FOR DEFECTIVE COILS

To perform this test, read from L_1 to the side of the coil connected to the start button contacts and the holding coil contacts. If a voltage reading is recorded by the voltmeter, the coil is assumed to be okay. To be sure, read the coil for continuity using an ohmmeter, after removing from circuit. **(See Figure 23-35)**

TESTING HOLDING CONTACTS

This test can be made by testing from L_1 to terminal L_2 of the holding coil with the start button energized. A voltage reading should be measured indicating that the contacts are closing when the coil circuit is made. If no voltage reading is recorded, the voltage to the coil is restricted by open contacts in the start button. As always, consider the possibility of loose or broken wires at the contacts of the stop or start buttons, etc. **(See Figure 23-36)**

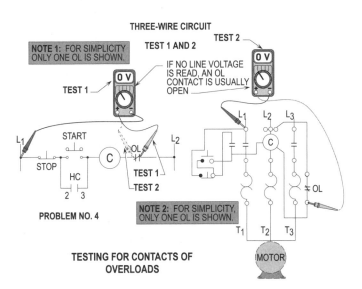

Figure 23-34. The above illustration shows the procedure for testing the contacts of the overloads in a three-wire control circuit.

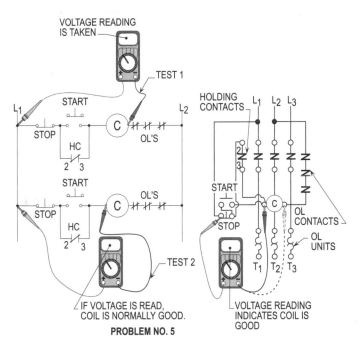

Figure 23-35. The above illustration shows the procedure for testing a coil in a three-wire circuit to determine if it's defective.

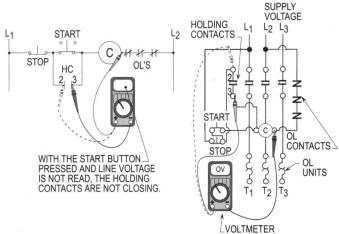

Figure 23-36. The above illustrates the procedure for testing the contacts of the holding circuits in a three-wire control circuit.

TROUBLESHOOTING VARIABLE FREQUENCY DRIVES (VFD)

First of all, the drive should be installed away from any high heat-producing apparatus. The air flow around the drive unit must be maintained at all times, with no other equipment or materials restricting air flow. Air conditioning should be provided to help cool the drive unit and its components.

Most troubleshooting procedures may be performed with the following tools: voltmeter, oscilloscope, digital voltmeter, AC ammeter, clamp-on ammeter, and standard hand tools.

Today's VFD systems have self-diagnostic readouts. It is imperative to have the manufacturer's manual for a particular unit and to follow the manufacturer's instructions on the use of the diagnostic system.

For basic troubleshooting:

(1) check incoming AC power supply;
(2) check all OCPD's;
(3) check output to motor;
(4) check tach feedback to controller if used;
(5) if need be, consult manufacturer's manual for correct troubleshooting procedure.

Figure 23-37 illustrates the troubleshooting procedures using a readout board to determine the cause of trouble. Note that within the adjustable speed drive system, there are low and high voltages of both AC and DC present.

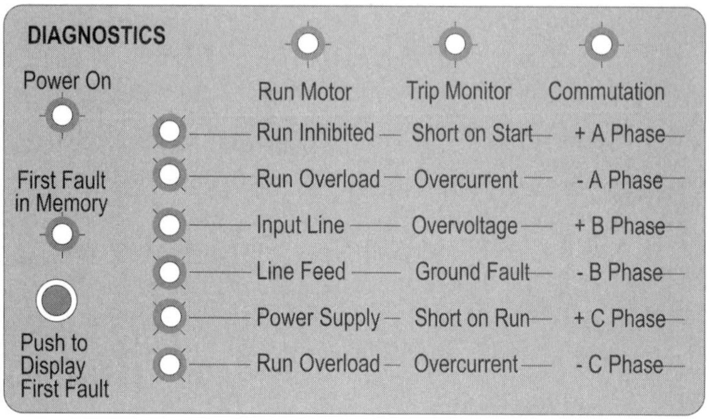

Figure 23-37. The above illustration shows a readout board being used to troubleshoot a problem with the operation of an adjustment speed drive unit. Note that this is just one of the readout solutions which can be performed when troubleshooting a drive unit. For troubleshooting tips, See Table 6 of the Appendix.

For safety, ALWAYS:
(1) Check all AC power sources.
(2) Check the DC bus for voltage.
(3) Allow sufficient time to discharge any power supply capacitors.

(4) The equipment ground and DC bus ground may not be the same potential. Most DC bus grounds float. Do not let the oscilloscope cabinet touch the chassis of the adjustable speed controller.
(5) Incorrectly connecting the leads to the SCR's will damage the SCR's.

Shorted diodes or SCR's can be easily determined by checking across their terminals with a volt-ohmmeter. Good rectifier cells have infinite resistance with reverse polarity and read approximately mid-scale with forward polarity. SCR's have resistance readings ranging from 12K to infinity. **(See Figure 23**

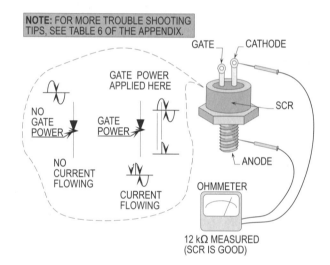

Figure 23-38. SCR's can be tested by using an ohmmeter. The ohmmeter will either measure 12,000 ohms, or the needle will peg, if the SCR is good.

TROUBLESHOOTING EDDY-CURRENT DRIVES

Eddy-current drive controllers usually consist of a voltage reference circuit, anti-hertz circuit, DC amplifier circuit, feedback circuit, and various potentiometers for controlling output speed of the magnetic drive system.

If a higher drive speed is designed, the speed potentiometer can be turned to a higher setting, which results in the following:

(1) The DC output voltage of the potentiometer is increased. (command voltage)

(2) When the command voltage is increased, an *error* voltage is produced between the command voltage and the *feedback* voltage coming from the tachometer generator (or magnetic pickup frequency to voltage converter). The error causes an increased voltage to be applied to the *magnetic drive coupling* field through a silicon controlled rectifier (SCR). This increased field excitation causes the magnetic drive coupling to accelerate until the feedback voltage signal again aligns to the command voltage signal. **Note:** The difference between the command voltage and the feedback voltage is the error voltage. When the drive is operating at a steady speed, the error voltage is practically zero. For more detailed information on troubleshooting Eddy-Current Drives, See Table 5 of the Appendix.

TESTING CERTAIN COMPONENTS

Testing a diode can be done by using an ohmmeter. If a diode (forward biased) measures a low resistance with the ohmmeter connected across it and a high resistance with the diode reversed (reversed biased), the diode, in all probability is okay. However, if the diode shows either high or low resistance in both directions, it is usually open or shorted and needs replacing **(See Figure 23-39)**
Triacs and diacs are much more difficult to test their diodes. When using an ohmmeter, both devices should normally show an open circuit in both directions. If they do not, they are almost always defective. **(See Figures 23-40 (a) and (b))**

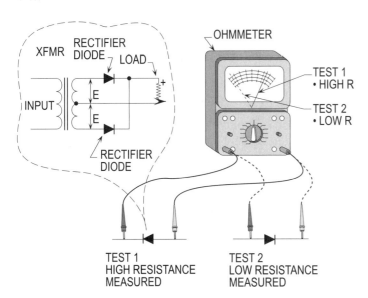

Figure 23-39. If a diode has a high resistance measurement in one direction and a low resistance reading in the reversed direction, the diode is normally good.

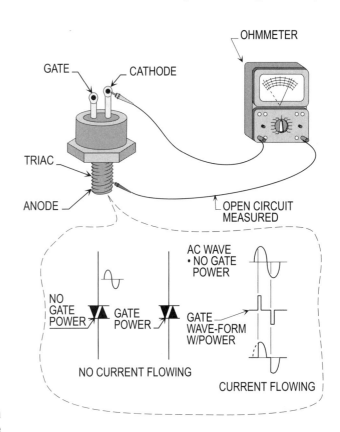

Figure 23-40(a). When testing a triac with an ohmmeter, if the measurement reads an open circuit, the triac is usually not defective.

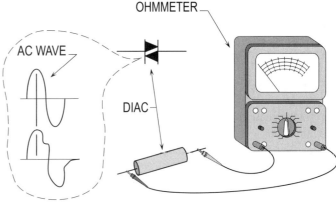

Figure 23-40(b). When testing a diac with an ohmmeter, if the measurement reads an open circuit, the diac is normally not defective.

TROUBLESHOOTING THE WINDINGS OF A WYE MOTOR FOR GROUNDS

To test for grounds in a wye-connected motor, connect one test lead to the frame of the motor and one test lead to the to one of the leads of the motor. If a resistance reading is measured, a winding is grounded. To ensure a more accurate test, move the test lead to each lead of the motor and take resistance measurements.

For checking each winding individually in a star-connected motor, disconnect the windings at the star point and test each winding for a ground. **(See A in Figure 23-41)**

See B in **Figure 23-41** which shows the procedure for checking all the windings for a ground to the frame of the motor.

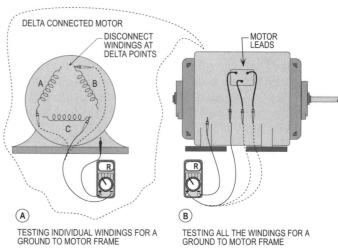

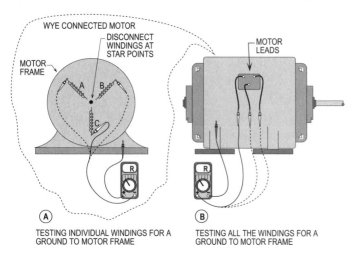

Figure 23-41. If a resistance is read on any winding or all the windings, a ground to the frame of the motor is usually present.

TROUBLESHOOTING THE WINDINGS OF A DELTA MOTOR FOR GROUNDS

To test for grounds in a delta connected motor, connect one test lead to the frame of the motor and one test lead to one of the leads of the motor. If a resistance reading is measured, a winding is grounded. To ensure a more accurate test, move the test lead to each lead of the motor and take resistance measurements.

For checking each winding individually in a delta-connected motor, disconnect the windings at the delta point and test each winding for a ground. **(See A in Figure 23-42)**

See B in **Figure 23-42** which shows the procedure for testing all the windings for a ground to the frame of the motor.

Note: For more troubleshooting tips, see Tables 1 through 11 of Appendix A in the back of this book.

Figure 23-42. If a resistance is read on any winding or all the windings, a ground to the frame of the motor is usually present.

TROUBLESHOOTING SOLID STATE CIRCUIT BOARDS

Assuming that a motor in a process machine will not start and run, the procedure for troubleshooting the circuitry to and from the solid state board(s) can be performed as shown in **Figure 23-43**.

TROUBLESHOOTING PROCEDURES FOR DETERMINING WHY THE MOTOR WILL NOT START AND RUN

1. Check to see if there is proper voltage going to L1 and L2 on the Input Board.

2. Check to see if there is proper voltage going to L1, L2, and L3 on the Output Board.

3. Check the output coming from the Feedback Board. If output is present, go to step 11.

4. Check the output coming from the Input Board. If output is present, go to step 8.

5. Check that the Start/Stop circuit works properly.

6. Check the Input Device for a proper signal going to Input Board.

7. Replace the Input Board. Problem should now be taken care of.

8. Check Feedback Device #1 for proper feedback.

9. Check Feedback Device #2 for proper feedback.

10. Replace the Feedback Board. Problem should now be take care of.

11. Check the output coming from the Output Board. if output is present, go to step 14.

12. Make sure there is power going to the Output Board.

13. Replace the Output Board.

14. Check if there is anything mechanically wrong with the system.

15. Replace the motor, if necessary.

Note that the problem can be the circuitry leading into or out of the solid state board. However, if the circuitry or the solid state board is not the cause of trouble, then it would be logical to conclude that there are mechanical problems or a defective motor.

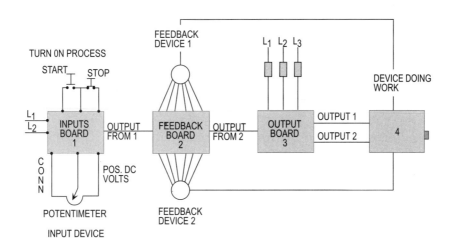

Figure 23-43. The above diagram shows the components that are checked to determine why the motor will not start and run. (See the Troubleshooting Procedures)

Chapter 23: Troubleshooting Motor Windings and Components

Section   **Answer**

_____   T   F   **1.** The running winding leads motors may be color coded, yellow for T_3 and black for T_4.

_____   T   F   **2.** A winding is grounded if it makes an electrical contact with the metal of the motor housing of a single-phase squirrel-cage induction motor.

_____   T   F   **3.** Two or more turns of wire that make contact with each other electrically are the cause of short-circuits.

_____   T   F   **4.** If a capacitor is shorted, the ohmmeter will normally read less than 25 ohms.

_____   T   F   **5.** The windings of shaded pole motors can be tested by using a voltmeter.

_____   T   F   **6.** If the field windings in a universal motor cannot be measured, an open-circuit is present.

_____   T   F   **7.** If there is a reading from one lead of the motor windings to the frame of a single-phase repulsion motor, there is no short-circuit involved.

_____   T   F   **8.** The windings of wye-connected motors can be tested for correct markings by applying 240 volts to leads T_1, T_2, and T_3.

_____   T   F   **9.** Delta-connected motors with nine leads have only three circuits with three leads each to find and mark.

_____   T   F   **10.** To test the windings for a wound-rotor motor to ground, connect one lead of the ohmmeter to the frame of the motor and the other lead to one of the motor leads.

_____ _____   **11.** Information on magnetic coils are normally listed in units of volt-amperes (VA) per manufacturers specifications.

_____ _____   **12.** At the line side of the magnetic starter, measure the voltage between L_1, L_2, and L_3 to verify if supply voltage is correct.

_____ _____   **13.** When testing the coil for control circuits, read between L_1 and the side of the coil connected to the overloads.

_____ _____   **14.** When performing a test on stop buttons, read between L_1 and the side of the stop button that is connected to the start button and terminal 3 of the holding contacts.

_____ _____   **15.** Holding contacts can be tested from L_1 to terminal 3 of the holding contacts with the start button deenergized.

_____ _____   **16.** New single-phase motors are tagged to identify the starting winding leads as _____ and _____.

_____ _____   **17.** To determine whether the winding is grounded, a _____ tester may be used.

_____ _____   **18.** Capacitors can become defective by _____ and _____ operation.

_____ _____   **19.** An open-circuit on an capacitor can be checked by using a _____ voltage tester.

_____ _____   **20.** The _____ of a motor consists of a steel frame and a laminated iron core and a winding formed of individual coils placed in slots for three-phase motors.

_____ _____   **21.** The stator for a three-phase motor can be connected _____ or _____.

_____ _____   **22.** When testing slip rings for synchronous motors, a_____ voltage is applied through brushes to slip rings and to the rotor.

_____ _____   **23.** When testing two-wire devices for control circuits, read between _____ and the side of the two-wire device connected to the coil.

_____ _____   **24.** When testing for a defective coil in control circuits, measure the voltage from L_1 to the side of the coil connected to the _____ button contacts and the terminals of the holding coil contacts.

_____ _____   **25.** When testing holding contacts for control circuits, a voltage reading should be measured indicating that the contacts are _____ when the coil circuit is energized.

_____ _____   **26.** Running windings are identified for older single-phase motors as:
(a) T_1 and T_2 (c) M_1 and M_2
(b) T_3 and T_4 (d) S_1 and S_2

_____ _____   **27.** To detect defects in a single-phase, split-phase induction motor, both the running and starting windings must be tested for:
(a) Grounds   (c) Short-circuits
(b) Open-circuits   (d) All of the above

_____ _____   **28.** When checking capacitors that are good, the resistance reading will be about _____ ohms.
(a) 10   (c) 25
(b) 20   (d) 50

_____ _____   **29.** Shaded pole motors are equipped with a short-circuited _____ winding that displaces in a magnetic position from the main winding.
(a) Auxiliary   (c) Running
(b) Starting   (d) All of the above

_____ _____   **30.** Three-phase motors with six leads are connected in a wye configuration by:
(a) L_1 to A; L_2 to B, L_3 to C(c) L_1 to C, L_2 to B, L_3 to A
(b) L_1 to B, L_2 to A, L_3 to C   (d) L_1 to C, L_2 to A, L_3 to B

_____ _____   **31.** Three-phase motors with nine leads are connected in a wye configuration by:
(a) L_1 to T_1; L_2 to T_2; L_3 to T_3 (c) L_1 to T_2; L_2 to T_3; L_3 to T_1
(b) L_1 to T_3; L_2 to T_2; L_3 to T_1 (d) L_1 to T_3; L_2 to T_1; L_3 to T_2

32. Three-phase motors with nine leads are connected in a delta configuration by:
      (a) T_4 to T_9; T_5 to T_8; T_6 to T_7 **(c)** T_4 to T_5; T_6 to T_7; T_8 to T_9
      (b) T_4 to T_7; T_5 to T_8; T_6 to T_9 **(d)** T_4 to T_7; T_5 to T_7; T_8 to T_9

33. The continuity between slip rings and brushes to the rotor can be interrupted in a wound-rotor motor by:
      (a) Slip rings that are dirty   **(c)** All of the above
      (b) Brushes not set properly   **(d)** None of the above

34. By removing the brushes to DC motors, ohmmeter readings can be taken and should zero when connected to the armature leads that are marked:
      (a) A_1 and A_2 **(c)** F_1 and F_2
      (b) M_1 and M_2 **(d)** S_1 and S_2

35. When testing for closed contacts of the start button in a control circuit, read from the side of the stop button connected to L_1 and to the side of the start button terminated to the coil and terminal _____ of the holding contacts.
      (a) 1   **(c)** 3
      (b) 2   **(d)** 4

36. Connect the leads for a wye configuration six lead motor.

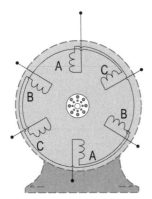

     ___ C

     ___ B

     ___ A

     ___ C L_1

     ___ B L_2

     ___ A L_3

SIX LEAD WYE MOTOR

37. Connect the leads for a delta configuration six lead motor.

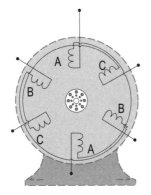

     ___ A   ___ L_1

     ___ B   ___ L_2

     ___ C   ___ L_3

SIX LEAD DELTA MOTOR

_____ _____

38. Number the leads for each winding in a delta configuration nine lead motor?

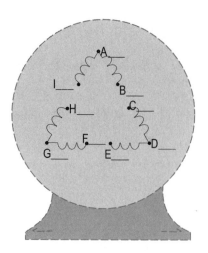

NINE LEAD DELTA MOTOR

_____ (A)_____

(B)_____

(C)_____

(D)_____

(E)_____

(F)_____

(G)_____

(H)_____

(I)_____

39. Number the leads for each winding in a wye configuration nine lead motor?

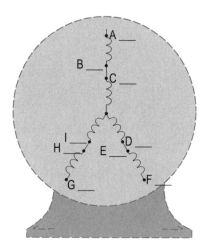

NINE LEAD WYE CONNECTOR

_____ (A)_____

(B)_____

(C)_____

(D)_____

(E)_____

(F)_____

(G)_____

(H)_____

(I)_____

40. Draw where the testing leads are to be placed when testing the voltage to the line side of the magnetic starter. (For simplicity, test each overload.)

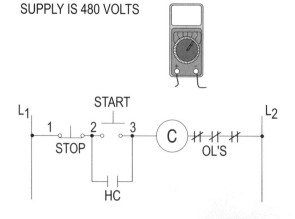

SUPPLY IS 480 VOLTS

41. Draw where the testing leads are to be placed when testing the voltage through the overloads and to the coil.

_____ _____

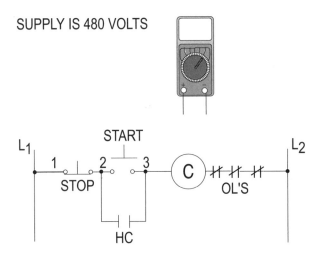

42. Draw where the testing leads are to be placed in two different tests to determine the common problems that could cause the interruption of voltage through the OL's to the coil terminals.

_____ _____

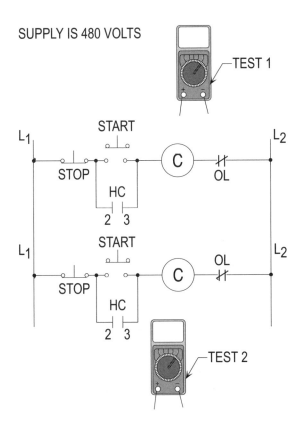

_____ _____

43. Draw where the testing leads are to be placed for two different tests to determine the voltage to a coil if it's defective. (Voltage is measured at Test 1 and Test 2.)

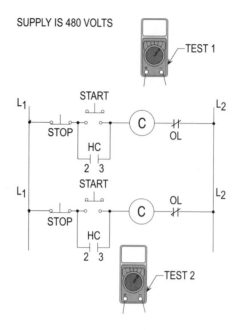

_____ _____

44. Draw where the testing leads are to be placed when testing the line voltage to the contacts (HC) of the stop and start buttons.

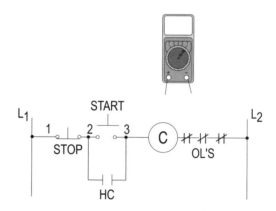

_____ _____

45. Draw where the testing leads are to be placed when testing the contacts of a stop button in a three-wire control circuit useing resistance.

46. Draw where the testing leads are to be placed when testing the contacts of a start button in a three-wire control circuit.

_____   _____

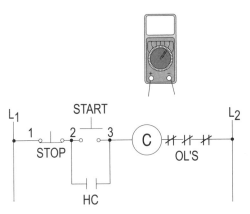

47. Draw where the testing leads are to be placed in two different tests for testing the contacts of the overloads in a three-wire control circuit.

_____   _____

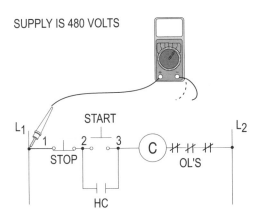

48. Draw where the testing leads are to be placed in two different tests for testing a coil in a three-wire circuit to determine if it's defective. (No voltage is measured at Test 1.)

_____   _____

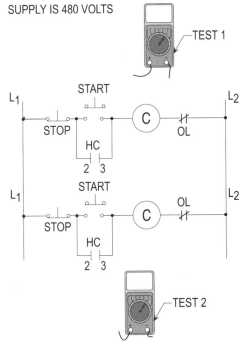

49. Draw where the testing leads are to be placed when testing the contacts of the holding circuit in a three-wire control circuit with the start button held closed..

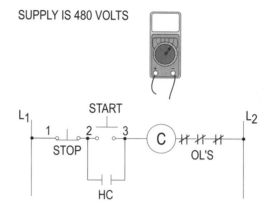

SUPPLY IS 480 VOLTS

Compressor Motors

Article 440 deals with individual or group installations having hermetically sealed motor compressors. The techniques for designing the proper size conductors, disconnecting means, and controllers are outlined and discussed.

The conductors supplying power to heating, air-conditioning, cooling and refrigeration (HACR) equipment are sized from the full-load amp (FLA) ratings of the compressor and condenser motor. To compensate for the starting periods and overload conditions, these FLA ratings are increased by 125 percent per **440.32.**

The overcurrent protection devices (OCPD's) protecting the branch-circuits from short-circuit and ground-fault currents are sized from the provisions listed in **440.22(A)**, which requires the FLA ratings to be increased from 175 up to 225 percent to allow the HACR equipment to start and run without tripping the OCPD ahead of the circuit.

The components used to supply the branch-circuits to HACR equipment may be required to be selected by the branch-circuit selection currents listed on the nameplate of the equipment per **440.4(C)** and **110.3(B)**.

Note, when the size of the circuit conductors and overcurrent protection device are listed on the nameplate of the equipment, these sizes must be selected, installed and used. **(See Figure 24-1)**

NAMEPLATE LISTING
440.1; 4

The overcurrent protection devices, running overload protection devices, conductors, disconnecting means, and controllers are sized and selected by the information provided on the nameplate listing for air-conditioning and refrigeration equipment. The information on the nameplate is very important to installers and service personnel, therefore, the nameplate should never be removed from the air-conditioner or refrigeration equipment.

MARKINGS ON HERMETIC REFRIGERANT MOTOR-COMPRESSORS AND EQUIPMENT
440.4

Hermetic refrigerant motor-compressors must be provided with the following information on the nameplate, giving the manufacturer's name, trademark, or symbol and designating identification, number of phases, voltage, and frequency. The information provided on the nameplate of the hermetic refrigerant motor-compressor is used to determine the ratings of branch-circuit conductors, ground-fault protection, short-circuits, disconnecting means, controllers, and other components of the electrical system.

MARKINGS ON CONTROLLERS
440.5

Controllers must be marked with information that lists the manufacturer's name, trademark or symbol, identifying voltage, phases, full-load current in amps, locked rotor current rating, or horsepower.

AMPACITY AND RATING
440.6

The full-load current rating in amps listed on the nameplate of the motor-compressor must be used to determine the branch-circuit conductor rating, short-circuit protection rating, motor overload protection rating, controller rating, or disconnecting means rating. The branch-circuit selection current (if greater) must be applied as shown instead of the full-load current rating. The full-load current rating must be used to determine the motor's overload protection rating. The full-load current rating listed on the compressor nameplate must be used when the nameplate for the equipment does not list a full-load current rating based upon the branch-circuit selection current (BCSC). For BCSC defined, see **440.2**. (**See Figure 24-1**)

HIGHEST RATED (LARGEST) MOTOR
440.7

When sizing the conductors for a feeder supplying A/C units and motors per **430.24**, the full-load current in amps of the largest motor is multiplied by 125 percent. The full-load current rating of the remaining motors are added to this total to derive the total FLA. See Figures 20-11 and 12 for an illustration pertaining to this rule.

When sizing the overcurrent protection device for two or more motors per **430.62(A)**, the full-load current in amps of the largest motor is multiplied by the percentages listed in **Table 430.52**. The full-load current rating in amps of the remaining motors are added to this total to derive the FLA to size the OCPD.

The full-load current ratings listed on the nameplate of the compressor-motor are used to determine the size conductors and overcurrent protection device using the same procedure. The larger of the two must be used.

See Figure 24-2 for feeder-circuit supplying motors and A/C units. Note that the A/C unit is the largest motor and not one of the motors in the group.

SINGLE MACHINE
440.8

Each motor controller must be provided with an disconnecting means. Air-conditioning and refrigeration systems are considered a single machine even through they consist of any number of motors. The number of disconnecting means to be provided are determined by applying **430.87, Ex.** and **430.112, Ex.**

DISCONNECTING MEANS - GENERAL
440.11

The full-load current rating of the equipment's nameplate or the nameplate branch-circuit selection current of the compressor, whichever is greater, must be used to size the branch-circuit conductors and the disconnecting means to disconnect air-conditioners and refrigeration equipment.

RATING AND INTERRUPTING CAPACITY
440.12

The full-load current rating of the equipment's nameplate or the nameplate branch-circuit selection current of the compressor, whichever is greater, must be sized at 115 percent to select the disconnecting means. A horsepower rated switch, circuit breaker, or other switches may be used as the disconnecting means per **430.109**. (**See Figure 24-3**)

Design Tip: A minimum load is derived when applying 115 percent for sizing the disconnecting means. Therefore, on larger units the 115 percent may not be of sufficient ampacity for opening the circuit under load.

The horsepower amperage rating may be selected from **Tables 430.247** through **430.250** when corresponding to the nameplate current rating or branch-circuit selection current of the motor-compressor or equipment if listed in amperage and not horsepower. The horsepower amperage rating for locked-rotor current must be selected from Tables

430.251(A) or (B) when the nameplate fails to list the locked-rotor current in amps. Note that the disconnecting means must be sized with enough capacity in horsepower that is capable of disconnecting the total locked-rotor current. **(See Figure 24-4)**

The full-load current rating in amps of the nameplate may be used to size a circuit breaker at 115 percent or more to disconnect a hermetically sealed motor from the power circuit. **(See Figure 24-5)**

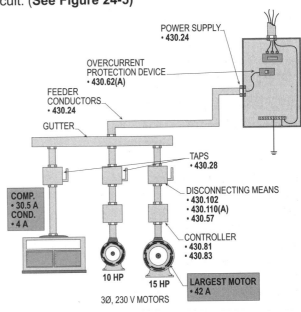

NEC 440.7

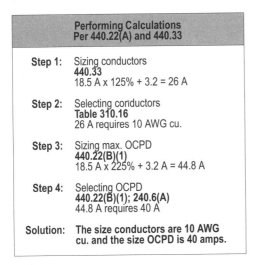

```
Performing Calculations
Per 440.22(A) and 440.33
```

Step 1:   Sizing conductors
440.33
18.5 A x 125% + 3.2 = 26 A

Step 2:   Selecting conductors
Table 310.16
26 A requires 10 AWG cu.

Step 3:   Sizing max. OCPD
440.22(B)(1)
18.5 A x 225% + 3.2 A = 44.8 A

Step 4:   Selecting OCPD
440.22(B)(1); 240.6(A)
44.8 A requires 40 A

Solution: **The size conductors are 10 AWG cu. and the size OCPD is 40 amps.**

NEC 440.6

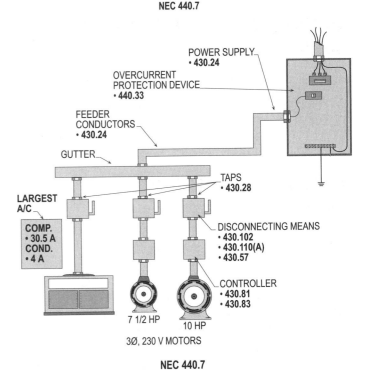

NEC 440.7

Figure 24-1. If the branch-circuit selection current (BCSC) on the nameplate calls for a certain circuit size and the OCPD size is specified, this rating must be used instead of actually calculating these values.

Figure 24-2. The above illustration shows the calculation procedure for sizing conductors and OCPD where the A/C unit or motor is the largest in the group of A/C units and motors.

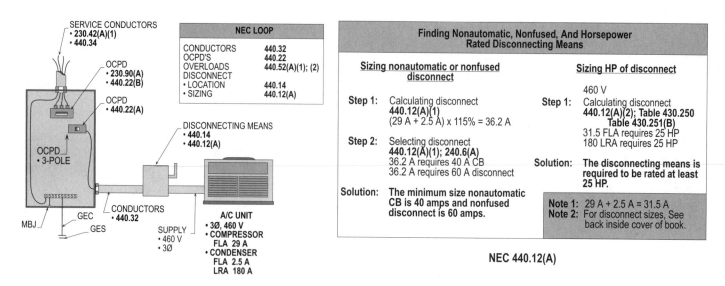

Figure 24-3. The full-load current rating in amps of the nameplate or the nameplate branch-circuit selection current of the compressor, whichever is greater, must be sized at 115 percent to size the disconnecting means.

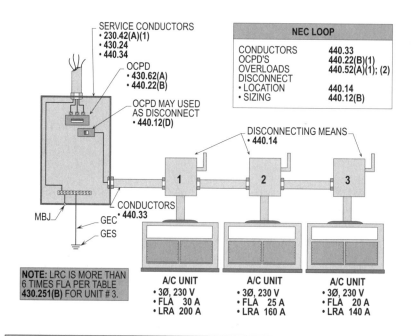

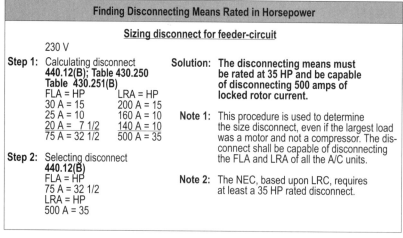

Figure 24-4. Sizing horsepower rating to select disconnecting means based upon the locked-rotor currents.

Design Tip: The circuit breaker must be sized at 115 percent or more of the branch-circuit selection current in amps, if this is greater in rating, so as to be capable of disconnecting the circuit safely. Comparison calculations should be made between the two.

Two or more hermetic motors or combination loads such as hermetic motor loads, standard motor loads, and other loads must have their separate values totaled to determine the rating of a single disconnecting means. This total rating must be sized at 115 percent of flux to determine the size disconnecting means required to disconnect the circuits and components in a safe and reliable manner. (**See Figure 24-6**)

coolers, and beverage dispensers, a cord, plug, or receptacle, which may be of the separable type, can be used to serve as a disconnecting means. (**See Figure 24-7**)

Design Tip: In some cases, room air-conditioners are not permitted to have a cord-and-plug connection serve as their disconnecting means. Such cases is where their unit switches for manual control are installed in air-conditioners mounted over 6 ft. above finished grade.

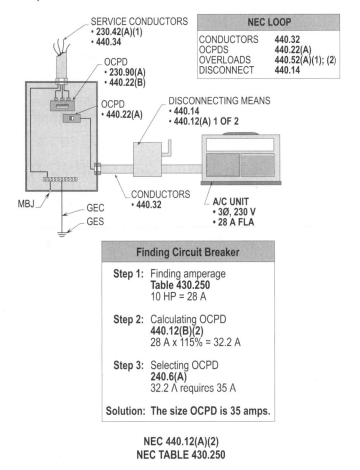

Figure 24-5. The full-load current rating of the nameplate may be used to size a circuit breaker at 115 percent or more to disconnect a hermetically sealed compressor motor and condenser fan motor from the power circuit based on total FLA.

CORD-CONNECTED EQUIPMENT 440.13

For cord-and-plug connected equipment such as room air-conditioners, home refrigerators and freezers, drinking water

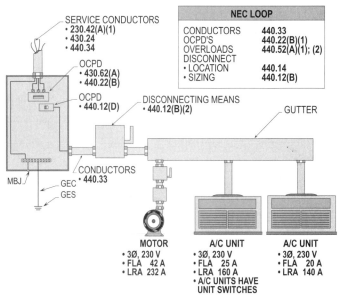

Figure 24-6. Two or more hermetic motors or combination loads such as hermetic motor loads, standard motor loads, and other loads must have their separate values totaled to determine the rating of a single disconnecting means.

LOCATION
440.14

The disconnecting means for air-conditioning or refrigeration equipment must be located within sight and within 50 ft. and be readily accessible to the user. An additional circuit breaker or disconnecting switch must be provided at the equipment if the air-conditioning or refrigeration equipment is not within sight or within 50 ft. The disconnecting means is permitted to be installed within or on the air-conditioning or refrigeration equipment. For the use of unit switches located in air-conditioning units review **422.34** per AHJ. **(See Figure 24-8)**

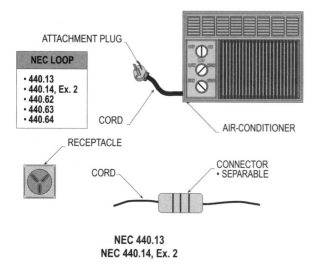

NEC 440.13
NEC 440.14, Ex. 2

Figure 24-7. Cord-and-plug connected equipment such as room air-conditioners, home refrigerators and freezers, drinking water coolers, and beverage dispensers are permitted to be disconnected by a cord and receptacle. A cord, plug, or receptacle, which may be of the separable type, can also be used to serve as such disconnecting means.

APPLICATION AND SELECTION
440.22

The branch-circuit fuse or circuit breaker ratings for hermetically sealed motors must be size with enough capacity to allow the motor to start and develop speed without tripping open the OCPD due to the momentary inrush current of the compressor and other components. Maximum protection is always provided by the ratings and settings of the OCPD being sized with values as low as possible. Hermetic refrigerant motor-compressors must be protected by properly sizing and selecting the ratings and settings of the overcurrent protection devices to protect the branch-circuit conductors and other components in the circuit from short-circuit and ground-fault conditions.

RATING AND SETTING FOR INDIVIDUAL MOTOR-COMPRESSOR
440.22(A)

The OCPD for hermetically sealed compressors must be selected at 175 percent (for minimum) or 225 percent (for maximum) of the compressor's FLA rating or the branch-circuit selection circuit current, whichever is greater. **(See Figure 24-9)**

OCPD's for hermetically sealed compressors may be selected up to 225 percent to allow the motor to start if the compressor will not start and develop speed when applying a lower rating selected at 175 percent or less.

> **Design Tip:** A normal circuit breaker must not be installed when the equipment is marked for a particular fuse size or HACR circuit breaker rating. The branch-circuit conductors must be protected only by that specified fuse size or HACR circuit breaker rating.

RATING OR SETTING FOR EQUIPMENT
440.22(B)

When sizing the overcurrent protection device, the rating or setting must be selected to comply with the number of hermetic motors, or combination of hermetic motors and standard motors installed on a circuit.

SIZING OCPD FOR TWO OR MORE HERMETIC MOTORS
440.22(B)(1)

The OCPD for a feeder-circuit supplying two or more air conditioning or refrigerating units must be sized to allow the largest unit to start and allow the other units to start at different intervals of time. The full-load current rating of the nameplate or the branch-circuit selection current rating in amps of the largest motor, whichever is greater, must be sized at 175 percent, if there are two or more hermetically sealed motors installed on the same feeder-circuit and other motor's FLC is added. **(See Figure 24-10)**

OCPD's for hermetically sealed motors may be selected up to 225 percent to allow the motor to start if the motor will not start and develop speed when applying a lower rating which is normally selected at 175 percent or less. **(See Figure 24-11)**

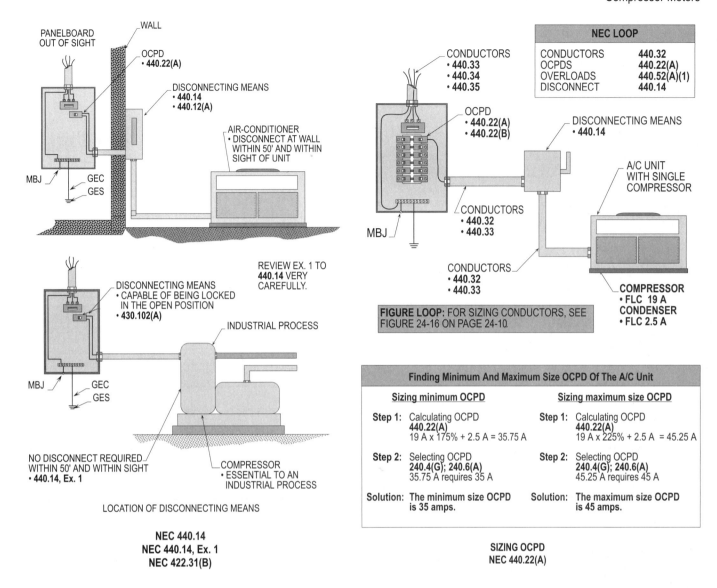

NEC LOOP

CONDUCTORS	440.32
OCPDS	440.22(A)
OVERLOADS	440.52(A)(1)
DISCONNECT	440.14

FIGURE LOOP: FOR SIZING CONDUCTORS, SEE FIGURE 24-16 ON PAGE 24-10.

Finding Minimum And Maximum Size OCPD Of The A/C Unit	
Sizing minimum OCPD	**Sizing maximum size OCPD**
Step 1: Calculating OCPD **440.22(A)** 19 A x 175% + 2.5 A = 35.75 A	**Step 1:** Calculating OCPD **440.22(A)** 19 A x 225% + 2.5 A = 45.25 A
Step 2: Selecting OCPD **240.4(G); 240.6(A)** 35.75 A requires 35 A	**Step 2:** Selecting OCPD **240.4(G); 240.6(A)** 45.25 A requires 45 A
Solution: The minimum size OCPD is 35 amps.	**Solution:** The maximum size OCPD is 45 amps.

SIZING OCPD
NEC 440.22(A)

NEC 440.14
NEC 440.14, Ex. 1
NEC 422.31(B)

Figure 24-8. The disconnecting means for air-conditioning or refrigeration equipment must be located within sight and within 50 ft. and be readily accessible to the user. An additional circuit breaker or disconnecting switch must be provided at the equipment if the air-conditioning or refrigeration equipment is not within sight or within 50 ft.

Figure 24-9. The OCPD for hermetically sealed compressors are selected at 175 percent (for minimum) or 225 percent (for maximum) of the compressor's FLA rating or the branch-circuit selection circuit current, whichever is greater.

SIZING OCPD FOR HERMETIC MOTORS AND OTHER LOADS WHEN A HERMETICALLY SEALED MOTOR IS THE LARGEST
440.22(B)(1)

When installing hermetically sealed motors and other loads such as motors on the same circuit, and the largest motor of the group is hermetic, the same procedure used for two or more hermetic motors on a feeder-circuit is used to size the overcurrent protection device. The full-load current rating of the equipment's nameplate or the branch-circuit selection current rating of the largest hermetic motor, whichever is

greater, must be sized at 175 percent and the sum of the full-load current ratings of the other motors and loads added to this largest hermetic motor load to size the OCPD.

SIZING OCPD FOR HERMETIC MOTORS AND OTHER LOADS WHEN A MOTOR IS THE LARGEST
440.22(B)(2)

When installing hermetically sealed motors and other loads such as motors on the same circuit, and the largest in the group is a motor, the overcurrent protection device must be sized and selected based on the percentages from **Table 430.52**. The maximum branch-circuit overcurrent protection device per **Table 430.52** must be used when the standard

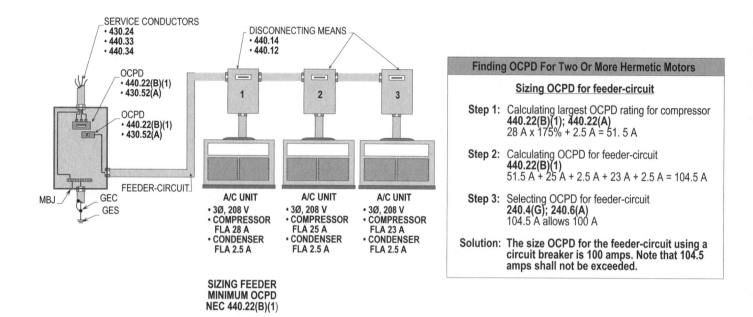

Figure 24-10. The full-load current rating in amps of the nameplate or the branch-circuit selection current rating of the largest motor, whichever is greater, must be size at 175 percent if there are two or more hermetically sealed motors with fan motors installed on the same feeder-circuit.

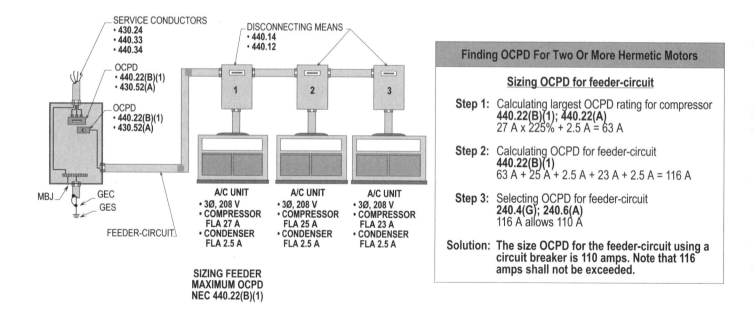

Figure 24-11. OCPD's for hermetically sealed motors may be selected up to 225 percent to allow the motor to start if the motor will not start and develop speed.

motor is the largest of the group and the sum of the full-load current ratings of the remaining hermetically sealed motors and other motors of the group added to this device rating. The next lower standard size overcurrent protection device below this total sum must be installed per **240.6(A)**. (**See Figure 24-12**)

Design Tip: The next larger, standard size OCPD is not permitted to be installed on the feeder-circuit, because there is not an exception to allow the next higher size per **440.22(B)(2) or 430.62(A)**.

USING 15 OR 20 AMP OCPD 440.22(B)(2), Ex. 1

When the equipment will start, run, and operate on 15 or 20 amp, 120 volt, single-phase branch-circuit, or 15 amp, 208 volt or 240 volt, a single-phase branch-circuit, the 15 or 20 amp overcurrent protection device may be used to protect the branch-circuit. However, the values of the overcurrent protection device in the branch-circuit must not exceed the values marked on the nameplate of the equipment. (**See Figure 24-13**)

Finding OCPD For Hermetic And Standard Motors	
Sizing OCPD for feeder-circuit	
Step 1: Finding FLA for motor **440.22(B)(2); Table 430.250** 15 HP = 46.2 A	**Step 5:** Selecting OCPD for feeder-circuit **440.22(B)(2); 240.6(A)** 158 A allows 150 A
Step 2: Calculating OCPD for motor **440.22(B)(2); Table 430.52** 46.2 A x 250% = 115.5 A	**Solution:** The size OCPD for the feeder-circuit using a circuit breaker is 150 amps.
Step 3: Selecting OCPD for motor **240.4(G); 240.6(A)** 115.5 A allows 110 A	**Note 1:** The conductors supplying power to a compressor with other motors are computed at 125% times the largest FLA plus the FLA of the other motors.
Step 4: Calculating OCDP for feeder-circuit **440.22(B)(2)** 110 A + 25 A + 23 A = 158 A	**Note 2:** The same rule applies where only compressors are supplied by such conductors.

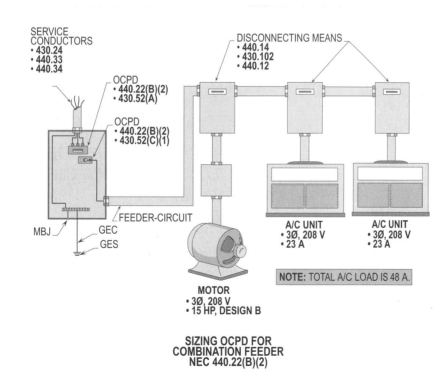

SERVICE CONDUCTORS
• 430.24
• 440.33
• 440.34

OCPD
• 440.22(B)(2)
• 430.52(A)

OCPD
• 440.22(B)(2)
• 430.52(C)(1)

DISCONNECTING MEANS
• 440.14
• 430.102
• 440.12

MBJ   GEC   GES

FEEDER-CIRCUIT

MOTOR
• 3Ø, 208 V
• 15 HP, DESIGN B

A/C UNIT
• 3Ø, 208 V
• 23 A

A/C UNIT
• 3Ø, 208 V
• 23 A

NOTE: TOTAL A/C LOAD IS 48 A.

SIZING OCPD FOR COMBINATION FEEDER NEC 440.22(B)(2)

Figure 24-12. When installing hermetically sealed motors and other loads such as motors on the same circuit, and the largest in the group is a motor, the overcurrent protection device is sized and selected based on the percentages from **Table 430.52**.

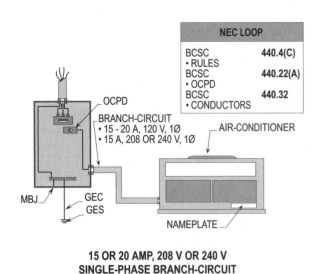

NEC LOOP	
BCSC	440.4(C)
• RULES	
BCSC	440.22(A)
• OCPD	
BCSC	440.32
• CONDUCTORS	

15 OR 20 AMP, 208 V OR 240 V
SINGLE-PHASE BRANCH-CIRCUIT
NEC 440.22(B)(2), Ex. 1

Figure 24-13. Where the equipment will start, run, and operate on 15 or 20 amp, 120 volt, single-phase branch-circuit, or 15 amp, 208 volt or 240 volt, a single-phase branch-circuit, with 15 or 20 amp overcurrent protection device must be used to protect the branch-circuit.

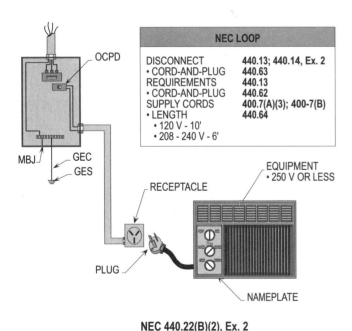

NEC LOOP	
DISCONNECT	440.13; 440.14, Ex. 2
• CORD-AND-PLUG	440.63
REQUIREMENTS	440.13
• CORD-AND-PLUG	440.62
SUPPLY CORDS	400.7(A)(3); 400-7(B)
• LENGTH	440.64
• 120 V - 10'	
• 208 - 240 V - 6'	

NEC 440.22(B)(2), Ex. 2

Figure 24-14. The rating of the overcurrent protection device must be determined by using the rating of the nameplate of the cord-and-plug connected equipment having a single-phase, 250 volt or less, hermetically sealed motor.

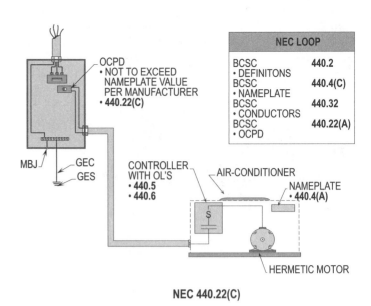

NEC LOOP	
BCSC	440.2
• DEFINITONS	
BCSC	440.4(C)
• NAMEPLATE	
BCSC	440.32
• CONDUCTORS	
BCSC	440.22(A)
• OCPD	

NEC 440.22(C)

Figure 24-15. The manufacturer's values marked on the equipment must not be exceeded by the overcurrent protection device rating, where the maximum overcurrent protective device ratings on the manufacturer's heater table for use with a motor controller are less than the rating or setting per **440.22(A) and (B)**.

USING A CORD-AND-PLUG CONNECTION NOT OVER 250 VOLTS
440.22(B)(2), Ex. 2

The rating of the overcurrent protection device must be determined by using the rating on the nameplate of the cord-and-plug connected equipment having a single-phase, 250 volt or less hermetically sealed motor. **(See Figure 24-14)**

PROTECTIVE DEVICE RATING NOT TO EXCEED THE MANUFACTURER'S VALUES
440.22(C)

The manufacturer's values marked on the equipment must not be exceeded by the overcurrent protection device rating. Where the maximum OCPD ratings listed by the manufacturer, this rating must not exceed the rating or setting per **440.22(A) and (B)**. **(See Figure 24-15)**

BRANCH-CIRCUIT CONDUCTORS
440.31

In general, to prevent conductors and motor components of the branch-circuit from over heating, the conductors must be sized with enough capacity to allow a hermetic motor to start and run. To ensure adequate sizing, a derating factor of 80 percent must be applied to the branch-circuit conductors or the conductors must be sized at 125 percent of the load served.

SINGLE MOTOR-COMPRESSORS 440.32

The conductors supplying power to an air-conditioning or refrigerating unit must be sized to carry the load of the unit plus an overload for a period of time that won't damage the components. The full-load current rating of the equipment's nameplate or branch-circuit selection current, whichever is greater, must be sized at 125 percent to size and select the conductors supplying hermetically sealed motors. **(See Figure 24-16)**

FIGURE LOOP. FOR SIZING OCPD, SEE FIGURE 24-9 ON PAGE 24-7.

Sizing Branch-Circuit Conductors

Sizing conductor based on controller

Step 1: Calculating FLA
440.32
19 A x 125% + 2.5 A = 26.25 A

Step 2: Selecting conductor
240.4(D)
26.25 A requires 10 AWG

Solution: **The size conductors 10 AWG THHN cooper.**

Sizing conductors based on branch-circuit selection current

Step 1: Selecting conductors
440.4(C)
30 A requires 10 AWG

Solution: **The branch-circuit selection current requires 10 AWG THHN copper conductors.**

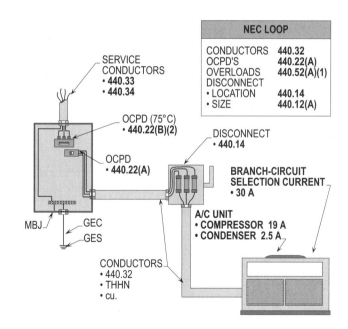

NEC LOOP	
CONDUCTORS	**440.32**
OCPD'S	**440.22(A)**
OVERLOADS	**440.52(A)(1)**
DISCONNECT	
• LOCATION	**440.14**
• SIZE	**440.12(A)**

SERVICE CONDUCTORS
• **440.33**
• **440.34**

OCPD (75°C)
• **440.22(B)(2)**

OCPD
• **440.22(A)**

DISCONNECT
• **440.14**

BRANCH-CIRCUIT SELECTION CURRENT
• **30 A**

A/C UNIT
• **COMPRESSOR 19 A**
• **CONDENSER 2.5 A**

MBJ

GEC

GES

CONDUCTORS
• 440.32
• THHN
• cu.

SIZING BRANCH-CIRCUIT CONDUCTORS
NEC 440.4(C)
NEC 440.32

Figure 24-16. The full-load current rating in amps of the equipment's nameplate or branch-circuit selection current, whichever is greater, must be sized at 125 percent to size and select the conductors supplying hermetically sealed motors.

TWO OR MORE MOTOR COMPRESSORS
440.33

Two or more compressors plus the other motor loads can be connected to a feeder-circuit. The largest compressor is computed at 125 percent of its FLA and the remaining compressor loads are added to this total at 100 percent of their FLA ratings. For units with a branch-circuit selection current, the circuit conductors are selected and based upon the equipment's nameplate values. **(See Figure 24-17)**

FIGURE LOOP: FOR SIZING OCPD, SEE FIGURES 24-10 AND 24-11 ON PAGE 24-8.

Sizing Feeder-Circuit Conductors

Step 1: Finding FLA of A/C Units
A/C unit #1 = 40 A + 3 A
A/C unit #2 = 25 A + 2.5 A
A/C unit #3 = 23 A + 2.5 A

Step 2: Calculating FLA
440.33
40 A x 125% + 3 A   =   53   A
25 A x 100% + 2.5 A = 27.5 A
23 A x 100% + 2.5 A = 25.5 A
Total load   = 106   A

Step 3: Selecting conductors
310.10(2); Table 310.16
106 A requires 2 AWG.

Solution: **The size THWN conductors are 2 AWG copper.**

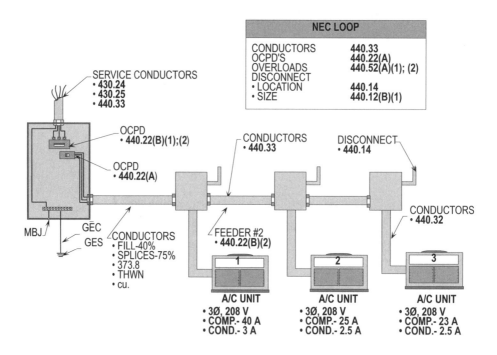

**SIZING FEEDER-CIRCUIT CONDUCTORS
NEC 440.33**

Figure 24-17. Two or more compressors plus the other motor loads can be connected to a feeder-circuit. The largest compressor must be computed at 125 percent of its FLA and the remaining compressor loads are added to this total at 100 percent of their FLA ratings.

COMBINATION LOADS
440.34

Two or more motor compressors with motor loads plus other loads may be connected to a feeder-circuit or service conductors. The largest compressor or motor load must be calculated at 125 percent plus 100 percent of the remaining compressors and motors. The other loads must be computed at 125 percent for continuous and 100 percent for noncontinuous operation per **215.2(A)(1)** and these total values are used to select conductors. **(See Figure 24-18)**

Finding Two Or More A/C Units With Other Loads

Step 1: Finding FLA of A/C Units
440.34
A/C unit #1 = 33 A + 2.5 A
A/C unit #2 = 27 A + 2.5 A
A/C unit #3 = 21 A + 2.5 A
Other loads = 65.6 A

Step 2: Calculating FLA
440.34; 215.2(A)(1)

33 A x 125% + 2.5 A	=	43.75 A
27 A x 100% + 2.5 A	=	29.5 A
21 A x 100% + 2.5 A	=	23.5 A
65.6A x 125%	=	82 A
Total load	=	178.75 A

Step 3: Selecting conductors
310.10(2); Table 310.16
178.75 A requires 3/0 AWG cu.

Solution: **The size THWN conductors for the feeder-circuit are 3/0 AWG copper.**

Note: The load of 178.75 amps requires the conductors and OCPD to be rated at 200 amps respectfully.

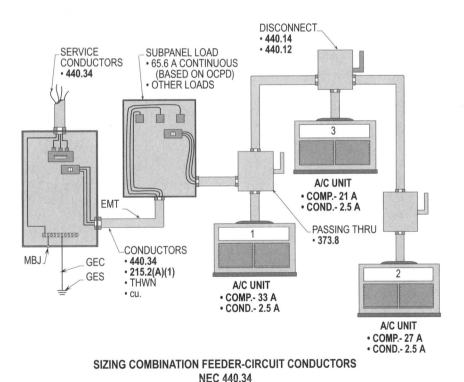

SIZING COMBINATION FEEDER-CIRCUIT CONDUCTORS
NEC 440.34

Figure 24-18. Two or more motor compressors with motor loads plus other loads may be connected to a feeder-circuit or service conductors. The largest compressor or motor load must be calculated at 125 percent plus 100 percent of the remaining compressors and motors plus the other loads.

MULTIMOTOR AND COMBINATION LOAD EQUIPMENT
440.35

The marking on the equipment's nameplate must be used when sizing the branch-circuit conductors for multimotor and combination load equipment. The conductors must be installed to have a rating equal to the minimum nameplate rating. Each individual motor or load contained in the unit is not required to be calculated individually to size and select the conductors to be installed.

CONTROLLERS FOR MOTOR COMPRESSORS
440.41

When sizing the circuit conductors to a controller, the full-load current rating and the locked rotor current rating of the compressor motor must be sized at continuous operation.

MOTOR COMPRESSOR CONTROLLER RATING
440.41(A)

The full-load current rating in amps on the equipment's nameplate or the branch-circuit selection current ratings, whichever is greater, must be used to size and select the motor controller. If necessary, the locked rotor current rating of the motor may be used to size and select the motor controller. (**See Figure 24-19**)

> **Design Tip:** The motor controller must be sized and selected by using the same procedure as was used for the sizing of the disconnecting means.

MOTOR COMPRESSOR AND BRANCH-CIRCUIT OVERLOAD PROTECTION
440.51

The overload (OL) protection for compressors may be accomplished by using OCPD's in separate enclosures, separate overload relays or thermal protectors which are an integral part of the compressor.

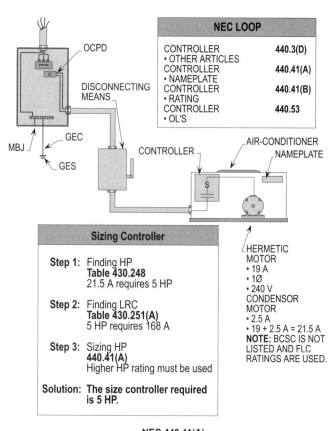

NEC 440.41(A)

Figure 24-19. The full-load current rating of the equipment's nameplate or the branch-circuit selection current ratings, whichever is greater, must be used to size and select the motor controller.

APPLICATION AND SELECTION
440.52

The overload relay for a motor compressor must trip at not more than 140 percent of the full-load current rating. If a fuse or circuit breaker is used for the protection of the motor compressor it must trip at not more than 125 percent of the full-load current rating in amps. (**See Figure 24-20**)

OVERLOAD RELAYS
440.53

Short-circuit and ground-fault protection is not provided by overload relays and thermal protectors. Overload relays and thermal protectors respond to any type of heat build up and open with a delay action which will not operate instantly, even on short-circuits or ground-faults. The branch-circuit overcurrent protection device for the circuit must operate and clear the circuit under short-circuit and ground-fault conditions.

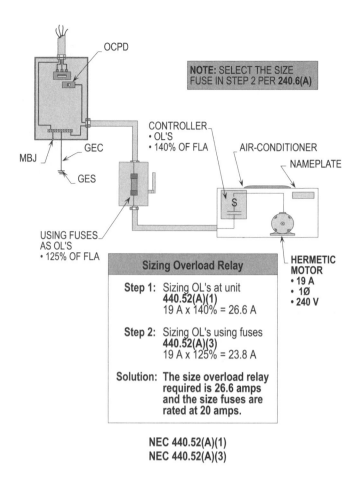

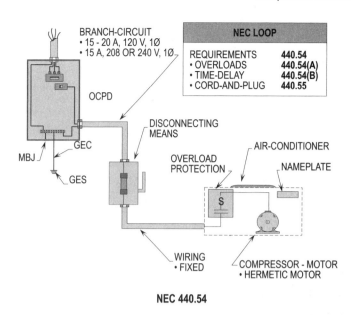

NEC 440.52(A)(1)
NEC 440.52(A)(3)

Figure 24-20. The overload relay for the motor compressor must trip at not more than 140 percent of the full-load current rating. If a fuse or circuit breaker is used for the protection of the motor compressor it must trip at not more than 125 percent of the full-load current rating.

MOTOR COMPRESSORS AND EQUIPMENT ON 15 OR 20 AMP BRANCH-CIRCUIT - NOT CORD-AND-PLUG CONNECTED 440.54

Overload protection must be provided for direct or fixed-wired motor compressors and equipment which is connected to 15 or 20 amp, 120 volt, single-phase branch-circuits. Note that 15 amp OCPD is required for 240 volt, single-phase branch-circuits.

Separate overload relays must be selected at not more than 140 percent of the compressor's FLA. Hermetic motors must be provided with fuses or circuit breakers that provides sufficient time delay to allow the motor to come up to running speed without tripping open the circuit due to the high inrush current. **(See Figure 24-21)**

Figure 24-21. Overload protection must be provided for direct or fixed-wired motor compressors and equipment that is connected to 15 or 20 amp, 120 volt, single-phase branch-circuits.

CORD AND ATTACHMENT PLUG CONNECTED MOTOR COMPRESSORS AND EQUIPMENT ON 15 OR 20 AMP BRANCH-CIRCUITS 440.55

When attachment plugs and receptacles are used for circuit connection they must be rated no higher than 15 or 20 amp, for 120 volt, single-phase circuits, or 15 amp, for 208 or 240 volt, single-phase branch-circuits. **(See Figure 24-22)**

ROOM AIR-CONDITIONERS-GENERAL 440.60

Room air-conditioners are usually cord-and-plug connected when installed on 120/240 volt, single-phase systems. However, they may be hard-wired. Air-conditioners are always hard-wired when installed on three-phase systems, or electrical supply systems over 250 volts.

GROUNDING 440.61

The following wiring methods when utilized to wire-in room air-conditioners must be grounded:

(1) Cord-and-plug connected

(2) Hard-wired (if within reach of the ground or grounded object)

(3) If in contact with metal

(4) Operating over 150 volts-to-ground

(5) Wired with metal-clad wiring

(6) Located in a hazardous location

(7) Installed in damp location (within reach of the user)

See Figure 24-23 for the wiring methods to be used to ground room air-conditioners.

BRANCH-CIRCUIT REQUIREMENTS 440.62

The full-load current rating of a room air-conditioner must be marked on the nameplate and must not operate at more than 40 amps on 250 volts, single-phase. The branch-circuit overcurrent protection device must be installed with a rating no greater than the circuit conductors ampacity or the rating of the receptacle serving the unit, whichever is less. The ampacity of a cord-and-plug connected air-conditioning window unit must not exceed 80 percent of the branch-circuit where no other loads are served. If other loads are served by the branch-circuit, the cord-and-plug connected air-conditioner unit must not exceed 50 percent of the branch-circuit. (**See Figures 24-24(a) and (b)**)

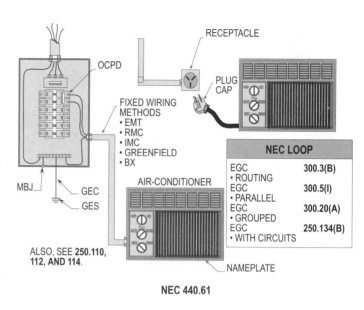

NEC 440.61

Figure 24-23. The above are wiring methods to be used to ground room air-conditioners.

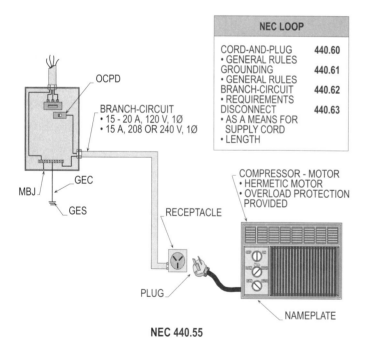

NEC 440.55

Figure 24-22. When attachment plugs and receptacles are used for circuit connection they must be rated no higher than 15 or 20 amp, for 120 volt, single-phase circuits, or 15 amp, for 208 or 240 volt, single-phase branch-circuits.

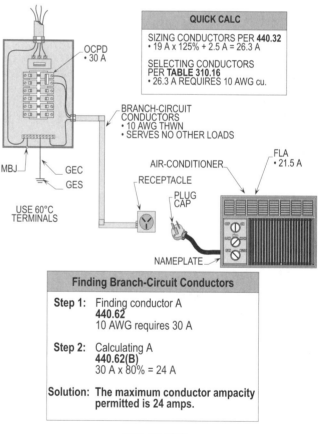

NEC 440.62(B)

Figure 24-24(a). The ampacity of a cord-and-plug connected air-conditioning window unit must not exceed 80 percent of the branch-circuit where no other loads are served. Note that 21.5 amps is less than 24 amps.

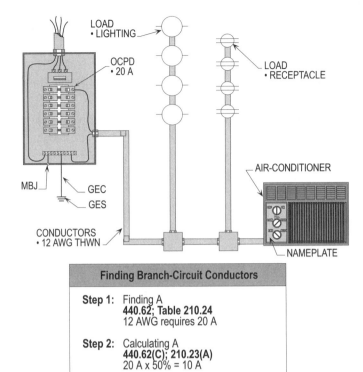

Finding Branch-Circuit Conductors

Step 1: Finding A
440.62; Table 210.24
12 AWG requires 20 A

Step 2: Calculating A
440.62(C); 210.23(A)
20 A x 50% = 10 A

Solution: The maximum A/C unit ampacity permitted is 10 amps.

NEC 440.62(C)

Figure 24-24(b). If other loads are served by the branch-circuit, the cord-and-plug connected air-conditioner unit must not exceed 50 percent of the branch-circuit.

DISCONNECTING MEANS
440.63

A cord-and-plug may serve as the disconnecting means for the room air-conditioner if all the following conditions are complied with:

(1) Operates at 250 volts or less
(2) Controls are manually operated
(3) Controls are within 6 ft. of the floor
(4) Controls are readily accessible to the user

Room air-conditioners may be hard-wired and located within sight of the service equipment or it may be wired so that it is readily accessible to a disconnecting switch for the user. However, such switch must be located within sight and or within the unit. **(See Figures 24-25(a) and (b))**

Design Tip: The rules for three-phase room air-conditioners are not be used for these type of units. Three-phase room air-conditioners must be hard-wired and provided with a disconnecting means that is readily accessible to the user.

SUPPLY CORDS
440.64

Room air-conditioners installed with flexible cords must be a length that is limited to 10 ft. for 120 volt circuits and 6 ft. for 208 or 240 volt circuits. Long cords must not be used for they are dangerous. Long cords can also be a shock or fire hazard. **(See Figure 24-26)**

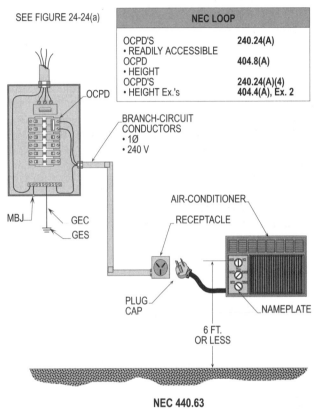

NEC LOOP	
OCPD'S • READILY ACCESSIBLE	240.24(A)
OCPD • HEIGHT	404.8(A)
OCPD'S • HEIGHT Ex.'s	240.24(A)(4) 404.4(A), Ex. 2

NEC 440.63

Figure 24-25(a). A cord-and-plug may serve as the disconnecting means for room air-conditioners if it operates at 250 volts or less, controls are manually operated, controls are within 6 ft. of the floor, and controls are readily accessible to the user.

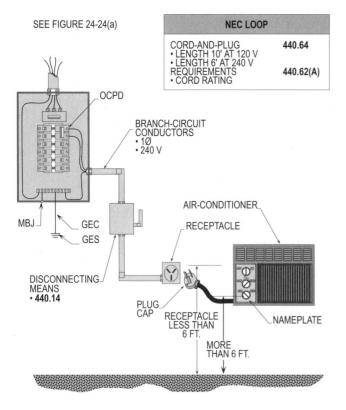

SEE FIGURE 24-24(a)

NEC LOOP	
CORD-AND-PLUG • LENGTH 10' AT 120 V • LENGTH 6' AT 240 V	440.64
REQUIREMENTS • CORD RATING	440.62(A)

NEC 440.63

Figure 24-25(b). A cord-and-plug may must not serve as the disconnecting means, if the room air-conditioner's manual controls are located 6 ft. above the finished grade.

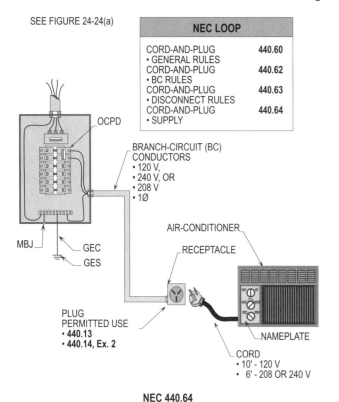

SEE FIGURE 24-24(a)

NEC LOOP	
CORD-AND-PLUG • GENERAL RULES	440.60
CORD-AND-PLUG • BC RULES	440.62
CORD-AND-PLUG • DISCONNECT RULES	440.63
CORD-AND-PLUG • SUPPLY	440.64

NEC 440.64

Figure 24-26. Room air-conditioners installed with flexible cords must be a length that is limited to 10 ft. for 120 volt circuits and 6 ft. for 208 or 240 volt circuits.

TROUBLESHOOTING COMPRESSOR MOTOR

To check the motor-compressor safely, turn off disconnect switch and disconnect all wiring from the motor terminals in the terminal box. The terminal at the right, when facing the compressor motor will be the starting terminal. Note that the center terminal is the common and the left terminal is the running winding terminal.

Using an ohmmeter, troubleshooting the windings for grounds as follows:

TESTING RUNNING WINDINGS

The first step in testing running windings is to disconnect all wires from the motor terminals. To check for resistance in the running winding, touch the ohmmeter leads to the "common" and "running" terminals and take a measurement. **(See Figure 24-27)**

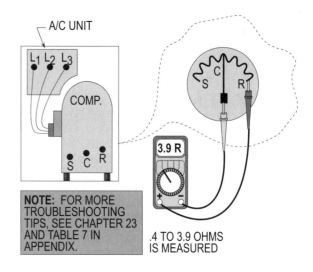

Figure 24-27. The running winding is usually not defective, if a resistance reading of .4 to 3.9 ohms is measured.

TESTING STARTING WINDINGS

The second step is to test the resistance in the starting winding, this test can be performed by touching the ohmmeter leads to the "common" and "starting" terminals. The resistance reading in the running winding from R to C will measure the lowest and the starting winding from S to C will be higher. Between R and S the reading is the total of the two, from 2.4 to 22.9 ohms. (.4 R +2 R = 2.4 R - 3.9 R + 19 = 22.9 R). **(See Figure 24-28)**

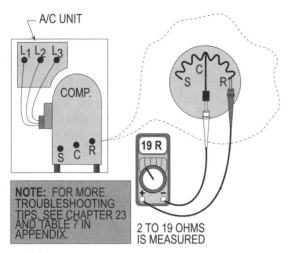

Figure 24-28. The starting winding is usually not defective, if a resistance reading of 2 to 19 ohms is measured.

TESTING FOR GROUNDS

The ohmmeter method can be used for testing grounds. One test lead is touched to the motor frame and the other is touched to each motor terminal. If the resistance measured is below 1,000,000 ohms, a ground from a winding is assumed. **(See Figure 24-29)**

Note: For more troubleshooting tips, see the illustrations in Chapter 23 and Table 7 of the Appendix.

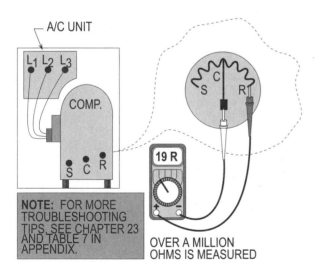

Figure 24-29. If a reading over one million ohms to ground is measured, there is usually no ground present.

Chapter 24. Compressor Motors

Section   **Answer**

_____   T   F   **1.** The overcurrent protection device is sized and selected by the information provided on the nameplate listing for air-conditioning and refrigeration equipment.

_____   T   F   **2.** Each motor controller for air-conditioning and refrigeration systems is not required to provided with an disconnecting means.

_____   T   F   **3.** A horsepower rated switch may be used as a disconnecting means.

_____   T   F   **4.** Room air-conditioners are not permitted to have a cord-and-plug serve as the disconnecting means for hermetic motors.

_____   T   F   **5.** An additional circuit breaker or disconnecting means must be provided at the equipment if an air-conditioning unit is not within sight or within 50 ft.

_____   T   F   **6.** An OCPD for hermetic seal compressors may be selected up to 250 percent to allow the motor to start and run.

_____   T   F   **7.** If a fuse or circuit breaker is used for the overload protection of the motor compressor it must trip at not more than 125 percent of the full-load current rating (in amps).

_____   T   F   **8.** Short-circuit and ground-fault protection for compressor motors may be provided (only) by overload relays and thermal protectors.

_____   T   F   **9.** The ampacity of a cord-and-plug connected air-conditioning window unit must not exceed 80 percent of the branch-circuit where no other loads are served.

_____   T   F   **10.** Room air-conditioners installed with flexible cords must be a length that is limited to 12 ft. for 120 volt circuits.

_____   _____   **11.** When sizing the conductors for a feeder supplying A/C units and motors, the full-load current in amps of the largest motor is multiplied by _____ percent.

_____   _____   **12.** The full-load current rating of the nameplate or the nameplate branch-circuit selection current of the compressor, whichever is greater, must be sized at _____ percent of the disconnecting means.

_____   _____   **13.** The disconnecting means for air-conditioning or refrigeration equipment must be located within sight and within _____ ft. and shall be readily accessible to the user.

_____   _____   **14.** Use full-load current rating of the nameplate or the branch-circuit selection current rating of the largest motor, whichever is greater to size the feeder OCPD at _____ percent if there are two or more hermetic sealed motors installed on the same feeder-circuit.

15. Two or more motor compressors plus other motor loads can be connected to a feeder-circuit with the largest motor compressor calculated at _____ percent of its FLA and the remaining compressor loads added to this total at _____ percent of their FLA ratings.

16. The overload relay for the motor compressor must trip at not more than _____ percent of the full-load current rating in amps.

17. When attachment plugs and receptacles are used for circuit connection they must be rated no higher than _____ amps, for 208 or 240 volt, single-phase branch-circuits.

18. The full-load current rating of a room air-conditioner must be marked on the nameplate and must not operate at more than _____ amps on 250 volts.

19. A cord-and-plug may not serve as the disconnecting means if the room air-conditioner manual controller is above _____ ft.

20. Room air-conditioners installed with flexible cord must be a length that is limited to _____ ft. for 208 or 240 volt circuits.

21. The OCPD for individual hermetically sealed compressors are selected at _____ percent (for minimum) of the compressor FLA rating or the branch-circuit selection current, whichever is greater.

    (a) 125   **(c)** 175
    (b) 150   **(d)** 225

22. The rating of the overcurrent protection device must be determined by using the rating of the nameplate of the cord-and-plug connected equipment having single-phase, _____ volt or less hermetic sealed motors.

    (a) 120   **(c)** 480
    (b) 250   **(d)** 600

23. Overload protection must be provided for direct or fixed-wired motor compressors and equipment which is connected to _____ or _____ amp, 120 volt, single-phase branch-circuits.

    (a) 15 or 20   **(c)** 30 or 40
    (b) 20 or 30   **(d)** 40 or 50

24. Room air-conditioners must be grounded per **440.61** for the protection of _____.

    (a) people   **(c)** all of the above
    (b) equipment   **(d)** none of the above

25. A cord-and-plug may serve as the disconnecting means for a room air-conditioner if it operates at _____ volts or less.

    (a) 120   **(c)** 250
    (b) 150   **(d)** 300

26. What size nonautomatic circuit breaker is required for an A/C unit with a compressor rated at 29 amps and the condenser fan rated at 2.5 amps?

_____ _____

27. What size nonfused disconnect is required for an A/C unit with a compressor rated at 29 amps and the condenser fan rated at 2.5 amps?

_____ _____

28. What size horsepower rated disconnect is required for an A/C with a compressor rated at 29 amps and the condenser fan rated at 2.5 amps with a 200 amp locked-rotor current rating? (Note that the supply voltage is 480 V, three-phase).

_____ _____

29. What size horsepower rated disconnect is required for the following loads on a three-phase, 230 volt system:

_____ _____

• A/C unit with a compressor rated at 29 amps with a 200 amp locked-rotor current rating
• A/C unit with a compressor rating 24 amps with a locked-rotor current of 150 amps
•A/C unit with a compressor rating of 20 amps with a locked-rotor current of 140 amps

30. What size circuit breaker (nonautomatic) is required to disconnect a hermetic sealed motor for an A/C unit rated at 29 amps? (Supply voltage is 280 V, three-phase).

_____ _____

31. What size horsepower rated disconnecting means is required to disconnect the following motor loads on a three-phase, 230 volt system:

_____ _____

• Motor rated at 38 amps with a 212 amp locked-rotor current rating
• A/C unit with a compressor rated at 28 amps with a 160 amp locked-rotor current rating
• A/C unit with a compressor rated at 24 amps with a 160 amp locked-rotor current rating

32. What is the minimum size OCPD, per **440.22(A)**, required for an individual A/C unit with a compressor rated at 20 amps and the condenser rated at 2.5 amps?

_____ _____

33. What is the maximum size OCPD, per **440.22(A)**, required for an individual A/C unit with a compressor rated at 20 amps and the condenser rated at 2.5 amps?

_____ _____

34. What is the minimum size OCPD, per **440.22(B)(1)**, required for a feeder-circuit with the following loads on a 230 volt, three-phase system:

_____ _____

• A/C unit with a compressor rated at 29 amps and the condenser fan rated at 2.5 amps
• A/C unit with a compressor rated at 26 amps and the condenser fan rated at 2.5 amps
• A/C unit with a compressor rated at 22 amps and the condenser fan rated at 2.5 amps

_____ _____

35. What is the maximum size OCPD, per **440.22(B)(1)** required for a feeder-circuit with the following loads on a 230 volt, three-phase system:

• A/C unit with a compressor rated at 29 amps and the condenser rated at 2.5 amps
• A/C unit with a compressor rated at 26 amps and the condenser rated at 2.5 amps
• A/C unit with a compressor rated at 22 amps and the condenser rated at 2.5 amps

_____ _____

36. What size OCPD is required for a feeder-circuit with the following loads on a 230 volt, three-phase system:

• A/C unit with a compressor rated at 26 amps and the condenser fan rated at 2.5 amps

• A/C unit with a compressor rated at 24 amps and the condenser fanrated at 2.5 amps
• 10 HP, 230 volt, three-phase, Design letter B motor

_____ _____

37. What size THHN copper conductors are required to supply an A/C unit with a compressor rated at 20 amps and the condenser fan rated at 2.5 amps?

_____ _____

38. What size THHN copper conductors are required to supply an individual A/C unit with a branch-circuit selection current of 30 amps?

_____ _____

39. What size THWN copper conductors are required for a feeder-circuit with the following loads on a 208 volt, three-phase system:

• A/C unit with a compressor rated at 30 amps and the condenser fan rated at 3 amps
• A/C unit with a compressor rated at 28 amps and the condenser fan rated at 2.5 amps
• A/C unit with a compressor rated at 24 amps and the condenser fan rated at 2.5 amps

_____ _____

40. What size THWN copper conductors are required for a feeder-circuit with the following loads on a 208 volt, three-phase system:

• A/C unit with a compressor rated at 30 amps and the condenser fan rated at 3 amps
• A/C unit with a compressor rated at 28 amps and the condenser fan rated at 2.5 amps
• A/C unit with a compressor rated at 24 amps and the condenser fan rated at 2.5 amps
• Other loads of 80 amps (continuous)

_____ _____

41. What size controller is required for an individual A/C unit with a compressor rated at 20 amps (BCSC) and the condenser fan rated at 2.5 amps on a 230 volt, single-phase system?

_____ _____

42. What size overload relay and fuses are required for an A/C unit with a compressor rated at 20 amps on a 240 volt, three-phase system?

_____ _____

43. What is the maximum value for a THWN copper conductor ampacity for a 10 AWG branch-circuit conductor supplying an air-conditioning window unit?

_____ _____

44. What is the maximum value for a THWN copper conductor ampacity for a 10 AWG branch-circuit conductor supplying an air-conditioning window unit with other loads such as lighting and receptacle loads?

Appendix

Troubleshooting Tables

Qualified personnel with proper testing equipment and tools may use the Tables in this appendix for troubleshooting problems related to motors, controls, adjustable speed drives, and eddy-current drives. However, these instructions do not cover all details or variations in equipment, nor do they provide for every possible condition to be met in actual practice.

The following Tables in this appendix can be used for troubleshooting tips:

TROUBLESHOOTING INDUCTION MOTORS

SYMPTOMS	AC Single Phase Motors				AC Three Phase Motors	Motors With Brushes
	Split Phase	Capacitor Start	Capacitor Start & Run	Shaded Poles		
	WHAT TO DO					
• Fails to start and run	A, B, F, C	A, B, F, H, C	A, B, H, I, Q	A, B, I, P, Q	A, B, E	A, B, L, M
• Motor does not always start, even without a load. Runs in forward or in reverse when started by manual means.	F, C	F, H, C	H, E		E	
• Starts and runs but heats very rapidly.	G, D	G, D	H, D	D	D	D
• Starts and runs but overheats.	D	D	H, D	D	D	D
• Sparking and arcing at brushes						J, K, L, M, N
• Severely high speed with sparking at the brushes.						O
• Increase in amps and motor over-heats.	D, P, Q	D, P, Q	D, P, Q	D, P, Q	D, P, Q	L, P, Q
• Motor blows OCPD and continues to operate when switch is off.	D, R	D, R	D, R	D, R	D, R	D, R
• Problems with vibration.						J, K, L, M, N, S

POSSIBLE CAUSES

A — Open circuit
B — Circuit open in motor winding (See Figure 15-4 in Chapter 15.)
C — Circuit open in starting winding
D — Winding is short-circuited or grounded
E — More than one winding is open
F — Open contacts in centrifugal switch
G — Centrifugal starting switch is not operating properly
H — Capacitor is defective
I — Motor is overloaded
J — Problems between mica and commutator

K — Commutator is dirty
L — Brushes are worn
M — Armature winding is shorted or open
N — Brushes not aligned and set
O — Shunt-winding is open
P — Problems with bearings
Q — Rotor Problems
R — Winding is grounded
S — Armature winding is shorted

TABLE 1

TROUBLESHOOTING REPULSION AND UNIVERSAL MOTORS

Symptoms	What To Do
• Failure to start	A, B, C, D, E, F, G, H, I J, K, L
• Excessive noise	M, N, O, P, Q
• Bearings overheating	P, Q, R, S, T, U, V, X, Y, Z, AA, BB
• Excessive brush wear	CC, DD, EE, FF, GG, HH
• Overheating of motor	II, JJ
• Commutator burned out	KK, LL, MM, NN, OO
• Governor problems	PP, QQ, RR, SS, TT, UU, VV, XX

Possible Causes

A — Blown fuse or
B — Circuit breaker open
C — Low voltage or
D — No voltage
E — Open circuit
F — Improper line connections
G — Excessive load
H — Brushes are worn or sticking
I — Brushes are incorrectly set
K — Excessive end play
L — Bearings are frozen
M — Unbalance conditions
N — Bent shaft or
loose parts
O — Faulty alignment or
Worn bearings
P — Dirt in air gap
Q — Uneven air gap
R — Motor needs oil
S — Dirty oil
T — Oil not reaching shaft
U — Excessive grease
V — Excessive belt tension
X — Rough bearing surface
Y — Bent shaft
Z — Misalignment of shaft
and bearing

AA — Excessive end thrust
BB — Excessive side pull
CC — Dirty commutator
DD — Improper contact with
commutator
EE — Excessive load
FF — Governor not acting
promptly
GG — High mica
HH — Rough commutator
II — Obstruction of ventilation
system
JJ — Overloading
KK — Worn bearings
LL — Moisture
MM — Acids or alkalies
NN — Harmful dust accumulation
OO — Overloading
PP — Dirty commutator
QQ — Governor mechanism
sticking
RR — Worn or sticking brushes
SS — Low frequency in
supply circuit
TT — Low voltage
UU — Incorrect connections or
Incorrect brush settings
VV — Excessive load
XX — Incorrect spring tension

TABLE 2

TROUBLESHOOTING WOUND-ROTOR MOTORS

Symptoms	What To Do
• Failure to start	A, B, C, D, E,
• Motor won't come up to speed	F, G, H, I, J, K, L, M, BB, CC
• Excessive noise	N, O, P, Q, R, S, T,
• Overheating of bearings	U, V, W, X, Y, Z, AA
• Overheating of motor	DD, EE
• Rotor or stator burned out	FF, GG, HH, II,

Possible Causes

A — Blown OCPD
B — Low or no a-c field supply
C — Bearings stuck or binding load
D — Open or shorted field
E — Open or shorted rotor
F — Broken slip rings
G — Open control device
H — Low a-c supply voltage
I — Binding load
K — Insufficient oil or grease
L — Low frequency in supply circuit
M — Dirt on slip rings
N — Vibration
O — Bent shaft
P — Loose parts
Q — Faulty alignment
R — Worn bearings

S — Dirt in air gap
T — Uneven air gap
U — Motor needs oil
V — Dirty oil
W — Oil not reaching shaft
X — Excessive grease
Y — Rough bearing surface
Z — Bent shaft
AA — Misalignment of shaft and bearing
BB — Broken or chipped brushes
CC — Improper brush contact
DD — Obstruction of ventilating system
EE — Overloading
FF — Worn bearings
GG — Moisture
HH — Dust accumulation

TABLE 3

TROUBLESHOOTING SYNCHRONOUS MOTORS	
Symptoms	**What To Do**
• Motor will not start and run	A, B, C, D, E, F, G, H, I, J, K, L,
• Motor will not accelerate to speed	J, K, L, N
• Motor fails to pull into step	M, L, N
• Motor pulls out of step or trips OCPD	O, P, Q, R, S, T, U, V
• Overheating	W, X, Y
Possible Causes	

A — Faulty connections	**O** — Exciter voltage low
B — Open circuit on one circuit	**P** — Open circuit in field and exciter circuit
C — Short circuit on one phase	**Q** — Short-circuit in field
D — Voltage falls too low	**R** — Reversed field
E — Friction too high	**S** — Load fluctuates widely
F — Field excited	**T** — Excessive torque peak
G — Too great of load	**U** — Power fails
H — Automatic field relay not working	**V** — Line voltage too low
I — Wrong direction of rotation	**W** — Overload condition
K — Low voltage	**X** — Over or under excitation
L — Field excited	**Y** — No field excitation
M — No field excitation	
N — Inertia of load excessive	

TABLE 4

TROUBLESHOOTING EDDY-CURRENT DRIVES

Symptoms	What To Do
• Motor does not start	A, B, C, D, E, F
• Motor runs but has no output	S, T, U, V, W, X
• Drive stops during operation	G, H, I, J ,K, L
• Unit overheats	Y, Z, AA, BB, CC
• Erratic operation	M, N, O, P, Q, R
• Runs at full speed only	DD, EE, FF
• Magnetic drive at stand-still or lower speed than expected with the speed potentiometer set at a higher speed	GG, HH, II
• Magnetic drive set 100% speed with no control	JJ, KK
• Magnetic drive has intermitten speed up or slow down	LL, MM, NN

Possible Causes

A — Loss of AC power.
B — Defective switch or breaker.
C — Blown fuse.
D — Motor starter not closing.
E — Overload or safety interlock open.
F — Loose or incorrect wiring or defective motor.
G — Controller malfunction, check controller.
H — Drive is overloaded.
I — Safety interlock.
K — Loose connection.
L — Open or defective clutch coil, check brushes first.
M — Controller malfunction.
N — Velocity feedback malfunction (Tach. Gen. or Mag. Pickup malfunction).
O — Electric noise or radio frequency interference.
P — Loose wiring connection.
Q — Contaminated slip rings.
R — Sticking or worn out brushes.
S — Check controller for input voltage.
T — Loose or incorrect wiring.
U — Open safety interlock.
V — Brushes not making contact.
W — Brake not releasing.
X — Open or defective clutch coil.
Y — Overload, check motor current.
Z — Operating below minimum speed.
AA — Air passages blocked on magnetic drive unit.
BB — Recirculating cooling air or ambient temperature too high.

CC — Brake not releasing or machine binding.
DD — Controller malfunction, check controller.
EE — Loss of velocity feedback signal (Tach. Gen. or Mag. Pickup).
FF — Mechanical lock up of clutch drum and rotor.
GG — Possible stall conditions or overload on the drive unit. Turn off control. Check driven load for restriction or fault condition.
HH — Possible open circuit in the drive unit fiield circuit. Check brushes and slip rings for continuity.
II — Check output voltage to the drive unit. If no or incorrect voltage check the silicon controller rectifiers (SCR's) or the gate pulse board is not sending pulse to the SCR's. Replace SCR's and pulse board as needed.
JJ — If voltage to drive unit is correct then check "Feedback" voltage from the tachometer generator (magnetic pickup).
KK — Check pulse board and SCR's.
LL — Check brushes and collector rings on drive unit.
MM — Check for loose or broken wires on the tachometer generator (or magnetic pickup).
NN — Check for proper adjustment of the tach generator or magnetic pickup.

TABLE 5

TROUBLESHOOTING ADJUSTABLE SPEED DRIVES

Symptoms	What To Do
• Overcurrent	A, B, C, D
• Tachometer loss	E, F, G, H, I, J, K, L, M, N
• Overspeed	O, P, Q, R, S
• Field current loss	T, U, V, W, X
• Sustained overload	Y, Z, AA
• Blower motor starter open	BB, CC, DD
• Open armature	EE, FF, GG
• Motor thermostat trip	HH, II, JJ, KK, LL, MM, NN, OO
• Controller thermostat trip	PP, QQ, RR, SS
•AC line synchronization fault	TT, UU, VV, WW, XX

Possible Causes

A — Incorrect armature current feedback scaling.
B — One or more thyristors not operating.
C — Improper current minor loop tuning.
D — Motor armature winding damaged.
E — Tach coupling failure.
F — Disconnected, loosely connected, or damaged tach wires .
G — Pulse tach supply voltage low.
H — Incorrect tach polarity.
I — Incorrect analog tach scaling.
K — Motor armature winding not connected or open circuit.
L — Blown inverting fault.
M — Inverting fault breaker tripped.
N — Tachometer failure.
O — Incorrect tach scaling.
P — Blown field supply.
Q — Improper speed loop tuning.
R — Pulse Tach Quadrature set to ON for a non-regenerative drive.
S — Incorrect pulse tach wiring.
T — Motor field wiring.
U — Blown field supply fuse(s).
V — Blown AC line fuse(s).
W — Field supply failure.
X — Disconnected, loosely connected or damaged wiring harness.
Y — Incorrect armature current feedback scaling.
Z — Blown field supply fuse(s).
AA — Mechanical binding preventing the motor shaft from rotating freely.
BB — Blown blower motor starter fuse(s).

CC — Disconnected, loosely connected or damaged blower motor starter wiring.
DD — Blower motor overload.
EE — Motor armature winding not connected or open circuit.
FF — Blown inverting fault (DC) fuse.
GG — Inverting fault breaker tripped.
HH — Damaged or disconnected motor thermostat wiring.
II — Inadequate ventilation.
JJ — Blower motor failure.
KK — Incorrect blower rotation.
LL — Blocked ventilation slots.
MM — Clogged filters.
NN — Excessive armature.
OO — One or more thyristors not operating.
PP — Inadequate heat sink ventilation.
QQ — Inadequate cabinet ventilation.
RR — Heat sink fan failure.
SS — Damaged or disconnected controller thermostat wiring.
TT — Blown AC line fuse(s).
UU — AC line frequency not within required range of 48-62Hz.
VV — Excessive AC line noise or distortion.
WW — Unstable AC frequency.
XX — Disconnected, loosely connected or damaged J6 ribbon cable.

TABLE 6

TROUBLESHOOTING COMPRESSOR MOTORS	
Symptoms	**What To Do**
• Compressor hums and fails to start.	A, B, C, D, E, F
• Compressor hums and cycles on overload protector, but fails to start.	G, H, I, J
• Starting winding remains in circuit after the compressor starts.	G, K, L, M, N
• Compressor starts but cycles on overload.	F, G, M, O, P
Possible Causes	

A — Disconnect switch is open.
B — Blown fuse or CB is open.
C — Defective wiring.
D — Overload protector is tripped.
E — Control contacts are open.
F — Overload protector is defective.
G — Low voltage.
H — Starting capacitor is defective.
I — Starting relay contacts are not closing.

J — Compressor motor is defective.
K — Starting relay is defective.
L — Starting capacitor is weak.
M — Running capacitor is defective.
N — Compressor motor is defective.
O — Compressor motor partially grounded.
P — Unbalanced line voltage (3Ø supply).

TABLE 7

TROUBLESHOOTING GENERATORS

Symptoms	What To Do
• Commutator	A
• Armature	B, C, G
• Brushes	D, E, F
• Overloaded	R
• Short-circuit	S
• Broken circuit	T
• Open-circuit	G, H, I, J, K, L

• Excessive current
  — in shunt winding
  — in series winding
	
	U, V

Possible Causes

A — Check for worn in grooves or ridges out of round
B — Check for short-circuit coils
C — Check for broken coils
D — Check setting at neutral points
E — Check and verify if they are in line
F — Check and verify if they are making good contact
G — Check for broken wires
H — Check for switch open
I — Check for safety fuses melted or broken
J — Check for faulty connections
K — Check for external circuit opening
L — Check for brushes not in contact
M — Check for excessive loading

N — Check for a ground and leak from short-circuit on line
O — Check for a dead short-circuit on line
P — Check for excessive current
Q — Check for eddy currents
R — Too many amps taken from machine
S — Usually caused by dirt etc. at commutator bars
T — Usually caused by a loose or broken band or wire etc..
U — Reduce speed and decrease voltage at terminals
V — By shunting, decrease current through field (remove some of field winding)

TABLE 8

TROUBLESHOOTING CONTACTORS AND RELAYS

Symptoms	What To Do
• Failure to energize	A, B, C, D, E, F, G, H, I, J,
• Failure to deenergize	K, O, P, Q
• Equipment fails to operate with contactor closed	L, M, N
• Pitted or discolored contacts	R, S, T
• Chatter or humming contactor	U, Z
• Coil has excessive temperature	V, W, X, Y

Possible Causes

A — Blown fuse, open line switch, or break in wiring

B — Line voltage is below normal

C — Overload relay is open or set too low

D — Control lever or start button is in OFF position

E — Pull-in circuit open, shorted or grounded

F — Contacts in protective or controlling circuit open or a pigtail connection is broken

G — Operating coil open or grounded circuit

H — Loose or disconnected coil wire

I — Test coil and replace if necessary

J — Normally closed contacts are welded together

K — Normally opened contacts are welded together

L — One contact is not closing

M — Contacts are burned

N — Contact pigtail connection is broken

O — Contacts in controlling or protective tripping circuits are closed, shorted or shunted

P — Tripping devices are defective: such as undervoltage relay plunger stuck or out of adjustment, defective stop button or defective time-relay

Q — Contact pressure spring or armature spring is too weak or improperly adjusted

R — Contacts are overheated from overload

S — Contacts are not fitted properly

T — Wiping action of contacts on closing is insufficient

U — Free movement of armature is hindered due to deformed parts, dirt, or lint

V — Excessive current or voltage is measured

W— Short-circuit is found in coil

X — Excessive eddy current and hysteresis is measured

Y — High room temperature is detected

Z — Voltage drop is measured at coil when closing

TABLE 9

TROUBLESHOOTING DRY TYPE TRANSFORMERS

Symptoms	What To Do
• Overheating	A, B, C, D, E
• Cable overheating	P
• Insulation is burned	R, S, T
• Failure of insulation	J, K, L, M, N
• Secondary voltage is too high	G
• CB is open or fuse is blown	O
• Excessive vibration and noise	Q

Possible Causes

A — Continuous overload problems

B — Wrong external connections are found

C — Poor ventilation is detected

D — High surrounding air temperatures are present

E — Clogged air ducts or inadequate ventilation problems

F — Loose connections to transformer terminal are found

G — Input voltage high or dirt accumulations on primary terminal leads are found

H — Terminal boards are not on correct tap position

I — Coils are short-circuited

J — Continuous overloads are measured

K — Dirt accumulations are found on coils

L — Mechanical damage is found

M — Lightning surges are detected

N — High core temperature due to high input voltage or low frequency is measured

O — Short-circuits, ground-faults, or overloads are detected

P — Improperly bolted connections are found

Q — Core clamps are loose or other loose hardware is found on enclosure

R — Lightning surge is found

S — Switching or line disturbance is detected

T — Broken bushings are found

TABLE 10

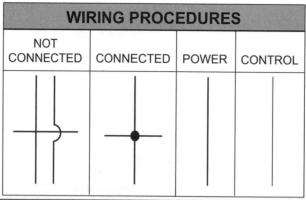

WIRING PROCEDURES

NOT CONNECTED	CONNECTED	POWER	CONTROL

CONTACT IDENTIFICATION

SPST - Single Pole Single Throw		SPST - Single Pole Single Throw		SPST - Single Pole Double Throw	
Single Break	Double Break	Single Break	Double Break	Single Break	Double Break
NO	NO	NC	NC	NC	NC

DPST - Double Pole Single Throw				DPDT - Double Pole Double Throw	
Single Break	Double Break	Single Break	Double Break	Single Break	Double Break
NO	NO	NC	NC	NC	NC

IDENTIFICATION OF POWER CONDUCTORS & TERMINALS

Phases	Single-phase	Single-phase	Three-Phase
Line Markings	L_1	L_1, L_2	L_1, L_2, L_3
Terminal Markings	T1	T1, T2	T1, T2, T3

WIRING METHODS

AC	Armored cable	NMC	Nonmetallic-Sheathed Cable
BX	Trade name for AC	NPLFA	Nonpower-Limited Fire Alarm Circuit
EMT	Electrical Metallic Tubing	OFNP	Nonconductive Optical Fiber Plenum Cable
ENT	Electrical Nonmetallic Tubing		
FMC	Flexible Metal Conduit	PLFA	Power-Limited Fire Alarm Circuit
ITC	Instrument Tray Cable	PLTC	Power-Limited Tray Cable
IMC	Intermediate Metal Conduit	PVC	Plastic Conduit
LTFMC	Liquidtight Flexible Metal Conduit	RMC	Rigid Metallic Conduit
LTFNMC	Liquidtight Flexible Non-Metallic Conduit	RNMC	Rigid Nonmetallic Conduit
		SE	Service-Entrance Cable
MC	Metal-clad Cable	SNM	Shielded Nonmetallic Sheathed Cable
MI	Mineral Insulated Metal-Sheathed Cable	TC	Tray Cable
		USE	Underground Service-Entrance Cable
NM	Nonmetallic-Sheathed Cable		

TABLE 11

DETERMINING FLC (in amps) OF TRANSFORMERS AND MOTORS

AC MOTORS				
SINGLE-PHASE				
FULL LOAD AMPERES				
HP	115 V	208 V	230 V	MIN. TRANSFORMER kVA
1/6	4.4	2.4	2.2	.53
1/4	5.8	3.2	2.9	.70
1/3	7.2	4.0	3.6	.87
1/2	9.8	5.4	4.9	1.18
3/4	13.8	7.6	6.9	1.66
1	16	8.8	8	1.92
1 1/2	20	11	10	2.4
2	24	13.2	12	2.88
3	34	18.7	17	4.1
5	56	30.8	28	6.72
7 1/2	80	44	40	9.6
10	100	55	50	12

THREE-PHASE					
FULL LOAD AMPERES					
HP	208 V	230 V	460 V	575 V	MIN. TRANSFORMER kVA
1/2	2.4	2.2	1.1	0.9	0.9
3/4	3.5	3.2	1.6	1.3	1.2
1	4.6	4.2	2.1	1.7	1.5
1 1/2	6.6	6	3	2.4	2.1
2	7.5	6.8	3.4	2.7	2.7
3	10.6	9.6	4.8	3.9	3.8
5	16.7	15.2	7.6	6.1	6.3
7 1/2	24.2	22	11	9	9.2
10	30.8	28	14	11	11.2
15	46.2	42	21	17	16.6
20	59.4	54	27	22	21.6
25	74.8	68	34	27	26.6
30	88	80	40	32	32.4
40	114	104	52	41	43.2
50	143	130	65	52	52
60	169	154	77	62	64
75	211	192	96	77	80
100	273	248	124	99	103
125	343	312	156	125	130
150	396	360	180	144	150
200	528	480	240	192	200
250	—	—	302	242	
300	—	—	361	289	
350	—	—	414	336	
400	—	—	477	382	
450	—	—	515	412	
500	—	—	590	472	

TRANSFORMERS				
SINGLE-PHASE				
	AMPERES			
kVA RATING	120 V	240 V	480 V	600 V
1	8.33	4.17	2.08	1.67
1 1/2	12..5	6.25	3.13	2.50
2	16.7	8.33	4.17	3.33
3	25.0	12.5	6.25	5.00
5	41.7	20.8	10.4	8.33
7 1/2	62.5	31.3	15.6	12.5
10	83.3	41.7	20.8	16.7
15	125	62.5	31.3	25.0
20	167	83.3	41.7	33.3
25	208	104	52.1	41.7
30	250	125	62.5	50
37 1/2	313	156	78.0	62.5
50	417	208	104	83.3
75	625	313	156	125
100	833	417	208	167
150	1,250	625	313	250
167	1,392	696	348	278
200	1,667	833	417	333
250	2,083	1,042	521	417
333	2,775	1,388	694	555
500	4,167	2,083	1,042	833

THREE-PHASE				
	AMPERES			
kVA RATING	120 V	240 V	480 V	600 V
3	8.3	7.2	3.6	2.9
6	16.6	14.4	7.2	5.8
9	25.0	21.6	10.8	8.7
15	41.6	36	18	14.4
20	55.6	48.2	24.1	19.3
25	69.5	60.2	30.1	24.1
30	83.0	72	36	28.8
37 1/2	104	90.3	45.2	36.1
45	125	108	54	43
50	139	120	60.2	48.2
60	167	145	72.3	57.8
75	208	180	90	72
100	278	241	120	96.3
112.5	312	270	135	108
150	415	360	180	144
200	554	480	240	192
225	625	540	270	216
300	830	720	360	288
400	1,110	960	480	384
500	1,380	1,200	600	480
750	2,080	1,800	900	720
1,000	2,780	2,400	1,200	960
1,500	4,150	3,600	1,800	1,440
2,000	5,540	4,800	2,400	1,920

TABLE 12

The following **Table** represents heater selection tables applicable to the overload relays used in Westinghouse Control Centers.

Compensated Ambient (Black reset rod)	Compensated Ambient (Black reset rod)	Heater Code Marking
.51 – .55	.48 – .51	FH10
.56 – .62	.52 – .57	FH11
.63 – .68	.58 – .63	FH12
.69 – .75	.64 – .70	FH13
.76 – .83	.71 – .77	FH14
.84 – .91	.78 – .85	FH15
.92 – 1.00	.86 – .93	FH16
1.01 – 1.11	.94 – 1.03	FH17
1.12 – 1.22	1.04 – 1.13	FH18
1.23 – 1.34	1.14 – 1.25	FH19
1.35 – 1.47	1.26 – 1.37	FH20
1.48 – 1.62	1.38 – 1.51	FH21
1.63 – 1.78	1.52 – 1.65	FH22
1.79 – 1.95	1.66 – 1.81	FH23
1.96 – 2.15	1.82 – 1.99	FH24
2.16 – 2.35	2.00 – 2.19	FH25
2.36 – 2.58	2.20 – 2.39	FH26
2.59 – 2.83	2.40 – 2.63	FH27
2.84 – 3.11	2.64 – 2.89	FH28
3.12 – 3.42	2.90 – 3.17	FH29
3.43 – 3.73	3.18 – 3.47	FH30
3.74 – 4.07	3.48 – 3.79	FH31
4.08 – 4.39	3.80 – 4.11	FH32
4.40 – 4.87	4.12 – 4.55	FH33
4.88 – 5.3	4.56 – 5.0	FH34
5.4 – 5.9	5.1 – 5.5	FH35
6.0 – 6.4	5.6 – 5.9	FH36
6.5 – 7.1	6.0 – 6.6	FH37
7.2 – 7.8	6.7 – 7.2	FH38
7.9 – 8.5	7.3 – 7.9	FH30
8.6 – 9.4	8.0 – 8.7	FH40
9.5 – 10.3	8.8 – 9.5	FH41
10.4 – 11.3	9.6 – 10.5	FH42
11.4 – 12.4	10.6 – 11.5	FH43
12.5 – 13.5	11.6 – 12.6	FH44
13.6 – 14.9	12.7 – 13.8	FH45
15.0 – 16.3	13.9 – 15.1	FH46
16.4 – 18.0	15.2 – 16.7	FH47
18.1 – 19.1	16.8 – 18.3	FH48
19.9 – 21.7	18.4 – 20.2	FH49
21.8 – 23.9	20.3 – 22.2	FH50
24.0 – 26.2	22.3 – 24.3	FH51
26.3 – 28.7	24.4 – 26.6	FH52
FOR STARTER SIZE 2		
28.8 – 31.4	26.7 – 29.1	FH53
31.5 – 34.5	29.2 – 32.0	FH54
34.6 – 37.9	32.1 – 35.2	FH55
38.0 – 41.5	35.3 – 38.5	FH56
41.6 – 45.0	38.6 – 42.3	FH57
FOR STARTER SIZES 3 & 4		
19.0 – 20.8	17.5 – 19.1	FH72
29.9 – 22.9	19.2 – 21.1	FH73
23.0 – 25.2	21.2 – 23.2	FH74
25.3 – 27.8	23.3 – 25.6	FH75
27.9 – 30.6	25.7 – 28.1	FH76
30.7 – 33.5	28.2 – 30.8	FH77
33.6 – 37.5	30.9 – 34.5	FH78
37.6 – 41.5	34.6 – 38.2	FH79
41.6 – 46.3	38.3 – 42.6	FH80
46.4 – 50.	42.7 – 46.	FH81
51. – 55.	47. – 51.	FH82
56. – 61.	52. – 56.	FH83
62. – 66.	57. – 61.	FH84
67. – 73.	62. – 67.	FH85
74. – 79.	78. – 73.	FH86
80. – 87.	74. – 80.	FH87
88. – 90.	81. – 87.	FH88
FOR STARTER SIZE 4		
88. – 50.	88. – 95.	FH88
96. – 105.	96. – 105.	FH89
106. – 116.	106. – 116.	FH90
117. – 128.	117. – 128.	FH91

Compensated Ambient (Black reset rod)	Compensated Ambient (Black reset rod)	Heater Code Marking
FOR STARTER SIZE GCA 5		
118. – 129.	110. – 119.	FH24
130. – 141.	120. – 131.	FH25
142. – 155.	132. – 143.	FH26
156. – 170.	144. – 158.	FH27
171. – 187.	159. – 173.	FH28
188. – 205.	174. – 190.	FH29
206. – 224.	191. – 208.	FH30
225. – 244.	209. – 227.	FH31
245. – 263.	228. – 247.	FH32
264. – 270.	248. – 270.	FH33
FOR STARTER SIZE GCA 6		
236. – 259.	219. – 239.	FH24
260. – 283.	240. – 263.	FH25
284. – 310.	264. – 287.	FH26
311. – 340.	288. – 316.	FH27
341. – 374.	317. – 347.	FH28
374. – 411.	348. – 381.	FH29
412. – 448.	381. – 417.	FH30
449. – 489.	418. – 455.	FH31
490. – 527.	456. – 494.	FH32
528. – 540.	495. – 540.	FH33

Note:

When selecting heating coils based on "other calculations", see pages 19-2 and 19-3 in this book.

TABLE 13

STAR-DELTA AND PART WINDING STARTERS

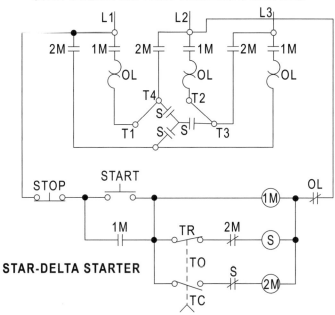

STAR-DELTA STARTER

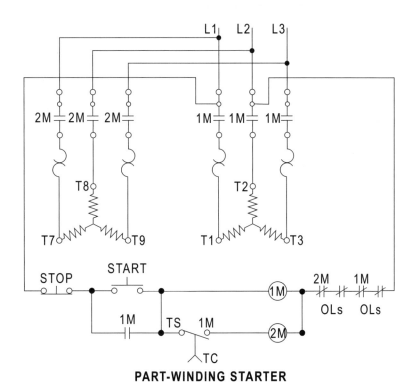

PART-WINDING STARTER

TABLE 14

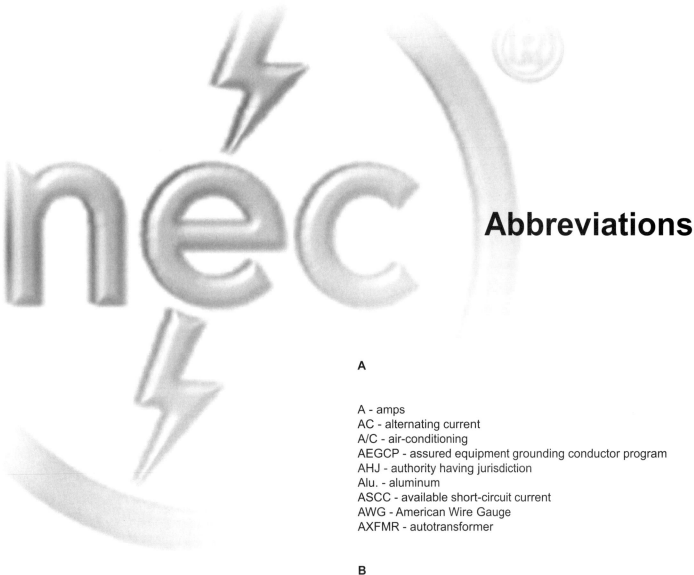

Abbreviations

A

A - amps
AC - alternating current
A/C - air-conditioning
AEGCP - assured equipment grounding conductor program
AHJ - authority having jurisdiction
Alu. - aluminum
ASCC - available short-circuit current
AWG - American Wire Gauge
AXFMR - autotransformer

B

BC - branch-circuit
BCSC - branch-circuit selection current
BJ - bonding jumper
BK - black
BL - blue
BR - brown

C

°C - Celsius
CB - circuit breaker
CEE - concrete-encased electrode
CL - code letter
CM - circular mills
CMP - Codemaking Panel
Comp. - compressor
Cond. - condenser
Cont. - continuous
cu. - copper
cu. in. - cubic inches

D

dia. - diameter
DC - direct-current
DPCB - double pole circuit breaker

E

EBJ - equipment bonding jumper
Eff. - efficiency
EGB - equipment grounding bar
EGC - equipment grounding conductor
EMT - electrical metallic tubing
ENT or ENMT - electrical nonmetallic tubing
Ex. - Exception
EExde - increase safety
Eexe - flameproof/increased safety components
Epf - explosionproof

F

°F - Fahrenheit
FC - feeder-circuit
FLA - full-load amperage
FLC - full-load current
FMC - flexible metal conduit
FPN - fine print note
FREQ. - frequency
Ft. - foot

G

G - ground
GE - grounding electrode
GEC - grounding electrode conductor
GES - grounding electrode system
GFL - ground-fault limiter
GFCI - ground-fault circuit interrupter
GFPE - ground-fault protection of equipment
GR - green
GRY - gray
GSC - Grounded service conductor

H

H - hot conductor
HACR - heating, air-conditioning, cooling, and refrigeration
HP - horsepower
Htg. - heating
Hz - hertz

I

I - amperage or current
IEC - International Electrotechnical Commission
IG - Isolated ground
in. - inches
INVT - inverse-time circuit breaker
INST. CB - instantaneous trip circuit breaker
IRA - inrush amps
ITSC - intrinsically safe circuits
INTP - interpole

K

kFT - 1000'
kV - kilovolts
kVA - kilovolt-amps
kvar - kilovar
kW - kilowatts
kWH - kilowatt-hour

L

L - length of conductor
LRA - locked rotor amps
Ld. - load
LPB - lighting panelboard
LRC - locked-rotor current
LTFMC - liquidtight flexible metal conduit
LTR - long time rated

M

MA - milli-amps
max. - maximum
MEL - maximum energy level
mf - microfarads
MGFA - maximum ground-fault available
min. - minimum
min. - minute
MR - momentary rated
Mt. - motor

N

N - neutral
NACB - nonautomatic circuit breaker
NB - neutral bar
NEC - National Electrical Code
NEMA - National Electrical Manufacturers Association
NFD - nonfused disconnect
NFPA - National Fire Protection Association
NLTFMC - nonmetallic liquidtight flexible metal conduit
NTDF - nontime-delay fuse

O

OCP - overcurrent protection
OCPD - overcurrent protection device
OSHA - Occupational Safety and Health Administration
OL - overload
OLP - overload protection
OL's - overloads
OR - orange

P

Ph. - phases (hots)
pri. - primary
PSA - power supply assembly
PF - power factor
PU - purple

R

R - ohms or resistance
REC. - receptacle
RD - red
RMC - rigid metal conduit

S

SBS - structural building steel
SIA - seal-in amps
SP - single-pole
S/P - single-phase
SCC - short-circuit current
sec. - secondary
SF - service factor
SPCB - single-pole circuit breaker
sq. ft. - square foot (feet)
SS - synchronous speed
STR - short-time rated
sq. in. - square inches
SWD - switched disconnect
SWG - switchgear
SDS - separately derived system

T

TDC - time-delay cycle
TDF - time-delay fuse
TDL - time-delay limiter
TP - thermal protector
TR - temperature rise
Tran. - transformer
TS - trip setting
TV - touch voltage

U

UF - underground feeder

V

V - volts
VA - volt-amps
VD - voltage drop

W

W - watts
WC - winding current
WT - white
WP - weatherproof
WV - winding voltage

X

XFMR - transformer

Y

YEL - yellow

Glossary of Terms

A

Active power is the true electrical power or real power supplying the load.

Across-the-line starter is a device consisting of contactor and overload relay which is used to start an electric motor by connecting it directly to the supply line.

Air gap is the air space between two electrically related parts such as the space between poles of a magnet or poles in an electric motor.

Alternating current (AC) is the current in an electrical circuit that alternates in flowing first with a positive polarity and then with a negative polarity.

Alternator is a rotating machine whose output is AC.

Ampere is an unit of measure for current flow. Note that one ampere equals a flow of one coulomb of charge per second.

Ambient conditions is the condition of the atmosphere adjacent to electrical equipment.

Ambient temperature is the temperature of the surrounding atmosphere cooling medium, which comes into contact with the heated parts of equipment.

Ambient temperature compensated is a device, such as an overload relay, which is not affected by the temperature surrounding it.

Ampacity is the current-carrying capacity in amperes.

Ampere is a unit of intensity of electrical current produced in a conductor by an applied voltage.

Armature is a special designed rotor.

Apparent power is the sum of active power and reactive power. It is determined by multiplying voltage times current.

Arc-chute is a cover around contacts designed to protect surrounding parts from arcing effects.

Armature reaction is the reaction of the magnetic field produced by the current on the magnetic lines of force which are produced by the field coil of an electric motor or generator.

Apparatus is a set of control devices used to help perform the intended control functions.

Automatic is a means of self acting that operates by its own mechanism, as such a change in pressure or temperature, etc.

Automatic controller is a motor or other control mechanism which uses automatic pilot devices as activating devices. These devices may be pressure switches, level switches or thermostats.

Autotransformer starter is equipped with an autotransformer which is designed to reduce the voltage to the motor terminals and reduce the starting current and still start the motor, etc.

Auxiliary contacts are contacts in addition to the main-circuit contacts and function with the movement of the latter.

Auxiliary device is any device other than motors and motor starters necessary to fully operate the machine or equipment.

Auxiliary interlock:

• Mechanical - A physical device or arm so arranged that it cannot close both starter circuits at the same time.

• Electrical - An additional contact mounted on the side of a magnetic starter.

B

Bearings is a device used to support the motor shaft, and allows it to rotate smoothly.

Bimetallic disc is a disc made up of two strips of dissimilar metals combined to form a single strip.

Breakdown torque is the maximum torque which a motor develops under increasing load conditions at rated voltage and frequency without an sudden drop in rotating speed.

Brushes are sliding contacts, usually made of carbon which are located between a commutator and the outside circuit in a generator or motor.

Branch-circuit is that portion of a wiring system extending beyond the final overcurrent device protecting the circuit.

C

Commutator is a device which reverses the connections to the revolving loops on the armature.

Capacitance is the ability to store electricity in an electrostatic field.

Capacitor is a device which is designed to introduce capacitance into an electric circuit.

Capacitor-start motor is an AC split-phase induction motor that has a capacitor connected in series with an auxiliary winding which provides a way for it to start. This auxiliary circuit is designed to disconnect to the motor when it reaches its speed.

Circuit is an electrical network of conductors that provides one or more paths for current.

Circuit breaker is a device designed to open and close a circuit either by a nonautomatic means or by an automatic means due to a predetermined overload of current.

Combination starter is a magnetic starter having a manually operated disconnecting means built into the enclosure housing the magnetic contactor or starter.

Compensating windings are the windings embedded in the main pole pieces of a compound DC motor.

Component is the smallest element of a circuit.

Contacts are connecting parts which co-act with other parts to connect or disconnect a circuit.

Contactor is a electro-mechanical device for connecting and disconnecting an electric power circuit.

Control is a device or group of devices which is used in some predetermined manner to govern the electric power delivered to an apparatus.

Control circuit is the control circuit of the control apparatus which carries the signals directing the performance of the controller.

Control, two-wire is a control function which utilizes a maintained contact type pilot device to provide undervoltage release.

Control circuit transformer is a control circuit transformer utilized to supply a reduced voltage suitable for the operation of control devices.

Control circuit voltage is the control circuit voltage providing the operation to the coils of magnetic devices.

Control, three-wire is a control function which utilizes a momentary contact pilot device and a holding circuit contact to provide voltage to the coil of a controller. The holding circuit maintains the control circuit voltage.

Controller is a device, or group of devices, which is used in some predetermined manner to connect and disconnect the electric power delivered to the apparatus.

Copper loss is the electrical power lost through the resistance of the coils due to the current flowing through the wire of the coils.

Core is the magnetic path through the center of coil or transformer.

Core losses is the loss of power in the coil (core) due to eddy currents and hysteresis.

Core transformer is an electrical transformer which has the core inside of the coils.

Counterelectromotive force is the voltage induced in the armature coil of an electric motor.

Counter torque is a repulsion force between two magnetic fields.

D

Delta-delta connected is a coil connection in which the primary and secondary coils are delta connected.

Delta-wye connected is a connection in which the primary is delta connected while the secondary windings are wye connected.

Diagram is a connection diagram showing the electrical connection between the parts of the control, and external connections.

Direct-current (DC) is a current that always flows in only one direction only.

Disconnect means is a motor circuit switch which is intended to connect and disconnect a circuit to a motor. It must be rated in horsepower and capable of interrupting the maximum current.

E

Efficiency is the ratio of output power (watts) to input power (watts).

Electric motor is a machine which converts electrical energy to mechanical energy.

Electromagnet is a magnet comprised of a coil of wire wound around a soft-iron core. When current is passed through the wire, a magnetic field is produced.

Eddy currents are the electrical currents circulating in the core of a transformer as the result of induction.

Electricity is the electrical charges in motion. Such movement is called current and is measured in amps.

Electrolytic capacitor is a capacitor which uses a liquid or past as one of its electrical storage plates.

Electromotive force is a voltage or force which causes free electrons to move in a conductor.

Electromagnet is a temporary magnet created by passing an electric current through a coil wound around a soft iron or other magnetic core.

Electromagnetism is a magnetic field which exists around a wire or other conductor if a current is passing through it.

Electromechanical is a device which uses electrical energy to create mechanical motion of force.

Electron is a negative electric charge.

Electron flow is the flow of electrons from a negative point to a positive point in a conductor.

Electrostatic charge is the electrical charge stored by a capacitor.

Electrostatic field is the stored electrical charges on the surface of an insulator.

Excitation is creating a magnetic field to be used to create electromagnetics when an electric current is passed through a coil.

F

Feeder-circuit is the conductors between the service equipment and the branch-circuit overcurrent device.

Float switch is a switch which is operated by a float and is responsive to the level of liquid.

Foot switch is a switch which is suitable for operation by an operators foot.

Frequency is the rate at which AC changes its direction of flow, it is normally expressed in terms of hertz (cycles) per second.

Fuse is an overcurrent protection device with a circuit opening fuseable member which opens when overheated by current passing through it.

G

Gate is one of the leads on a thyristor. This lead is the one that normally controls output when it is correctly biased.

Generator is a rotating machine which changes mechanical energy into DC.

Generator action is inducing voltage into a wire that is cutting a magnetic field.

Ground is any point on a motor component where the ohmic resistance between the component and the motor frame is one megaohm or less.

Guarded is covered, shielded, fenced, enclosed, or otherwise protected by means of suitable covers or casings, barriers, rails or screens, mats or platforms to remove the likelihood of dangerous contact or approach by persons or objects to a point of danger.

H

Hertz is a measurement of frequency and it actually means "cycles per second" of AC.

Hermetic refrigerant motor compressor is a combination consisting of a compressor and motor, both of which are enclosed in the same housing, with no external shaft or shaft seals, the motor operating in the refrigerant.

High side in a transformer, this marking indicates the high-voltage winding.

Horscpower is a unit of measure for power and it represents the force times distance times time. For example, one horsepower (HP) equals 746 watts, or 33,000 ft. lb. per minute, or 550 ft. lb. per second.

Hysteresis is the property of a magnetic substance that causes the magnetization to lag behind the magnetizing force.

I

Induced current is the current which flows in a conductor because of a changing magnetic field.

Inductance is electromotive force resulting from a change in magnetic flux surrounding a circuit or conductor.

Induction is the generation of electricity by magnetism.

Inductive reactance is the opposition in ohms to an AC as a result of induction. This is voltage resulting from cutting lines of magnetic force.

Interlock is an electrical or mechanical device actuated by the operation of a different device in which it is directly related.

Intermittent duty is a requirement or service that demands operations for alternate intervals of (1) load and no-load; or (2) load and rest; or (3) load, no-load and rest; such alternate intervals being definitely specified.

Isolating transformer is an isolating transformer used to electrically isolate one circuit from another.

Inverter is a circuit capable of receiving a positive signal and sends out a negative one or vice versa. It is a device which changes AC to DC or vice versa.

Impedance is the total opposition to current flow in a circuit and is measured in ohms.

J

Jogging is the rapid and repeated opening and closing of the circuit to start a motor from rest for the purpose of creating small movements of the motor or driven load.

K

kVA is the term used to rate transformers.

kvar is the reactive power in a circuit.

kW is used to rate the load of certain types of equipment, etc.

L

Lamination consist of sheet material which is sandwiched together to construct a stator or rotor of a rotating machine.

Limit switch is a switch which is operated by a part or motion of a power-driven piece of equipment. Such operation alters the electric or electronic circuits related to the equipment.

Locked-rotor current of a motor is the current taken from the line when the motor starts or the rotor becomes locked in place.

Low side in a transformer, this marking indicates which is the low-voltage winding.

M

Magnetic starter is a starter which is actuated by an electro-magnetic means.

Manual controller is a device which is manually closed or opened.

Manual reset is a device which requires manual action to re-engage the contacts after an overload.

Magnetic drive (magnetic clutch) is electromagnetic device that is connected between a three-phase motor and its load. Its main purpose is regulating the speed at load rotated speed.

Magnetic field is the invisible lines of force found between the north and south poles of a magnet.

Magnetic lines of force in a magnetic field, are imaginary lines which show the direction of the magnetic flux.

Maintained contacts closes the circuit when the push button is pressed and will open the circuit when the push button is pressed again.

Megaohm is one million ohms.

Motor action is the mechanical force that exist between magnets. Two magnets approaching each other will either pull toward or push away from the other. In other words, there is a pull and push action between the rotor and field poles of the motor.

Multi-speed motor is a motor which is capable of operating at two or more fixed speeds.

N

NEMA (National Electrical Manufacturers Association) is an organization which establishes certain voluntary standards relating to motors; such as operating characteristics, terminology, basic dimensions, ratings, and testing.

No-load speed is the speed reached by the rotor or armature when it rotates.

Non-automatic requires personal action and operation of devices for its control means.

Nonreversing is a control function which provides for operation in one direction only.

Normally open and normally closed is a term when applied to a magnetically operated switching device, signifies the position that the contacts are in.

Normally closed contacts are motor control contacts (set) that are open when the push button is depressed.

Normally open contacts are contacts (set) which are closed when the push button is depressed.

O

Ohm is a unit of electrical resistance of a conductor.

Out-of-phase is a condition where two or more phases of AC is changing direction at different intervals of time.

Over excited is a condition where a synchronous motor is equipped with a DC field which supplies more magnetization than is needed.

Overload protector is a device affected by an abnormal operating condition which causes the interruption of current flow to the device governed.

Overload relay is a device that provides overload protection for conductors and electrical equipment.

Overcurrent protective device (OCPD) is a device that operates on excessive current which causes the interruption of power to the circuit if necessary.

P

Parallel circuit is a circuit in which all positive terminals are connected at a common point and all negative terminals are connected to another point.

Periodic duty is a type of intermittent duty in which the load conditions are regularly recurrent.

Permanent-capacitor motor is a single-phase electric motor which uses a phase winding and capacitor in conjunction with the main winding. The phase winding is controlled by the capacitor which remains in the circuit at all times.

Permanent magnetism is magnet that will keep its magnetic properties indefinitely.

Permeability is a condition in which domains in a magnetic core can be made to line up to create magnetism.

Phase is the relationship of two wave forms having the same frequency.

Phase angle is the difference in angle between two sine wave vectors.

Phase shift is the creation of a lag or advance in voltage or current in relation to another voltage or current in the same electrical circuit.

Phase voltage is the voltage across a coil.

Polarity is a condition where a magnet has north and south poles which are positive and negative charge.

Polyphase is more than one phase, usually three-phase when related to generators, transformers, and motors.

Pounds force is an English unit of conventional measurement for force.

Power factor is the figure which indicates what portion of the current delivered to the motor is used to do work.

Primary coil is one of two coils in a transformer.

Prime mover is the primary power source which can be used to drive a generator.

Pull-in to torque is the maximum torque at which an induction motor will pull into step.

Pull-up torque is the minimum torque developed by an induction motor during the period of acceleration from rest to full speed.

Pull-out torque is the maximum torque developed by a motor for one minute, before it pulls out of step due to an overload.

Push button control is the control and operation of equipment through push buttons used to activate relays.

Push button switch is a switch utilizing a button for activating a coil and contact to open or close a circuit.

R

Rainproof is an enclosure so constructed, as to prevent rain from interfering with operation of the apparatus.

Raintight is an enclosure so constructed to exclude rain under specified test conditions.

Rated-load current is the load rated-load current for a hermetic refrigerant motor-compressor is the current resulting when the motor-compressor is operated at the rated load, rated voltage, and rated frequency of the equipment it serves.

Rating is a designated limit of operating characteristics based on conditions of use such as load, voltage, frequency, etc.

Rating, continuous is the rating which defines the substantially constant load which can be carried for an indefinitely long time.

Reactive power is the reactive voltage times the current, or voltage times the reactive current, in an AC circuit.

Rectifier is an electrical device which converts AC to DC by allowing the current to move in only one direction.

Relay is a device that operates by a variation of a condition which effects the operation of other devices in an electric circuit.

Relay contacts are the contacts that are closed or opened by movement of a relay armature.

Reluctance is the ratio between the magnetomotive force and the resulting flux.

Reset is to restore a mechanism or device to a prescribed state.

Reset, automatic is a function which operates automatically to re-establish certain circuit conditions.

Reset, manual is a function which requires a manual operation to re-establish certain circuit conditions.

Residual magnetism is the magnetism remaining in the core of a coil or an electromagnet after the current flow has been removed.

Resistance is a property of conductors which makes them resist the movement of current flow.

Resistance starting is a reduced-voltage starting method employing resistances which are short-circuited in one or more steps to complete the starting cycle of a motor, etc.

Resistors is an electrical-electronic device which is attached to a circuit to produce resistance to current flow.

Rheostat is a variable resistor with a fixed terminal and a movable contact.

Rotor is the rotating section which rotates within the stator of a motor.

Rotor impedance is the phasor sum of resistance and inductive reactance.

RPM is the revolutions per minute.

Running torque is the torque or turning effort determined by the horsepower and speed of a motor at any given point of operation.

S

Saturated is where an electrical or magnetic component can not receive any more electrical current or magnetism.

Saturation is a point at which a magnet will receive any more flux density.

Sealing, voltage or current is the voltage or current required to seat the armature of a magnetic circuit closing device to the make position.

Service factor is the number by which the horsepower rating is multiplied to determine the maximum safe load that a motor can carry continuously at its rated voltage and frequency.

Secondary coil is the coil that is connected to the load in the electrical circuit.

Self excitation is a condition of supplying excitation voltages by a device on the generator rather than from an outside source.

Self induction is a counterelectromotive force produce in a conductor when the magnetic field produced by the conductor collapses or expands after a change in current flow.

Separate excitation is a condition of producing generator field current from an independent source.

Short-time rating is referring to the motor load which can be carried for a short and definitely specified time.

Series circuit is a circuit in which all resistances and other components are connected so that the same current flows from point to point.

Series field is the total magnetic flux caused by the action of the series winding in a rotating piece of machine.

Series motor is a motor in which the field and armature circuits are connected in series.

Shaded-pole motor is a single-phase squirrel cage induction motor with stator poles slotted and used to create two sections in each pole.

Shading coil is a copper ring or coil that is set into a section of the pole piece and its function is to produce the lagging part of a rotating magnetic field for starting torque.

Short is any two points of a motor where there is zero, or extremely low, resistance between them or between two motor components.

Shunt field is a type of field coil designed for a DC motor which is connected in parallel with the armature.

Silicon controlled rectifier (SCR) is a semiconductor device which has the ability to block a voltage that is applied in either direction. On a signal applied to its gate, it is capable of conducting current even when the signal has been removed.

Single-phase is having only one AC or voltage in a circuit.

Slip is the difference between the synchronous speed of a motor and the speed at which it operates.

Slip ring motor has a rotor with the same number of magnetic poles at the stator.

Slip rings are equipped with circular bands on a rotor which are used to transmit current from rotor coils to brushes.

Slip speed is the difference between the rotor speed and synchronous speed in an induction motor.

Soft neutral position is a condition where the brushes of a repulsion electric motor are aligned with the stator field.

Solid state devices contain circuits and components using semiconductors.

Solid state controls are devices that control current to motors through semiconductors.

Solid state relay is a relay that uses semiconductor devices.

Split-phase (resistance-start) motor is a single-phase induction motor equipped with an auxiliary winding which is connected in parallel with the main winding.

Squirrel-cage rotor is designed with a rotor which is made up of metal bars which are short-circuited at each end.

Starter is a controller for accelerating a motor from rest to its running speed.

Starter, automatic is a starter which automatically controls the starting of a motor.

Starter, autotransformer is a starter which is provided with an autotransformer that provides a reduced voltage for starting.

Starter, part-winding applied voltage to partial sections of the primary winding of an AC motor.

Starter, reactor includes a resistor connected in series with the primary winding of an induction motor to provide reduced voltage for starting.

Starter, wye-delta connects the motor leads in a wye configuration for reduced voltage starting and reconnects the leads in a delta configuration for the run position.

Stall torque is the torque which the rotor of an energized motor produces when the rotor is not rotating.

Starting torque is the amount of torque produced by a motor as it breaks the motor shaft from standstill and accelerates to its running speed.

Stator is the portion which contains the stationary parts of the magnetic circuit with their associated windings when installed in a motor.

Static electricity is electricity at rest. It is also known in the industry as a static charge.

Stator field contains a magnetic field which is set up in the electric motor when the motor is energized and electric current is flowing.

Stator poles are the shoes on an electric motor stator which hold the windings and the magnetic poles of the stator.

Switch is a device for making, breaking or changing the connections in an electric circuit.

Switch, float is responsive to the level of a liquid.

Switch, foot is a switch that is operated by an operators foot.

Switch, general-use is a general-use type non-horse rated switch which is capable of interrupting the rated current at the rated voltage.

Switch, limit is operated by some part or motion which alters the electrical circuit associated with the equipment.

Switch, master controls the operation of contactors, relays, or other similar operated devices.

Switch, motor circuit is rated in horsepower, and is capable of interrupting the maximum operating current of the motor.

Switch, pressure is operated by fluid pressure, etc.

Switch, selector is a manually operated multiposition switch which is used for selecting an alternative control circuit.

Synchronous is a condition where the currents and voltages are in-step or in-phase.

Synchronous motor is an induction motor which runs at synchronous speed.

Synchronous speed is the constant speed to which an AC motor adjusts itself, depending on the frequency of the power source and the number of poles in the motor.

T

Tachometer is a device which is capable of measuring rotational speed of rotating machines.

Tap changer is a mechanical device which has the ability to change the voltage output of a transformer.

Taps are fixed electrical connections which are located at specific positions on a transformers coil.

Temperature, ambient is the temperature of the medium such as air, oil, etc. into which the heat of the equipment is dissipated.

Terminal is a point at which an electrical element may be connected to another electrical element.

Terminal board is an insulating base equipped with one or more terminal connectors used for making electrical connections.

Thermocouple is a device which consists of two unlike metals which are joined together and when heat is applied, a current will flow.

Thermal, cutout is an overcurrent protective device having a heater element which affects a fusible member that open the circuit due to an overload.

Thermal protector is a protective device that is an integral part of the motor which is designed to protect the motor windings from dangerous overloads.

Thermostat is an instrument which responds to changes in temperature to effect control over an operating condition.

Three-phase alternator is a rotating machine which generates three separate phases of AC.

Three-phase electric motor is a motor which operates from a three-phase power supply.

Timer is a device which is designed to delay the closing or opening of a circuit for a specific period of time.

Torque is a force that produces a rotating or twisting action

Torque, breakdown is the maximum torque which a motor develops with rated voltage when applied at rated frequency.

Torque, locked rotor is the minimum torque which a motor develops at standstill when rated voltage is applied at rated frequency.

Turns ratio is the ratio of the number of turns in the primary winding of a transformer to the number of turns in the secondary winding.

Two-capacitor motor is an induction motor which uses one capacitor for starting and one for running.

U

Under excited is a term used to describe the magnetizing power of synchronous motor.

Unity power factor is a power factor of 1 and this is the best PF that can be obtained in an electrical system.

Undervoltage protection is a device which operates on the reduction or failure of voltage, and has the ability to maintain the interruption of power.

Undervoltage release is a device which operates on the reduction or failure of voltage and has the ability to interrupt the power but not to prevent the re-establishment of the circuit.

Vector is an in-phasor diagram having lines with a specific length and direction.

Voltage is a force which, when applied to a conductor, produces a current in the conductor.

W

Watt is a unit of electrical power and is the product of voltage and amperage.

Wattmeter is an instrument used for measuring electrical power.

Wye or star connection is an electrical connection in which all of the terminals are joined at the neutral junction. After it is connected it resembles a wye connection.

Wye-wye connection is where the coil arrangement in which both the primary and the secondary coils are wye-connected.

Transformer is an device designed to change the voltage in an AC electrical circuit. Step-up transformers increase the voltage and lowers the current. Step-down transformers decrease the voltage and raises the current.

Transformer efficiency is the ratio of input to output power.